Artificial Intelligence in Pathology

Principles and Applications

Artificial Intelligence in Pathology
Principles and Applications

Second Edition

Edited by

Chhavi Chauhan
Director of Scientific Outreach, American Society for Investigative Pathology, Rockville, MD, United States

Stanley Cohen
Emeritus Founding Director, Center for Biophysical Pathology, Rutgers-New Jersey Medical School, Newark, NJ, United States; Adjunct Professor of Pathology, Northwestern University Feinberg School of Medicine, Chicago, IL, United States; Perelman School of Medicine, University of Pennsylvania, Philadelphia, PA, United States; Sidney Kimmell Medical College, Thomas Jefferson University, Philadelphia, PA, United States

Elsevier
Radarweg 29, PO Box 211, 1000 AE Amsterdam, Netherlands
125 London Wall, London EC2Y 5AS, United Kingdom
50 Hampshire Street, 5th Floor, Cambridge, MA 02139, United States

Copyright © 2025 Elsevier Inc. All rights are reserved, including those for text and data mining, AI training, and similar technologies.

Publisher's note: Elsevier takes a neutral position with respect to territorial disputes or jurisdictional claims in its published content, including in maps and institutional affiliations.

For accessibility purposes, images in this book are accompanied by alt text descriptions provided by Elsevier.

No part of this publication may be reproduced or transmitted in any form or by any means, electronic or mechanical, including photocopying, recording, or any information storage and retrieval system, without permission in writing from the publisher. Details on how to seek permission, further information about the Publisher's permissions policies and our arrangements with organizations such as the Copyright Clearance Center and the Copyright Licensing Agency, can be found at our website: www.elsevier.com/permissions.

This book and the individual contributions contained in it are protected under copyright by the Publisher (other than as may be noted herein).

Notices

Knowledge and best practice in this field are constantly changing. As new research and experience broaden our understanding, changes in research methods, professional practices, or medical treatment may become necessary.

Practitioners and researchers must always rely on their own experience and knowledge in evaluating and using any information, methods, compounds, or experiments described herein. In using such information or methods they should be mindful of their own safety and the safety of others, including parties for whom they have a professional responsibility.

To the fullest extent of the law, neither the Publisher nor the authors, contributors, or editors, assume any liability for any injury and/or damage to persons or property as a matter of products liability, negligence or otherwise, or from any use or operation of any methods, products, instructions, or ideas contained in the material herein.

ISBN: 978-0-323-95359-7

For information on all Elsevier publications
visit our website at https://www.elsevier.com/books-and-journals

Publisher: Stacy Masucci
Acquisitions Editor: Patricia Osborn
Editorial Project Manager: Matthew Mapes
Production Project Manager: Jayadivya Saiprasad
Cover Designer: Christian Bilbow

Typeset by STRAIVE, India

Contents

Contributors xv

Preface xix

Acknowledgments xxi

PART I Principles

CHAPTER 1 **The evolution of machine learning: Past, present, and future** **3**
Stanley Cohen

Introduction.. 3

Rules-based vs machine learning: A deeper look......................... 4

Varieties of machine learning 6

General aspects of machine learning 8

Deep learning and neural networks.. 9

The role of AI in pathology 11

Limitations of AI.. 12

General aspects of AI 13

References.. 14

CHAPTER 2 **The basics of machine learning: Strategies and techniques** **15**
Stanley Cohen

Introduction.. 15

Shallow learning 17

Geometric (distance-based) models 18

The K-means algorithm (KM) 21

Probabilistic models 22

Decision trees and random forests 24

The curse of dimensionality and PCA 26

Deep learning and the ANN... 27

Neuroscience 101 27

The rise of the machines 29

The basic ANN... 32

The weights in an ANN 33

Learning from examples: Backprop and stochastic
gradient descent 33

v

vi Contents

Convolutional neural networks 36
    Overview.. 36
    Detailed explanation... 36
Overfitting and underfitting... 39
Things to come 40
References... 41

CHAPTER 3   **Overview of advanced neural network architectures 43**
Benjamin R. Mitchell

Introduction.. 43
Network depth and residual connections 43
Autoencoders and unsupervised pretraining 46
Transfer learning.. 48
Generative models and generative adversarial networks 49
Recurrent neural networks.. 50
Reinforcement learning 51
Ensembles 53
Genetic algorithms... 55
References... 57

CHAPTER 4   **Complexity in the use of artificial intelligence in anatomic pathology... 59**
Stanley Cohen

Introduction.. 59
Life before machine learning 60
Multilabel classification 61
    Single object detection 62
Multiple objects 63
Advances in multilabel classification 64
Graphical neural networks.. 65
    Capsule networks.. 66
Weakly supervised learning 67
Synthetic data 69
N-shot learning 70
One-class learning 71
    Risk analysis 72
General considerations 73
Summary and conclusions.. 75
References... 76

Contents　vii

CHAPTER 5 **Dealing with data: Strategies of preprocessing data** 79
Stanley Cohen

Introduction.. 79
Overview of preprocessing.. 80
Feature selection, extraction, and correction 81
Feature transformation, standardization, and normalization 83
Feature engineering 83
Mathematical approaches to dimensional reduction................. 84
Dimensional reduction in deep learning 88
Imperfect class separation in the training set 89
Fairness and bias in machine learning 90
Summary.. 92
References... 92

CHAPTER 6 **Artificial intelligence in pathology: Easing the burden of annotation** 95
Benjamin R. Mitchell, Marion C. Cohen, and Stanley Cohen

Introduction.. 95
Artificial intelligence 101... 95
The human in the loop: Harvesting usable data 96
Reducing the need for annotated data............................ 97
Overview of unsupervised pretraining 98
Transfer learning.. 98
Unsupervised pretraining via clustering........................ 99
One class learning 100
Unsupervised pretraining via autoencoders 101
Unsupervised pretraining via generative adversarial
    networks.. 101
Reinforcement learning (RL) 102
Self-supervised learning 103
Zero shot learning... 104
Drowning in data: Quantum computing to the rescue............. 104
Summary and overview.. 106
References... 106

CHAPTER 7 **Digital pathology as a platform for primary diagnosis and augmentation via deep learning** 109
Anil V. Parwani and Zaibo Li

Introduction.. 109

viii Contents

Digital imaging in pathology 110
Telepathology 111
Whole slide imaging (WSI) 112
Whole slide image viewers 114
Whole slide image data and workflow management 115
Selection criteria for a whole slide scanner 116
Evolution of whole slide imaging systems 117
Infrastructure requirements and checklist for rolling out
    high-throughput whole slide imaging workflow solution 118
WSI and primary diagnosis 119
WSI and image analysis 122
WSI and deep learning 123
Conclusions 125
References 126

CHAPTER 8 **Artificial intelligence model development,
deployment, and regulatory challenges
in anatomic pathology 137**
Jerome Y. Cheng, Jacob T. Abel, Ulysses G.J. Balis,
Liron Pantanowitz, and David S. McClintock

Introduction 137
Development challenges 139
    Problem identification 139
    Dataset curation and annotation 139
Model development and training 141
Hardware and cost 142
Deployment challenges 142
Pathologist buy-in and transitioning to a digital workflow 143
IT infrastructure: Cloud computing vs. on-premises solutions 144
Lack of pathologist's experience with AI 144
    What is the right evidence standard for AI to be
        embedded in practice? 145
    What is required for clinical validation prior
        to using AI for diagnostic purposes? 145
    What is the ideal workflow when implementing
        AI in clinical practice? 145
    What model do pathology laboratories use to pay
        AI vendors? 147
    What is the business use case for deploying AI? 147
    Should residents or fellows be allowed to use AI,
        or is this "cheating?" 147

Contents   ix

Regulatory challenges... 148
    FDA.. 148
    European Union Conformité Européenne............................. 149
    CMS/CLIA... 149
Conclusion 150
Funding source... 150
Disclosures/conflicts of interest 150
References.. 151

CHAPTER 9   **Ethics of AI in pathology: Current paradigms and emerging issues.. 159**
Chhavi Chauhan and Rama R. Gullapalli

Introduction.. 159
Ethical AI study designs in pathology 163
    Inclusive AI design and bias.. 163
    Race in ethical AI design 165
    Stakeholder concerns: Consent and awareness..................... 166
Risks of AI in pathology and to pathologists—Real
    or imagined? 166
    Underestimating the risks of AI to pathology 167
    Overestimating the risks of AI to pathology 169
Institutional frameworks to enable ethical AI in pathology..... 170
    Transparency... 171
    Accountability... 172
    Governance 173
Recent developments in the use of AI in pathology 174
Conclusions.. 175
Acknowledgment 176
Disclosures... 176
Funding 176
References.. 176

PART II Applications

CHAPTER 10 Applications of artificial intelligence for image enhancement in pathology................... 183
Tanishq Abraham, Austin Todd, Daniel A. Orringer, and Richard Levenson

Introduction.. 183
Common machine learning tasks 184

Classification 184
Segmentation 184
Image translation and style transfer 185
Commonly used deep learning methodologies 185
Convolutional neural networks 185
U-nets... 187
Generative adversarial networks and their variants............. 187
Common training and testing practices................................... 188
Dataset preparation and preprocessing............................... 188
Loss functions.. 188
Metrics 189
Deep learning for microscopy enhancement
in histopathology 189
Stain color normalization 189
Mode switching 193
In silico labeling 196
Super-resolution, extended depth-of-field,
and denoising 199
Deep learning for computationally aided diagnosis
in histopathology 202
A rationale for AI-assisted imaging and
interpretation 203
Approaches to rapid histology interpretations 204
Future prospects.. 204
Acknowledgement 207
References.. 207

CHAPTER 11 Foundation models and information retrieval in digital pathology 211

H.R. Tizhoosh

Introduction... 211
Information retrieval.. 212
Image search 213
Validation of image search methods.. 216
Large deep models... 217
Foundation models 220
Generative AI 223
Information retrieval and foundation models 225
Conclusions.. 226
References.. 227

Contents **xi**

CHAPTER 12 Precision medicine in digital pathology via image analysis and machine learning............ 233
Peter D. Caie, Neofytos Dimitriou, and
Ognjen Arandjelović

Introduction.. 233
Precision medicine.. 233
Digital pathology 234
Applications of image analysis and machine
learning 235
Knowledge-driven image analysis 235
Machine learning for image segmentation............ 235
Deep learning for image segmentation 237
Spatial resolution 240
Machine learning on extracted data 241
Beyond augmentation 242
Practical concepts and theory of machine learning 243
Machine learning and digital pathology 243
Common techniques 244
Supervised learning 244
Unsupervised learning 248
Image-based digital pathology 249
Conventional approaches to image analysis 250
Deep learning on images 250
Regulatory concerns and considerations.................. 252
Acknowledgments 254
References... 254

CHAPTER 13 Generative deep learning in digital pathology...... 259
David Morrison, David Harris-Birtill, and Peter D. Caie

Introduction.. 259
Deep generative models 261
Generative models in the digital pathology pipeline................ 263
Color and intensity normalization......................... 263
Data adaptation 265
Data synthesis 266
Future directions 267
Conclusion 268
Acknowledgments 268
References.. 268

xii Contents

CHAPTER 14 Artificial intelligence methods for predictive image-based grading of human cancers 273

Gerardo Fernandez, Abishek Sainath Madduri, Bahram Marami, Marcel Prastawa, Richard Scott, Jack Zeineh, and Michael Donovan

Introduction 273
Tissue preparation and staining 275
Image acquisition 276
Stain normalization 277
Unmixing of immunofluorescence spectral images 278
Automated detection of tumor regions in whole-slide
    images 279
    Localization of diagnostically relevant regions
      of interest in whole-slide images 279
Tumor detection 280
Image segmentation 282
Nuclear and epithelial segmentation in IF images 283
Nuclei detection and segmentation in H&E images 284
Epithelial segmentation in H&E images 285
Mitotic figure detection 286
Ring segmentation 287
Protein biomarker features 288
Morphological features for cancer grading and prognosis 290
Modeling 294
Cox proportional hazards model 295
Neural networks 296
Decision trees and random forests 296
SVM-based methods: Survival-SVM, SVCR, and SVRc 297
Feature selection tools 298
Ground truth data for AI-based features 300
Conclusion 300
References 301

CHAPTER 15 Artificial intelligence and the interplay between cancer and immunity 309

Rajarsi Gupta, Tahsin Kurc, and Joel Haskin Saltz

Introduction 309
Immune surveillance and immunotherapy 310
Identifying TILs with deep learning 314

Spatial cancer biology with Pathomics,
immunohistochemistry, and immunofluorescence 327
Conclusion 332
References.. 333

CHAPTER 16 Overview of the role of artificial intelligence in pathology: The computer as a pathology digital assistant 343
John E. Tomaszewski

Introduction.. 343
Computational pathology: Background and philosophy........... 343
    The current state of diagnostics in pathology
      and the evolving computational opportunities:
      "why now?" 343
    Digital pathology versus computational pathology 345
    Data on scale 345
Machine learning tools in computational pathology:
    Types of artificial intelligence 346
The need for human intelligence-artificial intelligence
    partnerships... 348
Human-transparent machine learning approaches 349
    Explainable artificial intelligence 350
    Cognitive artificial intelligence....................................... 350
    Human-in-the-loop.. 351
    One-shot learning 352
Image-based computational pathology................................. 352
    Core premise of image analytics: What is a
      high-resolution image? 352
    The targets of image-based calculations 353
First fruits of computational pathology: *The evolving
    digital assistant* 354
    The digital assistant for quality control 354
    The digital assistant for histological object
      segmentation 355
    The digital assistant in immunohistochemistry 359
    The digital assistant in tissue classification....................... 359
    The digital assistant in finding metastases........................ 360
    The digital assistant in predictive modeling
      and precision medicine 361

xiv Contents

The digital assistant for anatomical simulation learning...... 362
The digital assistant for image-omics data fusion 362
Artificial intelligence and regulatory challenges 364
Educating machines-educating us: Learning how
to learn with machines 366
References.. 366

CHAPTER 17 Overview and coda: The future of AI.................... 369
Benjamin R. Mitchell and Stanley Cohen

Introduction.. 369
Transformers and attention... 371
Neuromorphic computing.. 373
Quantum computing 373
Summary and conclusions... 376
References.. 377

Index 379

Contributors

Jacob T. Abel
Department of Laboratory Medicine and Pathology, University of Washington, Seattle, WA, United States

Tanishq Abraham
Department of Biomedical Engineering, University of California, Davis, CA, United States

Ognjen Arandjelović
School of Computer Science, University of St Andrews, St Andrews, United Kingdom

Ulysses G.J. Balis
Department of Pathology, University of Michigan, Ann Arbor, MI, United States

Peter D. Caie
School of Medicine, QUAD Pathology, University of St Andrews, St Andrews, United Kingdom; Indica Labs, Albuquerque, NM, United States

Chhavi Chauhan
Director of Scientific Outreach, American Society of Investigative Pathology, Rockville, MD, United States

Jerome Y. Cheng
Department of Pathology, University of Michigan, Ann Arbor, MI, United States

Marion C. Cohen
CSAI, PA, United States

Stanley Cohen
Center for Biophysical Pathology, Rutgers-New Jersey Medical School, Newark, NJ; Perelman Medical School, University of Pennsylvania; Kimmel School of Medicine, Jefferson University, Philadelphia, PA, United States

Neofytos Dimitriou
School of Computer Science, University of St Andrews, St Andrews, United Kingdom

Michael Donovan
Department of Pathology, Icahn School of Medicine at Mount Sinai, New York, NY, United States

Gerardo Fernandez
Department of Pathology, Icahn School of Medicine at Mount Sinai, New York, NY, United States

Rama R. Gullapalli
Department of Pathology; Department of Chemical and Biological Engineering, University of New Mexico, Albuquerque, NM, United States

xvi Contributors

Rajarsi Gupta
Department of Biomedical Informatics, Stony Brook Medicine, Stony Brook, NY, United States

David Harris-Birtill
School of Computer Science, University of St Andrews, St Andrews, United Kingdom

Tahsin Kurc
Department of Biomedical Informatics, Stony Brook Medicine, Stony Brook, NY, United States

Richard Levenson
Professor and Vice Chair, Pathology and Laboratory Medicine, UC Davis Health, Sacramento, CA, United States

Zaibo Li
Department of Pathology, The Ohio State University Wexner Medical Center, Columbus, OH, United States

Abishek Sainath Madduri
Department of Pathology, Icahn School of Medicine at Mount Sinai, New York, NY, United States

Bahram Marami
Department of Pathology, Icahn School of Medicine at Mount Sinai, New York, NY, United States

David S. McClintock
Department of Laboratory Medicine and Pathology, Mayo Clinic, Rochester, MN, United States

Benjamin R. Mitchell
Department of Computer Science, Swarthmore College, Swarthmore, PA, United States

David Morrison
School of Medicine; School of Computer Science; Sir James Mackenzie Institute for Early Diagnosis, University of St Andrews, St Andrews, United Kingdom

Daniel A. Orringer
Associate Professor, Department of Surgery, NYU Langone Health, New York City, NY, United States

Liron Pantanowitz
Department of Pathology, University of Pittsburgh, Pittsburgh, PA, United States

Anil V. Parwani
Department of Pathology, The Ohio State University Wexner Medical Center, Columbus, OH, United States

Contributors **xvii**

Marcel Prastawa
Department of Pathology, Icahn School of Medicine at Mount Sinai, New York, NY, United States

Joel Haskin Saltz
Department of Biomedical Informatics, Stony Brook Medicine, Stony Brook, NY, United States

Richard Scott
Department of Pathology, Icahn School of Medicine at Mount Sinai, New York, NY, United States

H.R. Tizhoosh
Kimia Lab, Mayo Clinic, Rochester, MN, United States

Austin Todd
UT Health, San Antonio, TX, United States

John E. Tomaszewski
Pathology and Anatomical Sciences, University at Buffalo, State University of New York, Buffalo, NY, United States

Jack Zeineh
Department of Pathology, Icahn School of Medicine at Mount Sinai, New York, NY, United States

Preface

Artificial intelligence (AI) is a disruptive technology for pathology, as it has been for retail, banking, education, entertainment, social networking, and transportation, among other aspects of society. As noted in the preface to the first edition, pathology had been relatively slow to embrace this revolution. Digital pathology has really become established in the last decade or so. AI requires structured databases that only digital pathology can provide. In the short time since the previous edition, there has been an explosion of new techniques, new models, and new ways of combining diagnostic information from molecular to morphological biology and from single-cell analysis to spatial biology. The resulting explosion of the amount of data we can collect has made the implementation of AI for diagnostic, prognostic, and therapeutic purposes a necessity. The increasing use of AI in other fields, such as radiology, has also provided us with a wealth of relevant information to be incorporated. Additionally, we can use AI to look for important patterns in the massive data that has already been collected, that is, data mining both in defined datasets and in natural language sources.

A major barrier to the adoption of AI involves the need for physician buy-in and training and the education of a whole new generation of pathologists with the skills to understand and apply AI. In this regard, the evolution of molecular diagnostics was aided by the fact that molecular biology was already being routinely taught in medically oriented undergraduates and graduates. This is not the case for AI at present. There is therefore an unmet need for a book on AI that can serve as a text for both trainees and established pathologists. Unfortunately, texts on machine learning, deep learning, and AI (terms that are distinct but loosely overlap) tend to be either very general and superficial overviews for the layman or highly technical tomes assuming a background in linear algebra, calculus, and programming languages.

This book is designed to address these needs. It is the second edition of *Artificial and Deep Learning in Pathology*, which includes the original chapters with the addition of some reviews from a theme issue of *The American Journal of Pathology* on the same topic. This content has been updated where appropriate and integrated to provide a comprehensive introduction to the field. It is designed to (1) provide a working knowledge of the principles and methods of AI, (2) show how this can be implemented in the pathology laboratory, (3) provide a review of what has already been accomplished, and (4) point the way to the future. We cover the basic principles; advanced applications; challenges in the development, deployment, adoption, and scalability of AI-based models in pathology; the innumerous benefits of applying and integrating AI in the practice of pathology; and ethical considerations for the safe adoption and sustainable deployment of AI in pathology, as well as the emerging avenues for the readers to explore in the years to come. We touch on exciting new developments such as neuromorphic and quantum computing. Most importantly, a background in linear algebra, calculus, and computer programming is NOT necessary in order to understand the presented material.

Our target audience, in addition to trainees and newly minted pathologists, includes researchers, academicians, practitioners, policymakers, and administrators who need to know about the core AI principles and their applications in pathology. Because it covers the underlying principles of AI at a deep, though not mathematically explicit, level, it should be of use to individuals in other diagnostic imaging specialties such as radiology as well. The reader will find a significant amount of overlap in these chapters for the following reasons: each author reviewed some basic aspects of AI as they specifically relate to their assigned topic, each presents their own perspective on the underpinnings of the work they describe, and each topic considered can best be understood in the light of parallel work in other areas. The only place in which we attempted to strictly minimize overlap is in the early chapters that aim for a basic understanding of AI, with only limited references to the practical applications in pathology.

The ultimate insight that we wish to convey is that this transformative evolution of pathology will not lead to our digital replacements. Instead, there will be a partnership between man and machine, with the computer acting as a virtual pathology assistant. This will be a combination of both knowledge and wisdom in the service of both the clinical enterprise and translational research with the common goal of improving patient care.

Chhavi Chauhan
Stanley Cohen

Acknowledgments

This book is dedicated to the American Society for Investigative Pathology, and especially Bill Coleman, Mark Sobel, and Martha Furie, for supporting my efforts to proselytize on behalf of AI in pathology. Most importantly, it is also dedicated to my wonderful wife and scientific colleague Marion, children (Laurie, Ronald, and Ken), kids-in-law (Ron, Helen, and Kim, respectively), and grandkids (Jessica, Rachel, Joanna, Julie, Emi, Risa, Brian, Seiji, and Ava). One of the latter has already become a computer scientist, and another is well on her way. Marion, as both basic scientist and expert on institutional review, keeps me on track. I would also like to dedicate this book to Debbie Cohen, who ensures that I remain healthy enough to do this sort of thing. Last but not least, I want to acknowledge my computer, for without it, none of this could happen.

Stan Cohen

This book is dedicated to Dr. Stanley Cohen—a thought leader, pioneer, mentor, sponsor, well-wisher, and, most importantly, dear friend—who has been constantly guiding my path since I decided to embark on my journey to become an AI ethicist. Of course, despite his expert guidance, I could not have attained this phenomenal success had it not been for the unconditional love, constant support, and unimaginable inspiration that I have the unique privilege of receiving from my parents (Roop and Usha), my parents-in-law (Prashant and, especially, Umangi), my siblings (Bhavna and Harshita), my better half (Bhavin), and, most importantly, my guiding light, my daughter (Mohna). I look toward a hopeful future for humankind and thank each one of you readers for sharing your thoughts, wisdom, and knowledge to help us get there together.

Chhavi Chauhan

PART I

Principles

CHAPTER

The evolution of machine learning: Past, present, and future

1

Stanley Cohen

Center for Biophysical Pathology, Rutgers-New Jersey Medical School, Newark, NJ, United States; Perelman Medical School, University of Pennsylvania, Philadelphia, PA, United States; Kimmel School of Medicine, Jefferson University, Philadelphia, PA, United States

Introduction

The first computers were designed solely to perform complex calculations rapidly. Although conceived in the 1800s and theoretical underpinnings developed in the early 1900s, it was not until ENIAC was built that the first practical computer can be said to have come into existence. The acronym ENIAC stands for Electronic Numerical Integrator and Calculator. It occupied a 20×40 ft room and contained over 18,000 vacuum tubes. From that time until the mid-20th century, the programs that ran on the descendants of ENIAC were rule based in that the machine was given a set of instructions to manipulate data based on mathematical, logical, and/or probabilistic formulas. The difference between an electronic calculator and a computer is that the computer encodes not only numeric data but also the rules for the manipulation of these data (encoded as number sequences).

There are three basic parts to a computer: core memory, where programs and data are stored; a central processing unit that executes the instructions and returns the output of that computation to memory for further use; and devices for input and output of data. High-level programming languages take as input program statements and translate them into the (numeric) information that a computer can understand. Additionally, three basic concepts are intrinsic to every programming language: assignment, conditionals, and loops. In a computer, x = 5 is not a statement about equality. Instead, it means that the value 5 is assigned to x. In this context, though mathematically impossible, X = X + 2 makes perfect sense. We add 2 to whatever is in X (in this case, 5), and the new value (7) now replaces that old value (5) in X. It helps to think of the = sign as meaning that the quantity on the left is replaced by the quantity on the right. A conditional is easier to understand. We ask the computer to make a test that has two possible results. If one result holds, then the computer will do something. If not, it will do something else. This is typically IF-THEN statement. IF you are

Artificial Intelligence in Pathology. https://doi.org/10.1016/B978-0-323-95359-7.00001-7
Copyright © 2025 Elsevier Inc. All rights are reserved, including those for text and data mining, AI training, and similar technologies.

wearing a green suit with yellow polka dots, PRINT "You have terrible taste!" IF NOT print "There is still hope for your sartorial sensibilities." Finally, a LOOP enables the computer to perform a single instruction or set of instructions a given number of times. From these simple building blocks, we can create instructions to do extremely complex computational tasks on any input.

The traditional approach to computing involves encoding a model of the problem based on a mathematical structure, logical inference, and/or known relationships within the data structure. In a sense, the computer is testing your hypothesis with a limited set of data but is not using that data to inform or modify the algorithm itself. Correct results provide a test of that computer model, and it can then be used to work on larger sets of more complex data. This rules-based approach is analogous to hypothesis testing in science. In contrast, machine learning is analogous to hypothesis-generating science. In essence, the machine is given a large amount of data and then models its internal state so that it becomes increasingly accurate in its predictions about the data. The computer is creating its own set of rules from data rather than being given ad hoc rules by the programmer. In brief, machine learning uses standard computer architecture to construct a set of instructions that, instead of doing direct computation on some inputted data according to a set of rules, uses a large number of known examples to deduce the desired output when an unknown input is presented. That kind of high-level set of instructions is used to create the various kinds of machine learning algorithms that are in general use today. Most machine learning programs are written in PYTHON programming language, which is both powerful and one of the easiest computer languages to learn. A good introduction and tutorial that utilizes biological examples is Ref. [1].

Rules-based vs machine learning: A deeper look

Consider the problem of distinguishing squares, rectangles, and circles from each other. One way to approach this problem using a rules-based program would be to calculate relationships between area and circumference for each shape by using geometric-based formulas and then comparing the measurements of a novel unknown sample to those values for the best fit, which will define its shape. One could also simply count edges, defining a circle as having zero edges, and use that information to correctly classify the unknown.

The only hitch here would be the necessity to know pi for the case of the circle. But that is also easy to obtain. Archimedes did this somewhere about 250 BCE by using geometric formula to calculate the areas of two regular polygons: one polygon inscribed inside the circle and the other the polygon with which the circle was circumscribed (Fig. 1.1).

Since the actual area of the circle lies between the areas of the inscribed and circumscribed polygons, the calculable areas of these polygons set upper and layer bounds for the area of the circle. Without a computer, Archimedes showed that pi is between 3.1408 and 3.1428. Using computers, we can now do this calculation

Rules-based vs machine learning: A deeper look

FIG. 1.1

Circle bounded by inscribed and circumscribed polygons. The area of the circle lies between the areas of these bounds. As the number of polygon sides increases, the approximation becomes better and better.

Wikipedia Commons; Public domain.

for polygons with very large numbers of sides and reach an approximate value of pi to how many decimal places we require. For a more "modern" approach, we can use a computer to sum a large number of terms of an infinite series that converges to pi. There are many such series, the first of which have been described in the 14th century.

These rules-based programs are also capable of simulations; for example, pi can be computed by simulation. For example, if we have a circle quadrant of radius = 1 embedded in a square target board with sides = 1 unit and throw darts at the board so that they randomly land somewhere within the board, the probability of a hit within the circle segment is the ratio of the area of that segment to the area of the whole square, which by simple geometry is pi/4. We can simulate this experiment by computer by having it pick pairs of random numbers (each between 0 and 1) and , which is then using these as X-Y coordinates to compute the location of a simulated hit. The computer can keep track of both the ratio of the number of hits landing within the circle quadrant and the total number of hits anywhere on the board, which is then equal to pi/4. This is illustrated in Fig. 1.2.

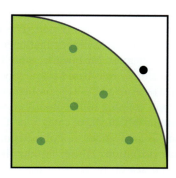

FIG. 1.2

Simulation of random dart toss. Here, the number of tosses within the segment of the circle is five, and the total number of tosses is six.

From Wikipedia Commons; public domain.

In the example shown in Fig. 1.2, the value of pi is approximately 3.3, based on only six tosses. With only a few more tosses, we can beat Archimedes, and with a very large number of tosses, we can come close to the best value of pi obtained by other means. Because this involves random processes, it is known as Monte Carlo simulation.

Machine learning is quite different. Consider a program that is created to distinguish between squares and circles. For machine learning, all we have to do is present a large number of known squares and circles, each labeled with its correct identity. From this labeled data, the computer creates its own set of rules, which the programmer does not give it. The machine uses its accumulated experience to ultimately correctly classify an unknown image. There is no need for the programmer to explicitly define the parameters that make the shape into a circle or square. There are many different kinds of machine learning strategies that can do this.

In the most sophisticated forms of machine learning, such as neural networks (deep learning), the internal representations by which the computer arrives at a classification are not directly observable or understandable by the programmer. As the program gets more and more labeled samples, it gets better and better at distinguishing the "roundness" or "squareness" of new data that it has never seen before. It uses prior experience to improve its discriminative ability, and the data itself need not be tightly constrained. In this example, the program can arrive at a decision that is independent of the size, color, and even regularity of the shape. We call this artificial intelligence (AI) since it seems more analogous to the way human beings behave than traditional computing.

Varieties of machine learning

The age of machine learning essentially began with the work of pioneers such as Marvin Minsky and Frank Rosenblatt, who described the earliest neural networks. However, this approach languished for several reasons, including the lack of algorithms that could generalize to most real-world problems and the primitive nature of input storage and computation during that period. The major conceptual innovations that moved the field forward were backpropagation and convolution, which we will discuss in detail in a later chapter. Meanwhile, other learning algorithms being developed also went beyond simply rule-based calculation and involved inferences from large amounts of data. These all address the problem of classifying an unknown instance based on the previous training of the algorithm with a known dataset.

Most, but not all, algorithms fall into three categories. Classification can be based on clustering of data points in an abstract multidimensional space (support vector machines), probabilistic considerations (Bayesian algorithms), or stratification of the data through a sequential search in decision trees and random forests. Some methods utilizing these techniques are shown in Fig. 1.3.

All of those latter approaches, in addition to neural networks, are collectively known as "machine learning." In recent years, however, neural networks have come

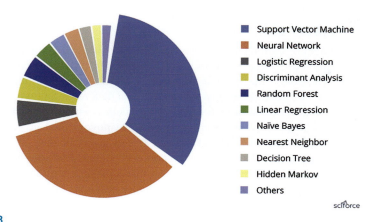

FIG. 1.3

Varieties of machine learning.
Reprinted from https://medium.com/sciforce/top-ai-algorithms-for-healthcare-aa5007ffa330, with permission of Dr. Max Ved, Co-founder, SciForce.

to dominate the field because of their near-universal applicability and ability to create high-level abstractions from the input data. Neural network-based programs are often defined as "deep learning." Contrary to common belief, this is not because of the complexity of the learning that takes place in such a network but simply due to the multilayer construction of a neural network as will be discussed below. For simplicity, in this book, we will refer to the other kinds of machine learning as "shallow learning."

The evolution of the neural network class of machine learning algorithms has been broadly but crudely based on the workings of biological neural networks in the brain. The neural network does not "think"; rather, it utilizes information to improve its performance. This is learning rather than thinking. The basics of all of these will be covered in this chapter. The term "Artificial Intelligence" is often used to describe the ability of a computer to perform what appears to be human-level cognition, but this is an anthropomorphism of the machine since, as stated above, machines do not think. In the sense that they learn by creating generalizations from inputted training data, and also because certain computer programs can be trained by a process similar to that of operant conditioning of an animal, it is easy to think of them as intelligent. However, as we shall see, machine learning at its most basic level is simply about looking for similarities and patterns in things. Our programs can create high levels of abstraction from building blocks such as these in order to make predictions and identify patterns. But "asking if a machine can think is like asking whether a submarine can swim" [2]. Nevertheless, we will use the term AI throughout the book (and in the title as well) as it is now firmly ingrained in the field.

General aspects of machine learning

The different machine learning approaches for chess provide good examples of the differences between rules-based models and AI models. To program a computer to play chess, we can give it the rules of the game, some basic strategic information along the lines of "it's good to occupy the center," and data regarding typical chess openings. We can also define rules for evaluating every position on the board for every move. The machine can then calculate all possible moves from a given position and select those that lead to some advantage. It can also calculate all possible responses by the opponent. This brute force approach evaluated billions of possible moves, but it was successful in defeating expert chess players. On the other hand, machine learning algorithms can develop strategically best lines of play based upon the experiences of learning from many other games played previously, much as prior examples improved a simple classification problem as described in the previous section. Early machine learning models took only 72 h of play to match the best rules-based chess programs in the world. Networks such as these learn to make judgments without implicit input regarding the rules to be invoked in making those judgments.

What most machine learning programs have in common is the need for large amounts of known examples as the data with which to train them. In an abstract sense, the program uses data to create an internal model or representation that encapsulates the properties of those data. When it receives an unknown sample, it determines whether it fits that model to "train" it. A neural network is one of the best approaches toward generating such internal models, but every machine learning algorithm does this in a certain sense. For example, a program might be designed to distinguish between cats and dogs. The training set consists of many examples of dogs and cats. Essentially, the program starts by making guesses, and the guesses are refined through multiple known examples. At some point, the program can classify an unknown example correctly as a dog or a cat. It has generated an internal model of "cat-ness" and "dog-ness." The dataset composed of known examples is known as the "training set." Before using this model on unlabeled, unknown examples, one must test the program to see how well it works. For this, one also needs a set of known examples that are NOT used for training but rather to test whether the machine has learned to identify examples that it has never seen before. The algorithm is analyzed for its ability to characterize this "test set."

Although many papers quantitate the results of this test in terms of accuracy, this is an incomplete measure. In cases where there is an imbalance between the two classes that are to be distinguished, it is completely useless. For example, overall cancer incidence for the period 2011–2015 was less than 500 cases per 100,000 men and women. I can create a simple algorithm for predicting whether a random individual has cancer or not by simply having the algorithm report that everyone is negative. This model will be wrong less than only 500 times, so its accuracy is approximately 99.5%! Yet this model is clearly useless, as it will not identify any patient with cancer.

As this example shows, we must not only take into account the accuracy of the model but also the nature of its errors. There are several ways of doing this. The simplest is as follows. There are four possible results for an output from the model: true positives (TP), false positives (FP), true negatives (TN), and false negatives (FN). There are simple formulas to determine the positive predictive value and the sensitivity of the algorithm. These are known as "precision" (Pr) and "recall" (Re), respectively. In some publications, an F1 value is given, which is the harmonic mean of Pr and Re.

Deep learning and neural networks

For many kinds of machine learning, the programmer determines the relevant features and their quantification. Features may include weight, size, color, temperature, or any parameter that can be described in quantitative, semiquantitative, or relative terms. The process of creating feature vectors is often known as feature engineering. There are some things that have too many features to do this directly. For example, a picture may have several megabytes of data at the pixel level. Prior to machine learning, image analysis was based upon specific rules and measurements to define features, and a rules-based program detected patterns based on edges, textures, shapes, and intensity to identify objects such as nuclei, cytoplasm, mitoses, necrosis, and so on, but interpretation of these was up to the human pathologist. Before neural networks, machine learning also needed humans to define features, a task often referred to as "feature engineering." These more powerful networks could make specific classifications and predictions from the image as a whole. Neural networks consisting of layers of simulated neurons took this a step further; a large part of the strength of these algorithms is that the network can abstract out high-level features from the inputted data all by itself. Those are based on combinations of initial data inputs and are often referred to as representations of the data. These representations are of significantly lower dimensionality than the original feature vectors of the input. The representations are formed in the network layers between input and output, referred to as hidden layers. Training refines these representations, and the program uses them to derive its own internal rules for decision-making based on them. Thus, the neural network learns rules by creating hidden variables from data.

Within the inner layers of the neural network, the process of abstraction leading to a representation is based on an iterative model often initiated by random variables and so is partially unexplainable to the human observer (which is why the term "hidden" is used for these inner layers). This lack of explainability is uncomfortable for some researchers in the field. However, if you were to ask an experienced pathologist how he or she arrived at a diagnosis based upon examination of a biopsy slide, they would find it difficult to explain the cognitive processes that went into that judgment, although if pressed, they could define a set of features consistent with that diagnosis. Thus, both neural networks and humans base their decisions on processes that are not readily explainable.

The simplest neural net has at least one hidden layer. Most neural nets have several layers of simulated neurons, and the "neurons" of each layer are connected to all of the neurons in the next layer. These are often preceded by a convolutional set of layers based loosely on the architecture of the visual cortex of the brain. These transform the input data into a form that can be handled by the following fully connected layer. There are many different kinds of modern, sophisticated machine learning algorithms based on this kind of neural network, and they will be featured predominantly throughout this monograph. Not all machine learning programs are based on these deep learning concepts. There are some that rely on different interconnections among the simulated neurons within them, such as so-called restricted Boltzmann networks. Others, such as spiking neural networks, utilize an architecture, the workings of which are similar to that of biological brains. While these represent examples of the current state of the art in AI, the basic fully connected and convolutional frameworks are simpler to train and more computationally efficient and thus remain the of choice as of 2019.

It should also be noted that it is possible to do unsupervised or weakly supervised learning where there is little to no a priori information about the class labels, and large datasets are required for these approaches as well. In subsequent chapters, we will describe these concepts in detail as we describe the inner workings of the various kinds of machine learning. Following these, we will describe how AI fits into the framework of whole slide imaging (WSI) in pathology. With this as a baseline, we will explore in detail examples of current state-of-the-art AI applications in pathology. This is especially timely in that AI is now an intrinsic part of our lives. The modern safety features in cars, decisions on whether or not we are good credit risks, targeted advertising, and facial recognition are some obvious examples. When we talk to Siri or Alexa, we are interacting with an AI program. If you have an iWatch or similar device, you can be monitored for arrhythmias, accidental falls, etc. AI's successes have included natural language processing and image analysis, among other breakthroughs.

AI plays an increasingly important role in medicine, from research to clinical applications. A typical workflow for using machine learning for integrating clinical data for patient care is illustrated in Fig. 1.4.

IBM's Watson is a good example of the use of this paradigm. Unstructured data from natural language is processed and outputted in a form that can be combined with more structured clinical information such as laboratory tests and physical diagnosis. In the paradigm shown here, images are not themselves interpreted via machine learning; rather, they are the annotation of those entities (diagnostic impression, notable features, etc.) that are used. As neural networks have become more sophisticated, they can take on the task of interpreting the images and, under pathologist and radiologist supervision, further streamline the diagnostic pathway. In Fig. 1.4, AI is the term used to define the whole process by which the machine has used its training to make what appears to be a humanlike judgment call.

The role of AI in pathology

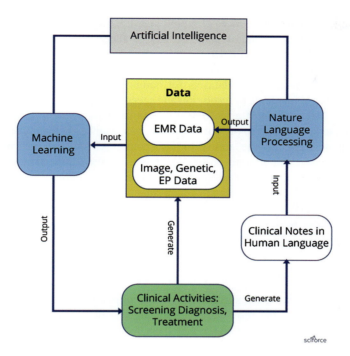

FIG. 1.4

The incorporation of artificial intelligence into clinical decision-making. Both precise, quantifiable data and unstructured narrative information can be captured within this framework.

Reprinted from https://medium.com/sciforce/top-ai-algorithms-for-healthcare-aa5007ffa330, with permission of Dr. Max Ved, Co-founder, SciForce.

The role of AI in pathology

It has been estimated that approximately 70% of a clinical chart is composed of laboratory data, and the pathologist is therefore an important interpreter of those data in the context of the remaining 30%. Even in the age of molecular analysis, the cognitive skills involved in the interpretation of image-based data by the pathologist remain of major importance in those clinical situations where diagnosis occurs at the tissue level. This has become increasingly difficult because of the complexities of diagnostic grading for prognosis, the need to integrate visual correlates of molecular data such as histochemical markers based on genes, gene expression, and immunologic profiles, and the wealth of archival material for comparative review. For these and many other reasons, research and implementation of AI in anatomic pathology, along with continuing implementation in clinical pathology, is proceeding at an exponential rate.

In many ways, the experiences of radiologists are helping to guide us in the evolution of pathology since radiologists continue to adopt AI at a rapid pace. Pathologists have been late in coming to the table. For radiologists, AI is mainly about analyzing images, and their images are relatively easy to digitize. In pathology, acquisition and annotation of datasets for computer input was not possible until WSI at the appropriate scale of resolution and speed of acquisition became available. Some of the drivers of WSI are efficiency, data portability, integration with other diagnostic modalities, and so on, but two main consequences of the move from glass microscopy to WSI are the ability to incorporate AI and the incorporation of computer-intensive imaging modalities. Although AI plays important roles in clinical pathology, in the interpretation of genomic and proteomic arrays, etc., where inference and prediction models are important, the focus of this book will be on the use of AI in anatomic pathology, with specific attention to the extraction of maximal information from images regarding diagnosis, grading, prediction, and so on.

Limitations of AI

While recognizing the great success of AI, it is important to note three important limitations. First, for the most part, current algorithms are one-trick ponies that can perform a specific task for which they have been trained, with limited capability of applying that training to new problems. It is possible to do so-called transfer learning on a computer, but it is still far removed from the kind of general intelligence that humans possess. Second, training an algorithm requires huge datasets that often number thousands of examples. In contrast, humans can learn from very small amounts of data. It does not take a toddler 1000 examples to learn the difference between an apple and a banana. Third, computation by machine is very inefficient in terms of energy needs compared with the human brain, even though the brain is much more complex than the largest supercomputer ever developed. A typical human brain has about 100 billion neurons and 100 trillion interconnections, which is much bigger than the entire set of neural networks running at any time across the globe. This has a secondary consequence. It has been estimated that training large AI models produces carbon emissions nearly five times greater than the lifecycle carbon emissions of an automobile.

There are several new approaches that have the ability to address these issues. The first is neuromorphic computing, which attempts to create a brainlike physical substrate in which the AI can be implemented instead of running solely as a software model. Artificial neurons, based upon the use of a phase-change material, have been created. A phase-change material is one that can exist in two different phases (for example, crystalline or amorphous) and easily switch between the two when triggered by laser or electricity. At some point, the material changes from a nonconductive to a conductive state and fires.

After a brief refractory period, it is ready to begin this process again. In other words, this phase-change neuron behaves just like a biological neuron. Artificial synapses have also been created. These constructs appear to be much more energy efficient than current fully software implementations such as both conventional neural networks and spiking net algorithms that more directly simulate neuronal behavior.

Much further in the future is the promise of quantum computing that will be orders of magnitude faster than the most powerful computers presently available. These are based on three-state "qubits" (0, 1, and an indeterminate superposition of these) rather than binary bits, and these states can hold multiple values simultaneously. A quantum computer can also process these multiple values simultaneously. In theory, this allows us to address problems that are currently intractable and could lead to major advances in AI. Although current quantum computers have only a small number of interacting qubits, it is conceivable that a quantum computer of the future could approach the complexity of the human brain with far fewer numbers of interacting qubits (unless, of course, the brain itself functions in part as a quantum computer, as some have speculated). This is not as far-fetched as it seems, as photosynthetic processes have been shown to exhibit quantum properties at physiological temperatures [3,4], thus demonstrating the feasibility of maintaining quantum coherence at room temperatures based on technologies that have had millions of years to evolve.

Even with current technology, however, both the potential and current applications of AI have demonstrated that it will play as powerful a role in pathology and medicine in general as it has already done in many other fields and disciplines. For both a good overview and historical perspective of AI, see Ref. [5].

General aspects of AI

For any use of AI in solving real-world problems, several practical issues must be addressed. For example:

a. For the choice of a given AI algorithm, what are the classes of learning problems that are difficult or easy, as compared with other potential strategies for their solution?
b. How does one estimate the number of training examples necessary or sufficient for learning?
c. What trade-offs between precision and sensitivity is one willing to make?
d. How does the computational complexity of the proposed model compare with that of competing models?

There is no current analytical approach to finding the best answer to such questions. Thus, the choice of algorithm and its optimization remains an empirical trial and error approach. Nevertheless, there are many aspects of medicine that have proven amenable to AI approaches. Some are "record-centric" in that they mine electronic health records, journal and textbook materials, patient clinical support systems, and so on, for epidemiologic patterns, risk factors, and diagnostic predictions. These all involve natural language processing. Others involve signal classification, such as the interpretation of EKG and EEG recordings. A major area of application involves medical image analysis. Just as they have for the digitization of images, our colleagues in radiology have taken the lead in this field with impressive results. Pathology has embraced this approach with the advent of WSI, which, in addition to its many other benefits, provides the framework in which to embed machine learning from biopsy specimens.

Because of the increasing importance and ubiquity of WSI, in the chapters to follow, there will be a focus on image interpretation, with some additional attention to the use of AI in exploring other large correlated data arrays such as those from genomic and proteomic studies. Interestingly, although genotype determines phenotype, using AI, we can sometimes work backward from image phenotype to genotype underpinning based on subtle morphologic distinctions that are not readily apparent to the naked eye. Natural language understanding, which is another major area of current AI research, is a vast field that requires its own monograph and is largely outside the scope of the present volume, though important for analyzing clinical charts, electronic health records, and epidemiologic information, as well as article and reference retrieval. By removing the mysticism surrounding AI, it becomes clear that it is a silicon-based assistant for the clinical and research laboratories rather than an artificial brain that will replace the pathologist.

References

[1] Jones MO. Python for biologists. CreateSpace Press; 2013.
[2] Dijkstra, EW. The threats to computing science (EWD-898) (PDF). Dijkstra, EW. Archive. Center for American History, University of Texas at Austin. [transcription].
[3] Fassioli F, Olaya-Castro A. Distribution of entanglement in light-harvesting complexes and their quantum efficiency. New J Phys 2010;12:8. http://iopscience.iop.org/1367-2630/12/8/085006.
[4] O'Reilly EJ, Olay-Castro A. Non-classicality of the molecular vibrations assisting exciton energy transfer at room temperature. Nat Commun 2014;5:3012.
[5] Sejnowski TJ. The deep learning revolution: artificial intelligence meets human intelligence. MIT Press; 2018.

CHAPTER

The basics of machine learning: Strategies and techniques

2

Stanley Cohen

Center for Biophysical Pathology, Rutgers-New Jersey Medical School, Newark, NJ, United States;
Perelman Medical School, University of Pennsylvania, Philadelphia, PA, United States; Kimmel
School of Medicine, Jefferson University, Philadelphia, PA, United States

Introduction

To begin, it is important to note that the term "artificial intelligence" is somewhat nebulous. It is usually used to describe computer systems that normally require human intelligence such as visual perception, speech recognition, translation, decision-making, and prediction. As we will see, pretty much all of machine learning fits under this broad umbrella. This chapter is a guide to the basics of some of the major learning strategies. To avoid repetition, I will use only hypothetical clinical examples that I am certain will not be covered elsewhere.

Machine learning refers to a class of computer algorithms that construct models for classification and prediction from existing known data, as compared to models based upon a set of explicit instructions. In this sense, the computer is learning from experience and using this experience to improve its performance.

The main idea of machine learning is that of concept learning. To learn a concept is to come up with a way of describing a general rule from a set of concrete examples. In abstract terms, if we think of samples being defined in terms of their attributes, then a machine language algorithm is a function that maps those attributes into a specific output. For most of the examples in this chapter, the output will involve classification, as this is one of the major building blocks for all the things that machine learning can accomplish.

In most cases, large datasets are required for optimal learning. The basic strategies for machine learning in general are (1) supervised learning, (2) unsupervised learning, and (3) reinforcement learning. In supervised learning, the training set consists of data that have been labeled and annotated by a human observer. Unsupervised learning does not use labeled data and is based upon clustering algorithms of various sorts. In both of these, improvement comes via sequential attempts to minimize the

---

Artificial Intelligence in Pathology. https://doi.org/10.1016/B978-0-323-95359-7.00002-9
Copyright © 2025 Elsevier Inc. All rights are reserved, including those for text and data mining, AI training, and similar technologies.

16 **CHAPTER 2** The basics of machine learning: Strategies and techniques

difference between the actual example (ground truth) and the predicted value. In reinforcement learning, the algorithm learns to react to its environment and finds optimal strategies via a reward function.

Although we tend to think of deep learning and neural networks as the newest forms of machine learning, this is not the case. The primitive ancestors of modern neural network architectures began with the work of Rosenblatt [1] in 1957. These neural networks were loosely based upon the model of neuronal function described by McCulloch and Pitt [2] about a decade earlier (see Ref. [3] for a review of the history of deep learning). Basically, it involves a sequence of simulated neurons arranged in layers, with the outputs of one layer becoming the input of the next. Each layer thus receives increasingly complex input from the layer beneath it, and it is this complexity that allows the algorithm to map an unknown input to an outputted solution. Note that the term "deep learning" does not refer to the sophistication of the program or to its capacity to mimic thought in any meaningful way. Rather, it simply refers to the fact that it is built of many layers of nonlinear processing units and their linear interconnections. At its most fundamental level, deep learning is all about linear algebra with a little bit of multivariate calculus and probability thrown in.

Meanwhile, a number of other models of machine learning were being developed, which are collectively referred to, by many though not all, as shallow learning. Just as is the case of deep learning, in all of these models, the program is presented with a large dataset of known examples and their defining features. In the simplest case, there are two classes of objects in the training set, and the program has to construct a model from these data that will allow it to predict the class of an unknown example. If the two classes are bedbugs and caterpillars, the useful features might be length and width. To distinguish between different faces, a large number of different morphometric features might have to be determined. The development of these architectures progressed in parallel with that of neural networks. A major advantage of neural networks is that they can derive their own features directly from data (feature learning), while shallow learning requires the programmer to explicitly create the feature set (feature engineering). Another advantage of deep learning is that it can be used for reinforcement learning in which the algorithm learns to react to an environment by maximizing a reward function.

In spite of this obvious advantage, the field of deep learning languished for many years, partially because the early models did not live up to the hype (they were billed as thinking machines) but mostly because the computational power and storage capacities available were insufficient until relatively recently. Deep learning tends to be more computationally "needy" than shallow learning.

In all forms of machine learning, the computer is presented with a training set of known examples. Once the program has been trained, it is presented with a validation set. This is another set of known examples. However, this is withheld until training is complete. At that point, the program is tested with the examples in the validation set, to determine how well it can identify these samples that it has never seen. It is important to note that accuracy alone is not a valid measure of performance. As an example, if I know that the incidence of a disease is 1%, I can design a simple program

Shallow learning   **17**

with an accuracy of 99%. Simply call every example healthy. Obviously, this is not very useful. Since accuracy does not give a complete picture, two other measures are usually used: precision and recall. For a typical classification problem, "precision" is defined as the number of true positives divided by the sum of true and false positives. It is also known as the positive predictive value (PPV) and is essentially a measure of a classifier's exactness. In other words, what proportion of positive identifications was actually correct? "Recall" is the number of true positives divided by the sum of true positives and false negatives. This is also known as sensitivity. There are a number of other ways of determining how good a given machine learning model is, including (1) the F1 score, which is twice the product of precision and recall divided by the sum of precision and recall, and (2) AUC (area under the curve) which estimates that a classifier will rank a randomly chosen positive instance higher than a randomly chosen negative instance. Finally, a confusion matrix allows visualization of the performance of an algorithm. Each row of the matrix represents the instances in a predicted class, while each column represents the instances of an actual class. These concepts are important in allowing one to evaluate results in the published literature and compare the efficacy of different models of machine learning applied to the same dataset.

With a bit of simplification, we can say that the basic aim of a machine learning strategy is to discover differences and/or similarities as well as patterns among sets of things. This can lead to high-level behaviors that mimic biological intelligence such as finding associations and correlations among patterns, interpreting and acting upon sensory information, processing language, and so on. While, in general, the datasets for machine learning must be large, it is impossible to overstate the importance of the quality of that data. For example, a study of patients with sepsis at a Chicago hospital (see Ref. [4]) concluded that patients with low pH levels in their blood were less likely to return to the hospital soon after being discharged! This is counter to the known fact that metabolic acidosis in sepsis is associated with increased risk. It turned out that the data included patients who died during their hospital stay and who were obviously not candidates for readmission. The patients least likely to be readmitted were the ones who had already been discharged to the funeral home. When the deceased were excluded, it turned out that patients with low pH were in significantly greater danger than those with pH closer to normal, as expected. This was a classic example of data bias.

Shallow learning

Shallow learning algorithms can be broadly classified as geometric models, probabilistic models, and stratification models. Since much of machine learning is based on linear algebra, it makes sense that the data itself is best represented as a matrix (Fig. 2.1). This is also a data representation for deep learning.

In this example, there are nine samples representing two distinct classes. Each sample is defined in terms of its feature set (in this example, [4]). In machine learning

18 **CHAPTER 2** The basics of machine learning: Strategies and techniques

Sample	Feature 1	Feature 2	Feature 3	Feature 4	CLASS
1	a1	b1	c1	d1	Class 1
2	a2	b2	c2	d2	Class 1
3	a3	b3	c3	d3	Class 2
4	a4	b4	c4	d4	Class 1
5	a5	b5	c5	d5	Class 2
6	a6	b6	c6	d6	Class 2
7	a7	b7	c7	d7	Class 1
8	a8	b8	c8	d8	Class 1
9	a9	b9	c9	d9	Class 1

FIG. 2.1

A dataset in matrix form. Each feature set is written as a letter with a number that shows which sample it belongs to. Note that these need not be unique; for example, a1, a3, and a7 can be identical. In fact, there *must* be shared feature values for learning to occur. In this dataset, four attributes define each sample. When samples in a training set are presented to a machine learning program, they are generally arranged in random order.

jargon, samples (or examples) are also known as "instances," and features are also known as "attributes." The set of features for a given sample is called its "feature vector." In this context, a vector is defined as an ordered collection of n elements rather than as a directed line segment. The features (a1, a2, etc.) can be anything that can be amenable for computer manipulation. If one feature is color, we would have to define that feature by assigning each color to a different number (red would be coded as 1, blue as 2, and so on). We can refer to the dimensionality N of the feature vector (here 4) or, in some cases, the dimensionality of the entire dataset. In order for this to be a usable training set, no sample can have a completely unique feature set since if that were the case, the computer would have no basis for generalization to new unknown samples; all it could do would be to memorize the list.

Geometric (distance-based) models

These fall into the category of clustering algorithms. They will be discussed in detail as to illuminate many of the concepts involved in more complicated forms of machine learning.

The best way to begin to talk about clustering is with an example. Bloodletting was a widely accepted therapeutic practice for a variety of diseases beginning about 3000 years ago and still is practiced for conditions such as polycythemia vera and hemochromatosis [5]. Back in the good old days of medicine, this was accomplished by the use of bloodsucking leeches. There are both bloodsucking leeches and non-bloodsucking leeches (I assume the latter are vegetarians). You are given the unenviable task of distinguishing between these two different species when presented with an unknown sample. Luckily, the leeches can be identified morphologically as well as functionally, and the two classes are of different sizes. You accept the challenge, with the proviso that you also will be given a list consisting of a number of different samples of each, with their respective weights and lengths as known examples. This will be the training set. We will ignore the validation step in this example.

You are then presented with the unknown leech. Unfortunately, you do not remember how to calculate standard deviations, so high school statistics is out. A good approach might be to plot the lengths and weights of your known samples (instances) as seen in Fig. 2.2, where reds and blues correspond to the two different classes. The colored shapes on the graph are the examples, and their coordinates represent their attributes. In this example, it is possible to draw a straight boundary line exactly separating the classes. All that is now necessary is to weigh and measure the unknown leech and see on which side of the line it falls, which tells you which of the two leech classes it belongs to. This is a geometric model since the distance between two points tells you how similar they are. It has learned from data how to predict the class of an unknown sample from its attributes. While this may seem needlessly complicated for a situation with samples such as this that have only a few attributes and fit into only two classes, consider the more general case where there may be multiple distinct classes, and the number of attributes may run into the hundreds or thousands.

Unfortunately, life is not as simple as has just been described. There are obviously many possible lines that can separate the two classes, even in this simple example. How do we treat novel samples that are on or close to the line? In fact, how do we know where to draw the line? What happens if we cannot find a simple line, i.e., there might be a nonlinear boundary? Finally, is it possible to extract information from the data if we do not have a priori knowledge of the classes of our samples (so-called unsupervised learning)?

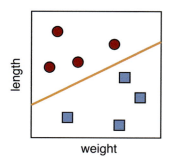

FIG. 2.2

Plot of length and weight of the two different leech species.

Copyright © 2018 Andrew Glassner under MIT license.

To treat novel samples on or close to the line, we look for a point in the training space closest to the novel point and assume that they are in the same class. However, the points may be exactly halfway between two points in two different classes, the test point itself might be inaccurate, and the data might be "noisy." The solution is the **K-nearest neighbor algorithm (KNN)**, which, though simple, is a strong machine learning strategy. Just use several (K) nearest neighbors instead of one and go with the majority, or use an average or weighted average.

It has already been noted that this may, at first glance, seem no better than a rules-based algorithm involving statistical analysis. However, consider the situation where the number of attributes (feature vector) has hundreds or thousands of attributes rather than just two.

Even with more complex but still linearly separable plots, there are often many different ways to draw the boundary line, as illustrated in Fig. 2.3A. In fact, there are an infinite number of linear separators in this example. The best line is the one that provides the greatest separation of the two classes. For this purpose, we obviously need to consider only the closest points to the line, which are known as support vectors. Using these, the algorithm creates a line that is farthest away from all the points in both the clusters (Fig. 2.3B). The algorithm that does this is known as a **support vector machine (SVM)**.

An SVM can do much more than finding optimal linear boundaries. It can even transform nonlinear data to a linear form by transforming the data in a new coordinate system in a higher dimensional space by means of a kernel function, which itself involves only linear operations. This is known as the "kernel trick."

Fig. 2.4 shows, on the right, an example of two classes without a linear boundary that can separate them. In this example, there are two classes, each with two defining attributes. Clearly, there is no linear boundary that can separate them. In this case, a circle (nonlinear boundary) will do the job. However, if you imagine pulling one class up into the third dimension, as shown on the left of Fig. 2.4, then it is easy

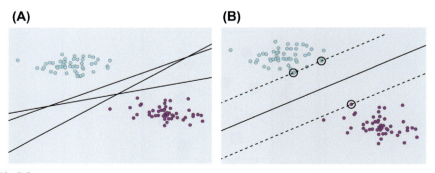

FIG. 2.3

(A): Possible linear separators for the clusters shown. (B): Support vectors are used to construct the line that best separates them.

Copyright © 2018 Andrew Glassner under MIT license.

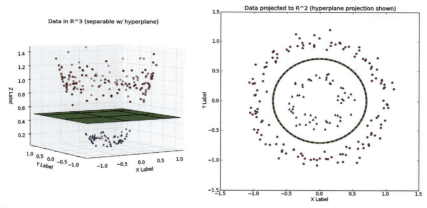

FIG. 2.4

Creating a linear separator for the data by adding a dimension to it. The separating plane shown in the 3D space on the left is analogous to a line drawn in 2D space. In general, for any N-dimensional space, a linear construct such as this is defined as a hyperplane.

Reprinted with permission from Eric Kim (www.eric-kim.net).

to see that a flat plane can divide them. In general, an SVM can raise the dimensionality of the data space until an $(N-1)$ hyperplane can separate the classes. In this example, the two-dimensional plot is transformed into a three-dimensional space such that an $N = 2$ hyperplane can provide a boundary.

The K-means algorithm (KM)

This is also a classic clustering algorithm, but it is discussed in a separate section because it is important not to confuse it with KNN, and so they are presented separately for pedagogical rather than conceptual reasons. This is another important distinction, in that unlike the KNN, it involves unsupervised learning. As such, it has important applications not only on its own but also in conjunction with some deep learning methods.

The K means algorithm attempts to find clustering in unlabeled data. One starts by making a guess as to how many clusters there are (this will be refined later). A centroid is the arithmetic mean position of all the points in a cluster. Next make a random guess as to where the central points of each cluster are so as to create k pseudo-centers. These should be reasonably distant from each other but need not have relationship to the actual clusters and can be chosen randomly (unless there is some prior knowledge about the data structure). Next assign each point of the dataset to its nearest pseudo-center. We now form clusters with each cluster comprising all of the data points associated with its pseudo-center. Next update the location of each cluster's pseudo-center so that it is now indeed in the center of all its members. Do this recursively until there are no more changes to cluster membership.

We can determine the best choice for K by plotting within-cluster scatter as a function of the number of clusters. As an extreme example, if K equals the number of data points, there is obviously no variance within each cluster, as they each consist of one element. The smaller the number of clusters, the greater the variance within the clusters. When variance is plotted as a function of K, the equation is reminiscent of exponential decay, and it is easy to pick a value of K such that larger values do not appreciably change the overall variance. This is analogous to the so-called "scree plots" used in principal component analysis (PCA), described in a separate section below.

Probabilistic models

In the previous section, the basic building blocks of clustering and distance were used to learn relationships among data and to use those relationships to predict class labels. We can also approach this problem by asking the following question: given the feature set for an unknown example, what is the probability that it belongs to a given class? More precisely, if we let X represent the variables we know about, namely our instance's feature values, and Y the target variables, namely the instance's class, then we want to use machine learning to model the relationship between X and Y.

Since X is known for a given instance, but Y is not (other than in the training set), we are interested in the conditional probabilities: what is the probability of Y given X. This is another way of saying that given a sample's feature set (X), predict the class it belongs to (Y). In mathematical notation, this is written as $P(Y|X)$, which is read as "the probability of Y given X." As a concrete example, X could indicate the phrase "my fellow Americans," and Y could indicate whether the speaker is a member of the class politician (thus using the words "class" and "politician" in the same sentence, possibly for the first time).

The classification problem is thus simply to find $P(Y|X)$. The probabilities of several kinds of occurrence can be calculated directly from the training set. In the data matrix in Fig. 2.1, for any given feature X, we can simply count the number of times its value appears divided by the total number of variables, giving $P(X)$. For any given value of Y, we can count similarly the ratio of the number of times that class appears relative to the overall number of classes. In this manner, we can calculate values for the sets X,Y to get $P(Y)$ and $P(X)$.

There is one more thing we can do, and that is to calculate $P(X|Y)$, the probability of the feature, given the class, which is simply based on looking only at that class and determining how many times the feature value X appears in samples of that class in the training set. We now know $P(X|Y)$. Given this information, we can calculate the probability of the class given our knowledge of its feature set, $P(Y|X)$, by means of the easily proven Bayes theorem (the only equation used in this chapter):

$P(Y|X) = (P(X|Y)*P(Y))/P(X)$ which, if a Venn diagram is drawn, becomes intuitively obvious.

In practice, this formula is too simple since in order to make predictive assignments from attributes to class, each sample must have more than one attribute, and in real-world problems, many more than one. The procedure for dealing with multiple attributes is more complicated [6]. In any case, the Bayes theorem is a formulation of one of the basic rules for conditional probabilities. However, the problem with computing probabilities in this way is that some of the features in a feature set may be dependent on each other (for embryos, mass and cell number are dependent variables), which makes the calculations difficult. In practice, this complication is ignored, and the features are treated as if they were independent. This algorithm is known as **Naïve Bayes,** where naïve means that we are foolish enough to assume independence. Surprisingly Naïve Bayes works remarkably well in spite of this limitation, and so it has become an important machine learning model. There are much more complex models that deal with dependency relationships; these are known as Bayesian optimal classifiers. There is also a middle ground that is based on the idea that some things are conditionally dependent but some things are not.

These approaches fall into the framework of probabilistic graphical models that are known as Bayesian Belief Networks or Directed Graphical Networks. These are networks that represent a set of variables and their conditional dependencies via the use of a directed graph. A simple network of this kind is shown in Fig. 2.5. Rain influences whether the sprinkler is turned on, and both rain and the sprinkler influence whether we get wet. Bayesian Belief Networks are beyond the scope of this chapter, but additional information may be found in Ref. [7]. All

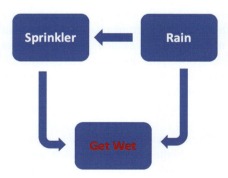

FIG. 2.5

Relationships among three objects. Whether it is raining or not determines whether the sprinkler gets turned on, but not vice versa, so this is a directed graph. Similarly, either rain or the sprinkler will get something wet, but the state of wetness does not cause either rain or the sprinkler to happen, so these arrows are also directional.

24 **CHAPTER 2** The basics of machine learning: Strategies and techniques

these algorithms learn from probability distributions within their sample space rather than spatial metrics, as was the case for SVMs and other geometrical models.

All machine learning algorithms have problems of some sort. For example, KNN does not handle large sets of multivariate data very well. Naïve Bayes works on discrete attributes, but only if there are enough examples to get good estimates for all the probabilities that have to be estimated from the training dataset.

Decision trees and random forests

The Decision Tree algorithm is a simple classic supervised learning model that works surprisingly well. The classifiers discussed above all expect attribute values to be presented at the same time. To make a diagnosis, we usually begin with symptoms, then obtain historical information and look for signs based on physical examination. Based upon this sequence of events, laboratory tests and radiologic examinations are ordered. Within the laboratory there may be reflexive testing; an initial test result may influence the choice of the next set of tests to be performed. Thus, information may be accumulated sequentially. A decision tree will be given its data all at once but utilize it in a sequential manner such as that described above. The game of "twenty questions" is a good example of a binary decision tree.

A decision tree uses a treelike graph that represents a flow-chart-like structure in which the starting point is the "root." Each internal node of the tree represents a test on an attribute or subset of attributes. Each branch from the node represents the outcome of the test, and the final node is a "leaf" that represents a class label. One can construct a simple decision tree by hand. However, it is easy to design an algorithm that learns the tree from data. As is the case for other kinds of machine learning, in the decision tree approach supervised learning takes labeled examples and builds a classifier by computing the sequence of branch choices. Decision trees, as the name implies, can be used for decision-making as well as classification. In that case, each of the classes represents different decision choices; for example, should I complete this chapter or watch the political debate? An example of a decision tree is shown in Fig. 2.6, which predicts an output based on a set of binary decisions. The tree proceeds from left to right.

There are lots of different ways that a decision tree can be constructed from the dataset, based upon which attributes to choose for each node, and what conditions are to be used for splitting at that node. The problem is to look at the available attributes and pick whichever is best to make a decision fork at each branch of the tree. The algorithm chooses "best" as that choice that maximizes the information gain for that

Decision trees and random forests

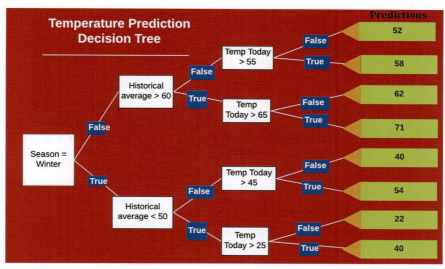

FIG. 2.6

A decision tree for predicting tomorrow's maximum temperature. Seattle, 12/27/2017. From Koehrsen, Random forest simplified. In this tree, there is a binary decision made at each branch, so each parent node gives rise to two children nodes.

https://medium.com/@williamkoehrsen/random-forest-simple-explanation-377895a60d2d.

step. It does so by the use of equations from information theory that are based on the concept of entropy and attempts to minimize entropy. Unlike the situation in physics, entropy in information theory is a measure of how random a set of things are. A set of things that are all the same has low entropy, and a set of things that are completely random has high entropy. Trying to minimize entropy is a good way of generating decision trees. However, measures other than entropy can also be used; in practice, they tend to perform similarly. The process is terminated when there are no attributes left to test on, or when the remaining examples are of a single class. When it gets to that node the program returns a class label.

One can construct decision trees for a given dataset in a variety of ways: (a) using different subsets of the data for different trees, (b) using different decision criteria, (c) limiting the depth of each tree, and so on. For (a) a technique called bagging is commonly used. In bagging, multiple new training subsets are taken by sampling from the main dataset with replacement after each sample is generated. Each subset is used in a different tree. A **Random Forest** consists of a set of several of such decision trees, with majority voting or some other consensus criterion determining final output values (Fig. 2.7).

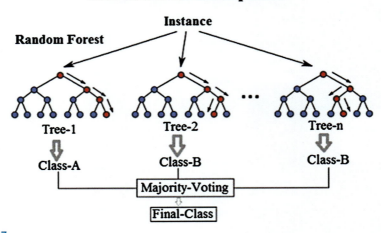

FIG. 2.7

A random forest composed of three decision trees. Koehrsen, Random forest simplified.
https://medium.com/@williamkoehrsen/random-forest-simple-explanation-377895a60d2d.

The curse of dimensionality and PCA

Intuitively, adding more features to the data would seem to improve our ability to classify it, but in many applications, this can make things worse [8]. As the dimensionality of the feature set increases, the space in which the samples reside increases with the power of the dimension, and so the sample density at any given point falls below that which is needed for an accurate prediction. The obvious solution is to get more data, but this may not always be possible in practice. There are techniques for dimensional reduction that minimize the loss of information that comes from this reduction. The classic example from statistical models is known as PCA, which was first described in the 1930s, way before the onset of machine learning. PCA operates by finding the direction of maximum variance in the data. This is the dimension along which the data are most spread out and is the first principal component. The next one is the line perpendicular to this component that has the next best spread of the data, and so on. This process is repeated until there are as many principal components as the original data have dimensions. In essence, it is simply a linear transformation of the data from one n-dimensional space to another n-dimensional space that retains the information in the original set. In order to reduce the dimensionality of this transformed set, we must discard some of these components. Since the n components have already been ordered according to importance, if we want to project down the k dimensions ($k < n$), we just keep the first k components and discard the rest. We can plot the loss of information by a graph showing variance as a function of component number (in mathematical terms, the curve shows the eigenvalues of the covariance matrix on the y-axis and the component number on the x-axis). This construct is known as a scree plot.

Deep learning and the ANN

Shallow learning works well in those situations where one can explicitly define the feature vectors of the problem. Problems involving image interpretation and classification, as well as those requiring natural language processing, are examples where these models perform poorly because of the complexity of the data. The human brain is very good at this because it makes use of a series of processing levels or "layers," each one doing a more complex analysis than the previous one. Deep learning computational approaches patterned on the brain deal with this issue very well. Although not the only form of deep learning, neural network-based models are the ones most currently employed.

As we shall see, ANNs develop increasingly complex internal representations of the data and, in the process, automatically extract the relevant salient feature vectors of the dataset. In this chapter we will cover the basics of fully connected and convolutional networks. More advanced architectures will be covered in the next chapter. The aim is to present the inner workings of ANNs in detail (the second-floor view), rather than merely providing the descriptive 30,000 feet overview that appears in most of the medical literature, while avoiding the underlying mathematics and programming details that become apparent at ground level. There is only one equation in the entire exposition, and it is easily derived. It should be noted that there are many other deep learning algorithms that are not based on connected layers in linear sequences such as those described here (e.g., Boltzmann machines). As these have only rarely been employed in pathology applications, they will not be covered here.

In the examples provided, the emphasis will be placed on image interpretation, but the principles can also be applied to genomics and expression arrays, as well as natural language processing. Natural language processing and semantic understanding are not covered; currently, state-of-the-art programs such as IBM's Watson and similar efforts have proven disappointing in the clinical setting. The focus will be entirely on supervised models. Just as there are shallow network models for unsupervised learning, there are a number of unsupervised deep learning strategies, and some of these will be covered in later chapters.

Neuroscience 101

As indicated above, models for deep learning are (very) loosely based upon the neural network architecture of the mammalian brain and how the neurons within the networks comprising the brain process signals both inside the neuron and from one neuron to another. The first mammalian neural nets came into being about 200 million years ago. Through a process of "continuous performance improvement," we now have the cellular structure shown in Fig. 2.8.

The neuron receives signals at its dendritic end, usually from multiple other neurons. Those signals are integrated within the cell, and if they rise past a threshold value, they are passed through the axon as an action potential. Signals are propagated via electrical pulses or spikes generated via the modulation of ion channels at the cell

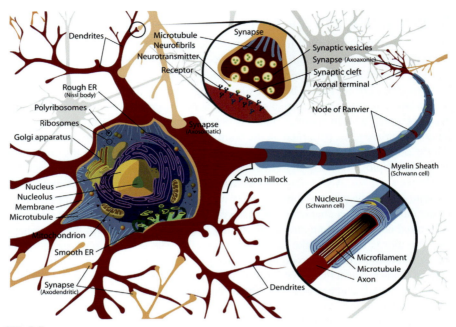

FIG. 2.8

Schematic of a typical neuron. The cell can receive signals from many other neurons via its dendrites. If the sum of these exceeds a threshold value, the neuron passes the integrated signal onto other neurons. The human brain has approximately 100 billion neurons. LadyofHats, Creative Commons.

membrane and are transduced across the synapses that separate the neurons via electrochemical reactions involving a variety of chemical mediators. Each action potential is a simple on-off burst of electrical energy. The only exception is for sensory neurons that terminate in a transducer. For example, in the Pacinian corpuscle, the action potential varies with stimulus intensity.

Since each dendritic input interacts with the others via multiple interacting electrochemical events, there may be a nonlinear sum of inputs, especially since the spatial separation of the inputs affects their interaction. For this reason, the neuronal threshold for firing can be an all-or-none event (in mathematical jargon, this is a Heaviside ramp function), while the total process is nonlinear. Moreover, information can move backward in a neuron through current flow. In other words, the neuron has an internal feedback system. A single backpropagating action potential can activate slow dendritic voltage-gated ionic currents, which in turn flow back toward the spike initiation zone in the axon, resulting in additional action potentials. Effectively this gives us a two-layered neural network that resides in a single neuron. There are experiments that show that directional selectivity and coincidence detection may be implemented even at this basic computational level.

The pulses of electrical activity that mediate interneuron signaling can be modulated on the basis of pulse frequency, pulse width, or pulse arrival time in relation to other incoming signals. A neuron can receive either excitatory or inhibitory signals. Additional modulation of signals comes at the synapses between two brain cells. A synapse between two neurons is strengthened when the neurons on either side of the synapse (input and output) have highly correlated outputs. This is known as Hebbian learning. In essence, this means that if one neuron repeatedly excites another, then the connection between them is facilitated. Hebbian learning allows modification of the flow of information within the network. In the brain, the synaptic connections are constantly being remodeled based upon the integrated set of signals they are responding to at any given moment. Remodeling is not well understood but appears to be based on the accumulation and depletion of chemical gradients. Some synapses can be completely eliminated via interaction with astrocytes or glia, presumably via cytokines.

This picture is further complicated in that the neuron, taken as a whole, can show bistable behavior. Bistability is a property of a dynamical system that exhibits more than one stable point. A model proposed by Lowenstein and Sompolinsky [9] involves bistability with respect to the dendritic concentration of Ca^{++}. If it is forced to be in the up state at one end of the dendrite, and in the down state at the other, then a standing wave of Ca^{++} concentration is created instead of a gradient. As long as no further input occurs, the wave front will maintain its position. This model is very important. Although the process of activating and firing a neuron occurs at very short time scales, working memory requires longer time scales, and this model is one simple way of integrating transient effects into a persistent trace, which is essentially a definition of working memory.

Since each neuron receives inputs from many other neurons and transmits its output to many other neurons, the neurons essentially work as aggregators; the total amount of output is controlled by the total input from all its dendritic connections. Since the axon can also connect to multiple neurons, this allows a complex, multilayer structure to develop. The human brain has approximately 100 billion neurons. Even though arranged as subsystems, there are huge numbers of possible interconnections.

The rise of the machines

After approximately 1.9 million years and 3 months (author's note: It was 1.9 million when this chapter was begun, 3 months ago) of brain evolution, human scientists began to model the brain. Rosenblatt's single-layer perceptron [1] was based upon the first mathematical model of a neuron by McCullough and Pitts [2], the MCP neuron. This was inspired by the Hebbian concept of synaptic plasticity based on the adaptation of brain neurons during the learning process. Applying different relative weights to the inputs affects the output, so Rosenblatt developed a supervised learning algorithm for this modified MCP neuron that enabled it to figure out the correct

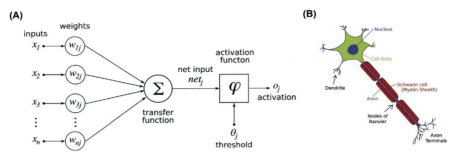

FIG. 2.9

An artificial neuron (which is an abstraction in a computer program rather than a physical entity) is diagrammed in (A). Each of the inputs *x* is modified by a weight *w* and then summed just as the synaptic inputs would be in a real neuron. This sum Σ is passed to an activation function, ϕ and if it exceeds the value of the function's threshold θ it is passed on as an output o. In this schematic (Σ and ϕ) together are the analogs to a biological neuron. Chrislb, Creative Commons. For comparison, a biological neuron is shown in (B).

(NickGorton-commonswiki).

weights by learning from data all by itself. Such an artificial neuron, which is a mathematical abstraction rather than a physical construct, is diagrammed in Fig. 2.9 and compared to a simple representation of a biological neuron.

This artificial neuron, which is the basic building block of most of the artificial networks of deep learning, is a great oversimplification in the following ways. Since the mathematical analysis of pulse-based coding is difficult, the model neuron receives as input a single steady value at each dendrite that is connected to an upstream neuron (there are spiking artificial neural networks (ANNs), but they are not in widespread use). The signal variability during learning is represented by a set of weights associated with the signals, and these weights change as the neural network learns. Another important difference between a biological and a modern ANN is that in the former, the threshold is based upon a simple off-on ramp, whereas, in the latter, a smooth curve such as a sigmoid or similar kind of function is used.

Initially, sigmoid functions were used as the transfer function. The curve for a typical sigmoid function is shown in Fig. 2.10. The sigmoid's value ranges from 0 to 1. The sigmoid is a continuous, nonlinear, differentiable function. Note that, unlike the Heaviside, the activation threshold can be any value we choose in its range. Current ANNs often use other functions, such as the hypertangent function or the rectified linear function (ReLU) as well. As we shall see, in a neural network weights are adjusted by a process that involves finding the minimum of a loss function through the use of calculus, and so the threshold must be a continuous function. Additionally, the function must allow the network to deal with nonlinear relationships. One of the criticisms of the earlier neural networks shortly after they were

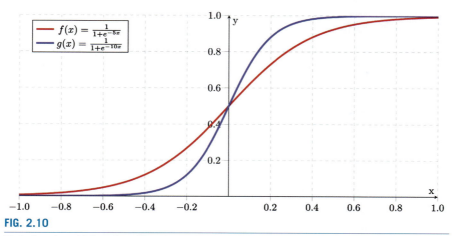

FIG. 2.10

Plot of a sigmoid function. The slope of the curve can be adjusted by the choice of the multiplier for x in the equation. The curve's bounds on the x-axis are from 0 to 1, thus making it a reasonable substitute for a ramp activation.

M. Thoma, Wikimedia Commons.

published was related to their inability to deal with nonlinear relationships. Although there was a great deal of hype surrounding the Rosenblatt perception, in 1969, Minsky [10] published a book that not only showed how limited Rosenblatt's perception (and indeed, any single-layer perceptron) actually is but also led to an intellectual climate change among the "artificial intelligentsia," namely the so-called AI-winter in the 1980s. However, major contributions by many investigators brought back ANNs into the forefront of AI research, based on the multilayer concept, nonlinear activations, backpropagation, and convolution. The beauty of a modern ANN and its many advanced variants that will be covered in the next chapter is that it can represent the output of any continuous bounded function and thus can mimic the output of any algorithmic structure that uses mathematics for its underlying implementation (i.e., most, if not all of them). Thus, in addition to its own unique abilities, an ANN can do anything a shallow learning network can do, but for data that are not too complex, shallow learning can be more computationally efficient. However, these modern artificial networks, in evolving beyond these humble beginnings, have become truly sophisticated models employing concepts from linear algebra, multivariate calculus, mathematical statistics, game theory, reinforcement learning, and evolutionary theory. In order to emphasize how much further we have to go to achieve true general intelligence in an artificial substrate, it is humbling to note that the human brain contains many billion neurons, and rather than being merely passive devices, each functions as an independent small computer in the way that it processes data. If we think of the brain as a set of giant interacting networks, the number of possible connections is literally astronomically large.

The basic ANN

A typical neural net is composed of several layers of nodes, as shown in Fig. 2.11.

Each neuron in the net has the properties described for perceptrons. However, in addition to the input and output layers, there are a variable number of "hidden" layers that connect the input to the output layers. They are called hidden layers for the following reason: since we know the inputs and receive output data, both these layers are known to us. In contrast, the layers in between are not readily accessible to the observer. Since the computer has no cognition, each run of the algorithm is, in essence, a double-blind experiment. This has raised philosophical issues about the explainability of deep learning models.

The basic features of an ANN are as follows:

- Each input neuron gets only one input, directly from outside; those of the hidden layers get many.
- Neurons in a layer do not connect to each other.
- Each neuron in one layer connects to all the neurons in the next layer.

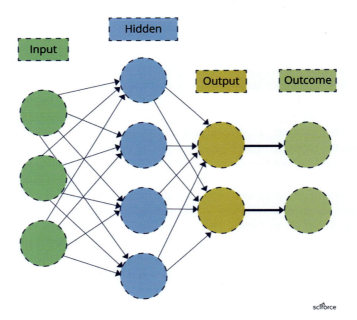

FIG. 2.11

The multilayer construction of a neural network. Each *colored circle* represents a neuron (which passes the aggregated signal through an activation function) and the lines are the inputs and output connecting them. In this example, only a single hidden layer is shown. In practice, multiple hidden layers are used [1–4].

From SciForce: "White paper on Top Algorithms for Health," by permission of M. Ved, "Top Algorithms for Healthcare" SciForce-Medium.

- The ANN is a synchronous model; all nodes in a given layer receive their inputs and calculate their activities at the same time. This means that all the calculations in the n layer must be calculated before we can start the calculations for the (n+1) layer.

Thus, the output of one layer becomes the input of the next. Because of the summing process, each layer receives increasingly complex inputs. At any given layer, the network creates an internal representation of the data that incorporates relationships among the feature vectors of each sample. Basically, a neural network utilizes multiple levels of representation to model both simple and complex relationships among data. They are powerful classifiers and can extract relations, predict outcomes, and make inferences from data, all without the need to manually define the feature vectors for that set. For example, instead of inputting the eye-snout ratios and other defining characteristics of dogs versus cats to classify an unknown image as to its dog-ness or cat-ness, one can simply present multiple known examples of each to the program.

The weights in an ANN

As was the case for the Rosenblatt perception, every input within an ANN is associated with a weight that represents the relative importance of that input with respect to its contribution to the sum of signals received by the neuron. For a node receiving input from two nodes of the previous layer, the total input is the sum of the output of the first node times its weight plus the output of the second node times its weight. This is the product of two vectors. In general, one can set up the computation of each layer of nodes as a single matrix-vector product. Basically, there is a matrix of weights multiplied by the input vector, which produces the aggregated output when acted upon by the activation function. When these calculations are generalized to multilayer networks with many nodes, they are seen to be the result of simple multiplicative steps in linear algebra. In some ANNs a bias term is also added; this is a scalar that is added to the input to ensure that a few nodes per layer are activated regardless of signal strength. It is because the weighted inputs summed is a linear function that a nonlinear activation function must act upon them so that the ANN is not restricted to linear problems. Both weights and biases are modified throughout the training of an ANN.

Learning from examples: Backprop and stochastic gradient descent

The traditional learning method for feed-forward networks such as the ANN is to take a fixed network architecture and then use the examples to find appropriate values for the weights that lead to correct performance. The fact that each node in a layer has an incoming connection from every node in the previous layer does not interfere with this adjustment since any unnecessary connection can be set to zero during this process.

The network learns the proper weights from the training set using a gradient descent algorithm based upon The error backpropagation algorithm (backprop). The basics of backpropagation were derived in the context of control theory by multiple researchers in the 1960s. Subsequently, in the 1980s, a number of investigators proposed that it could be used for neural networks. Thus, it is difficult to assign priority to this fundamental technique that is the basis for neural network training (see Ref. [3] for historical details).

The way Backprop works is as follows: We initially pick a training instance and set the activity of the input nodes to the values from that instance. Activities are then propagated through the ANN in its initial state (the initial weights, within certain constraints, are arbitrary and often are randomly chosen) The output is then examined. At this point, the output values are compared to the correct answer, which we know since the example is from the labeled training set. The labels are known as "ground truth" and the object is to generate a result that matches this ground truth. If this turns out to be the case, then a miracle occurred. If not, the difference between ground truth and the prediction is quantified by a so-called loss or cost function. This is often, but not always, a mean square error function. The task is now to minimize the cost function by adjusting the weights in an iterative manner. This is not a straightforward process since each weight contributes to more than one output, and so we cannot simply apply corrections evenly among all the weights. In practice, we work backward from the output layer through the hidden layers rather than starting at the beginning of the net. This is known as gradient descent.

Fig. 2.12 shows a simple diagram of an error curve, where J(w) is the loss function and w is a weight. The gradient, or slope, shows the direction of change in the weight needed to reach either a minimum or a maximum. That is the point at which

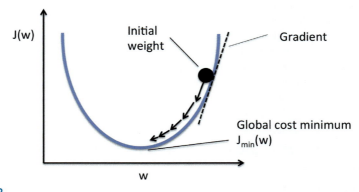

FIG. 2.12

The gradient is given by the slope of the curve directly above the point we are interested in. Lines that go up as we move right have a positive slope, and otherwise, they are negative. The weight is adjusted in a stepwise manner as shown.

Mixtend, Creative Commons. License.

the gradient (slope) is zero. At each step of the gradient descent method, the iterative steps are proportional to the negative of the gradient of the function at the current point. The gradient is calculated from the partial derivatives of the loss function with respect to the weights (multivariate calculus to the rescue).

The actual situation is not that simple. As the dimensionality of the space increases, the loss function curve can get complex. In Fig. 2.13 it can be seen that there is no single minimum; instead there are local minima that can be reached by different starting trajectories. There is no computationally efficient way of calculating the true global minimum. However, in practice, local minima are good enough for excellent algorithmic performance. One can always take different starting points for the weights and compare outputs as a way of sampling the local minima.

This process is refined over the set of training examples. Backpropagation is one of the reasons that training sets for deep learning must be large. Interestingly, if the network has too many layers, training can suffer. As we proceed further and further backward down the net, the weighting corrections get smaller and smaller, and they will become closer and closer to 0. This is known as the vanishing gradient problem. It can be overcome in a specialized ANN known as a residual neural network, which will be covered in the next chapter.

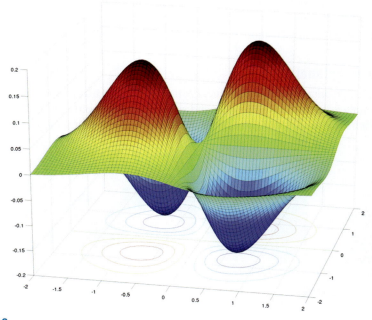

FIG. 2.13

A complex 3D curve showing multiple local minima and saddle points. The situation is even more complex in multidimensional space.

Wikimedia Commons.

Finally, it is important to note that this algorithm assumes that tweaking all the weights independently and simultaneously actually leads to a reduction in this error. This is not intuitive since changing one weight could cause ripple effects throughout the rest of the network. It turns out, however, that if we make the changes to the weights small enough the assumption generally holds true, and the weights do, in fact, reach good error reduction.

Convolutional neural networks
Overview
Convolutional neural networks (CNNs) are sufficiently complex, so a brief overview will be helpful prior to a more complete discussion. In brief, images are too large to be analyzed at the input pixel level. Arguably, the first CNN was "AlexNet" in 2012 by A. Krizhevsky et al. [11]. The CNN extracts image features to create an increasingly complex representation. It uses feature extraction filters on the original image and uses some of the hidden layers to move up from low-level feature maps to high-level ones. CNNs have two kinds of layers: convolutional and pooling (subsampling). Convolutional filters are small matrices that are "slid" over the image. The matrix is combined with the underlying image piece and the maximum value from each convolution is obtained. In CNN, the convolutional layers are not fully connected, but their output is usually passed to one or more fully connected layers that perform the final classification. An overview of the process is seen in Fig. 2.14.

Detailed explanation
CNNs take advantage of the hierarchical pattern in data such as images and assemble more complex patterns using smaller and simpler patterns before moving these representations into traditional fully connected networks. They are loosely based upon

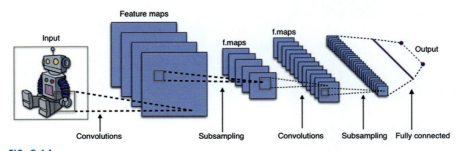

FIG. 2.14

Schematic of a typical convolutional network diagramming convolutions and subsampling. The image representation at the final layer is passed to fully connected layers and a classification prediction is made.

Creative Commons.

Convolutional neural networks 37

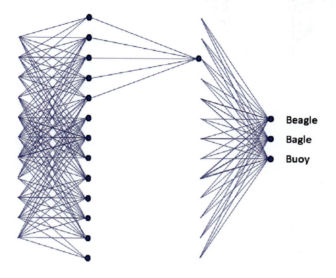

FIG. 2.15

This figure illustrates the beginning and end of a CNN. Not all the neurons of one hidden layer will connect to each neuron in the next layer. However, in all cases, there are neurons that share connections from the previous layer so as to create overlapping fields (not shown). The output of neuronal fields is down-sampled to further reduce the dimensionality of the sample space.

Wikimedia Commons.

the organization of the visual cortex of the brain. Individual cortical neurons respond to stimuli only in a restricted region of the visual field known as the receptive field. The receptive fields of different neurons partially overlap such that they cover the entire visual field. The CNN mimics this in that its layers are not fully connected. This is shown in Fig. 2.15.

As noted above, in CNN a convolutional matrix (also called filter or kernel) is "slid" across the image and applied at each position. The resulting value then becomes the value for that pixel in the result. This is shown in Figs. 2.16 and 2.17.

The convolution is then followed by a pooling layer that down-samples the feature maps by summarizing the presence of features in patches of the feature map. Max pooling simply involves using the largest value rather than averaging them. This makes the resulting representations more robust with respect to changes in the position of the feature in the image (local translation invariance). CNNs trained on similar data produce similar internal representations, and so they are useful for transfer learning, which will be discussed in subsequent chapters.

As seen above, the filters themselves are just matrices of numbers. By proper choice of values, they can detect specific features in the images such as horizontal and vertical lines, edges, shapes, and so on. These numbers can be modified by back-propagation. The filters in Photoshop are convolution filters and can be used to

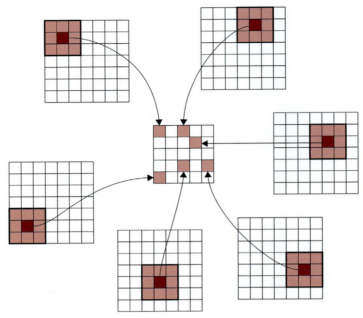

FIG. 2.16
To convolve an image with a filter, the filter is moved across the image and applied at each position. In practice, blank rows and columns are added at each end so that there is no loss of coverage when the filter reaches the farthest position in the row or column. If padding (blank grids around the perimeter) was added to the grid being scanned by the 9×9 convolution filter, there would have been no dimensional reduction at this step.

Copyright © 2018 Andrew Glassner under MIT license.

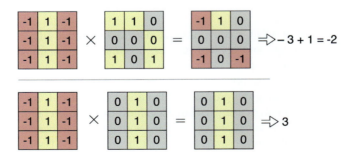

FIG. 2.17
Applying the filter to an image. The filter is on the left. The section of the image currently overlaid by the filter is in the middle. The result of multiplying each pixel by its corresponding filter value is shown on the right. Adding up the nine values gives the final sum on the right.

Copyright © 2018 Andrew Glassner under MIT license.

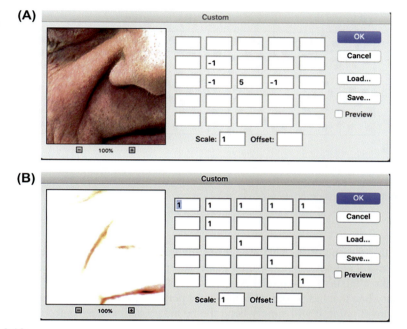

FIG. 2.18

Closeup of part of my face. (A) mild convolution (most detail remains). (B) after more convolution. In (B), the filter was chosen so as to detect line boundaries. In (A), the filter was chosen so as to simply make me look better. Your results may vary.

experiment with these concepts. Under "other" in the Filter tab there is an entry labeled Custom. Fig. 2.18 is a screenshot showing a convolution filter applied to a portrait of my face.

Overfitting and underfitting

In all of the examples of machine learning discussed above, their central purpose is to generalize. Generalization is the model's ability to give correct outputs to a set of inputs that it has never seen before. Underfitting simply means that the algorithm has not learned well enough from the data. The concept of overfitting is a bit more complicated because it means that the algorithm learns too well! A model so powerful that it "memorizes" the data (overfitting) would cause each sample to define a separate class, while a model that is too weak cannot find any reliable class distinctions (underfitting). For a given model, finding the balance between overfitting and underfitting is more art than science.

Instead of these extremes, consider an actual scenario, as shown in Fig. 2.19, which depicts the results of a clustering model. In case A, there is a smooth boundary

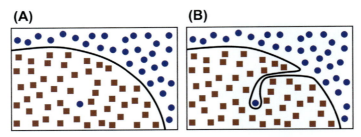

FIG. 2.19

(A) There is a smooth boundary line that does a good job of separating the two classes, with only one error. This is a good fit from which we can generalize well. (B) The boundary curve distorts itself to accommodate a single aberrant point (overfitting).

Copyright © 2018 Andrew Glassner under MIT license.

line with only one error. This model has high accuracy. Moreover, we can easily place a novel unknown point on either side, accepting the fact that we will occasionally make a mistake. The model generalizes well. In contrast, in case B, we have a boundary line that has separated both classes perfectly. In this case, when the model sees a novel point, it will simply bend the boundary again so as to accommodate it rather than generalize it for prediction. Perfection is the enemy of the truth. The subject of overfitting is a complex one, and the details are beyond the scope of this chapter. For a complete discussion of overfitting and mechanisms for dealing with it, see Ref. [12].

Things to come

While the many models for deep learning are impressive in their sophistication and power, we are only at the beginning of the age of artificial intelligence. Our human brains still have a number of obvious advantages over ANNs as presently implemented. For example, we can learn complex patterns from a very small training set of examples, and we are much better than machines at avoiding both overfitting and the increase of errors with multiple processing layers due to vanishing gradients during backpropagation. In fact, backpropagation itself seems biologically unlikely, based in part on measures of latency during biologic learning, and because very few neurons have taken courses in multivariate calculus. For these reasons, there has been an increasing interest in models for deep learning that more closely mimic our current understanding of biological neural function. This includes both the development of spiking neural networks and attempts to replace backpropagation in more traditional models, the latter via synthetic gradients, random feedback, or direct feedback alignment. With respect to spiking networks, the notion of a set of weighted inputs across layers is replaced by an integrate-and-fire model that integrates incoming pulses based upon pulse coding. Such models lend themselves well to synaptic plasticity, which means that there is a correlation between pre- and postsynaptic neurons based upon long-term depression and potentiation effects. They have intrinsic temporal information processing activity and are amenable to implementation of unsupervised learning strategies. Current spiking models are relatively simplistic and are not yet competitive with traditional ANNs for practical applications, but there is a great deal of ongoing research on optimizing them.

Of great potential significance is the creation of artificial neurons and synapses that will allow for the construction of spiking networks in hardware as well as software. For example, Tuma et al. [13] have described an artificial stochastic phase-change neuron in which membrane potential is represented by a reversible amorphous-to-crystal phase transition. More recently, Schneider et al. [14] have described an artificial synapse in which the strength of an electrical current flowing across it can be precisely controlled analogous to the way ions flow between neurons. This "synapse" is based on nanotextured magnetic Josephson junctions. Advances in the design of spiking network architecture implemented in hardware such as this may lead to a true artificial brain, assuming that we can also model the hierarchical architecture, modular structure, and cascading networks of the real thing, as will be done in 2338 for Chief Operations Officer Data in Star Trek TNG using an alloy of platinum and iridium (Wikipedia).

At the current state of the art, ANN and CNN are one-trick ponies that are trained to do specific tasks, and they find it difficult to generalize this learning to novel tasks. They can only do this in a limited manner. If one trains a CNN to recognize prostatic cancer and another to recognize breast cancer, they will only work on prostate and breast cancer, respectively. Using so-called transfer learning it is possible to use the weights of the early layers of the trained prostate CNN to replace the initial random weightings in the training of the breast CNN. A human, however, can generalize across problems to a much greater degree. We use the term "artificial general intelligence" (AGI) to mean the representation of generalized human cognitive abilities in software (or ultimately in hardware) so that, faced with an unfamiliar task, the AI system could find a solution. What is currently known as AI is more correctly called weak AI. Most of the artificial intelligentsia feel that AGI will come about within this century. We will recognize the arrival of AGI in academic circles when it asks for (and is granted) tenure.

References

[1] Rosenblatt F. The perception-a perceiving and recognizing automaton. Cornell Aeronautical Laboratory; 1957. Report 85-460-1.
[2] McCulloch WS, Pitts W. A logical calculus of the ideas immanent in nervous activity. Bull Math Biophys 1943;5:115–33.
[3] Sejnowski TJ. The deep learning revolution. Cambridge, MA: MIT Press; 2018.
[4] Smith G, Cordes J. The 9 pitfalls of data science. Oxford University Press; 2019.
[5] Greenstone G. The history of bloodletting. BC Med J 2010;12–4.

[6] Wasilewska A. Bayesian classification, https://www3.cs.stonybrook.edu/~cse634/19Bayes2.pdf.

[7] Koller D, Friedman N. Probabilistic graphical models. Cambridge, MA: MIT Press; 2009. See also https://ai.stanford.edu/~koller/Papers/Koller+al:SRL07.pdf.

[8] Carter B. Typical sets and the curse of dimensionality Stan blog., 2017, https://mc-stan.org/users/documentation/case-studies/curse-dims.html Refs. [For PCA, overfitting and underfitting].

[9] Loewenstein Y, Sompolinsky H. Temporal integration by calcium dynamics in a model neuron. Nat Neurosci 2003;6:961–7.

[10] Minsky M, Papert SA. Perceptions: an introduction to computational geometry. Cambridge, MA: MIT Press; 1969.

[11] Krizhevsky A, Sutskever I, Hinton GE. Imagen et classification with deep convolutional neural networks. In: Advances in neural information processing systems; 2012. p. 1106–1114 https://arxiv.org/pdf/1803.01164.pdf.

[12] Glassner A. Chapter 9: overfitting and underfitting. In: Deep Learning: from basics to practice; 2018. p. 334–49. Kindle ASIN: B079XSQNRX.

[13] Tuma T, et al. Stochastic phase-change neurons. Nat Nanotechnol 2016;11:693–9.

[14] Schneider ML, et al. Ultra-low power artificial synapses using nano-textured magnetic Josephson junctions. Sci Adv 2018;4(1), e1701329. https://doi.org/10.1126/sciadv.1701329.

CHAPTER

Overview of advanced neural network architectures

3

Benjamin R. Mitchell

Department of Computer Science, Swarthmore College, Swarthmore, PA, United States

Introduction

The field of deep learning has expanded rapidly over the past few years, and modern practitioners now use a wide range of methods and techniques to achieve state-of-the-art performance. Many of the best performing systems now use significantly more complex architectures than the fully connected feed-forward networks used in the past. These techniques draw inspiration from a range of different fields, including statistical methods, time series analysis, game theory, information theory, operant conditioning, genetic evolution, and modern neuroscience (though for the last few it is worth a reminder that "inspired by" is not the same as "accurately modeling").

This chapter will cover some of the most popular variations, including residual connections (ResNet), unsupervised pretraining (autoencoders), transfer learning, generative models (GANs), recurrent networks (LSTMs), reinforcement learning, ensemble methods, and evolutionary algorithms. When reading this chapter, keep in mind that the methods presented are largely ways of modifying or expanding on the basic techniques presented in the previous chapter. This is done to address problems where those basic techniques are insufficient. The techniques described here can be used to address problems such as computational efficiency, classifier accuracy, and lack of large amounts of labeled training data.

Network depth and residual connections

The "depth" of a deep neural network can be highly variable depending on the task. In the earliest days of artificial neural networks, most networks were "shallow," meaning they did not have hidden layers. This was not because the early network designers did not want to have deeper networks; rather, it was because early

Artificial Intelligence in Pathology. https://doi.org/10.1016/B978-0-323-95359-7.00003-0
Copyright © 2025 Elsevier Inc. All rights are reserved, including those for text and data mining, AI training, and similar technologies.

43

44 CHAPTER 3 Overview of advanced neural network architectures

researchers had no reliable method for training these deeper networks. As time progressed, developments in the field have allowed for the successful training of deeper and deeper networks. Today, some networks may have hundreds of hidden layers.

The primary advantage of adding more hidden layers is that it allows the network to encode more complex functions. Since each layer takes the output of the previous layer as its input, each layer can be thought of as computing some sort of abstract "features" of the previous layer. Thus, each additional layer can compute a more abstract and complex function of the original data. A deeper network has more computational power than a shallower one, at least in theory.

In practice, this theoretical power may not always be easy to realize. There are several practical problems that can arise from adding extra hidden layers. The first and simplest is that adding more hidden layers means adding more connections. Those extra connections give the network more representational power, but they also represent extra parameters for which we must propagate information through and learn good values. In other words, more connections mean the network will operate more slowly and may take more epochs to train.

Even if we are willing to pay the computational cost of adding more connections, another phenomenon can arise with too many hidden layers: the *vanishing gradient* problem. Since the standard method for learning weights in a neural network uses error backpropagation via the chain rule, the more hidden layers, there are the more times we must "chain" the derivatives together. This can lead to the gradient becoming flattened and "diffused" as we get farther and farther away from the output units. This problem is particularly severe when using sigmoid activation functions since the sigmoid function will always have a derivative between 0 and 1, and multiplying small numbers yields even smaller numbers. This is the primary reason that most modern neural networks use a rectified linear (ReLU) activation function [1]; the ReLU activation is less prone to gradient diffusion, allowing deeper networks to be used successfully.

However, even with the use of ReLU units, there are practical limits to how deep a simple feed-forward network can be successfully trained. To push depth beyond that limit, more advanced network architectures are required. One method that has become very popular in recent years is using *residual* connections [2]. This involves creating connections that "skip" one or more layers (also sometimes called "skip connections") (Fig. 3.1). In typical modern practice, these residual connections do not have tunable weights but rather are a way to simply copy the activation from a lower level.

The key mathematical insight behind making these connections "skip" a layer is that it makes it much easier for that layer to "learn" to replicate the input of the previous layer; with the addition of the residual connections, simply outputting 0 will result in the output value being equal to the node the residual connection originates from. In effect, any nonzero output of the node is now *added* to the value from the earlier layer rather than *replacing* it (as would be typical in a standard feed-forward network).

Network depth and residual connections

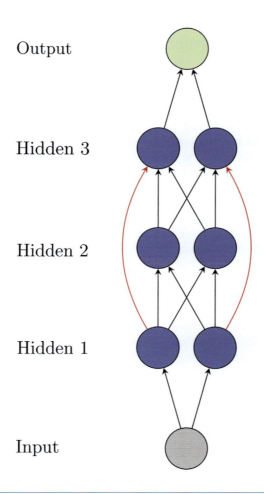

FIG. 3.1

Many modern deep networks make use of skip connections, which allow for easier optimization in very deep networks. Here, the units in the first layer have direct connections to corresponding units in the third hidden layer, in addition to their "normal" connections to units in the second hidden layer.

The term "residual" means the leftover or remaining error. In the case of residual connections, the idea is that each additional layer of the network now only has to try to fit the residual error, simplifying the learning process. The result is that networks that make use of residual connections can be made much deeper than are possible otherwise without falling prey to gradient diffusion or other optimization difficulties. The term ResNet may be used to refer to the original network proposed using this type of connection [2], but the term is also sometimes used to refer to any network using residual connections.

Autoencoders and unsupervised pretraining

One of the known drawbacks of deep neural networks is that they require very large amounts of data to train. This is a natural consequence of the power of the model; the more connections a network has, the more "power" it has to encode complex functions. However, it has long been known that the more parameters we need to estimate, the more data necessary to estimate good values for them. Since deep neural networks can have millions of free parameters, it is unsurprising that very large datasets may be needed to train them. As a few examples, the MNIST data set is a classic benchmarking set that is been around for over 20 years [3,4]. These days, it is considered a "toy" problem due to its comparatively small scale and the ease of achieving high performance. However, even this "small" data set has 60,000 images for training. The ImageNet data set, a more modern benchmarking set, consists of over 15 million high-resolution images [5]. In practice, many authors use only a subset of this full set, but even the subsets tend to have over a million training images.

This requirement for very large training sets would already be enough to potentially limit the problems to which deep neural networks can be applied, but the true scenario is even worse. Not only do we need very large amounts of data, but most deep networks are trained using backpropagation, which is a supervised learning method. This means that we need a human annotation giving the correct label (i.e., desired output) for each and every one of our millions of training examples.

For some tasks, this process can be crowdsourced; for instance, the ImageNet data set was created using Amazon's Mechanical Turk to pay large numbers of unskilled annotators to label images that had been scraped from the web [5]. Unfortunately, for many problems of interest, relying on unskilled annotators is not possible. In fields like medical imaging, it is common that humans require a significant amount of training to be able to correctly and reliably annotate images. Thus, even if a large set of images can be collected, it may not be practical to annotate them all. In this scenario, where a large amount of data is available but only a small portion can be annotated, pure supervised learning is unlikely to succeed. Fortunately, we do have other options.

One method for handling this scenario is called *unsupervised pretraining*. The idea behind unsupervised pretraining is that, prior to applying a supervised learning technique, we will first apply an unsupervised learning technique in an attempt to make the final supervised optimization problem easier.

A classic example of this is to first train one or more layers of a neural network as an *autoencoder* [6]. The autoencoder is trained simply to reconstruct its input; that is, the desired output is merely a copy of the input (Fig. 3.2). If the network has enough hidden nodes, this task becomes trivial. Still, if the network is restricted in capacity then it is forced to learn some low-dimensional encoding. This can be thought of as analogous to a lossy compression scheme; the "compressed" data are encoded by the activations of the hidden nodes.

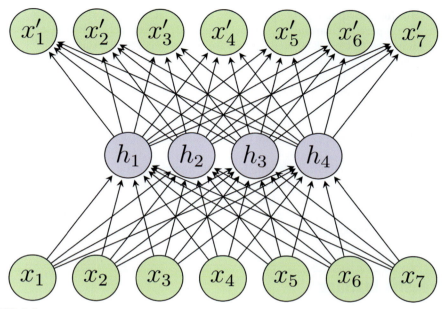

FIG. 3.2

An autoencoder network is trained to reproduce its inputs with as little reconstruction error as possible. Since the hidden layer has fewer units than the input and output layers, the network learns to produce a reduced dimensionality encoding in the hidden layer. This is an "unsupervised" training process since it requires no additional human annotation to the raw data.

A single autoencoder layer is trained using backpropagation, which is a supervised learning technique; however, the "correct output" values are just a copy of the input values, meaning no human annotation or labeling is required. Thus, the method as a whole is considered an unsupervised technique. If an even more abstract and compressed representation is desired, we can use the output of one autoencoder layer as the input for another. The result is called a *stacked autoencoder*.

A similar network style can be generated using a restricted Boltzmann machine (RBM) instead of a feed-forward network; the result of stacking RBMs is sometimes called a *deep belief network*. The training method for RBMs is quite different (see Ref. [7] for details), but for the purposes of unsupervised pretraining, the end result is functionally similar.

Once the unsupervised model has been trained, it is then used as the basis for a supervised model, typically by simply adding one or two extra fully connected layers to the top. At that point, the entire network is trained using a standard supervised approach. The intuition behind the use of unsupervised pretraining is that the "unsupervised" step is essentially finding a feature-based representation of the data suitable for reconstructing the original inputs. So long as a similar set of features

48 CHAPTER 3 Overview of advanced neural network architectures

is useful for classification, pretraining the network in an unsupervised setting can allow the supervised learning stage to achieve good convergence even in cases where the amount of labeled data would otherwise be too small to allow it.

It should be noted that unsupervised pretraining is not a magic bullet; in particular, it can reduce the amount of labeled data required to a point, but there will still be a point at which there is simply too little labeled data to learn a model that generalizes well. Additionally, it is nearly always the case that unsupervised pretraining will be outperformed by fully supervised training using the same amount of data (i.e., if you have labels for all your data, you are better off simply doing supervised training from the beginning).

That said, this is still a potentially useful method, particularly in cases where acquiring large amounts of labeled data is considerably more expensive than acquiring large amounts of unlabeled data. Additionally, there are several more recent variations on the autoencoder idea that may have advantages on certain tasks. Variational autoencoders [8] and sparsity [9] are some examples.

Transfer learning

Another potential way around the problem of not having enough data to train a deep neural network is the process of *transfer learning* [10]. Similar to unsupervised pretraining, transfer learning is a method to "start" learning on your data set using a network with weights that have already been pretrained. The difference is that in transfer learning, the "pretraining" is achieved by simply doing regular, supervised learning on a different but related problem. Here, the key insight is that if two problems involve similar data, then the low-level features needed to solve one problem are likely to be useful for the other as well.

To perform transfer learning, the first step is to train a deep neural network on a large data set that is as similar as possible to the task you ultimately want to do. In particular, the similarity of the data is the most important. What this means in practice is that if you want to train a digit classification network, MNIST is a reasonable data set to start out with. However, if you want to classify natural images of the world, a network trained on MNIST is likely to transfer poorly. Instead, a network trained on the ImageNet data set would likely give a much better result since that data set is already composed of natural images of the world.

One potential issue here is that training a large, complex network on a large, complex data set requires a large amount of computational hardware (and/or a lot of training time). Fortunately, in many instances, it is possible to use a pretrained network created by someone else. Several prominent "pretrained" networks are available to download for a variety of deep learning toolkits; check the information for the deep learning library you are using to find out what is available.

The actual transfer learning process involves taking a pretrained network, removing the fully connected "topper" on the network that does the classification, and replacing it with a new fully connected "topper" (with randomly initialized weights).

At this point, the network is trained as usual using your data set. Again, the primary advantage is that the training can succeed with a much smaller labeled training set than would be required to train the same network "from scratch" (i.e., starting with randomly initialized weights).

It is also possible to selectively update only part of the network. This is primarily useful if you are using a large pretrained network and have limited computational power available, as it will speed up training but can result in lowered accuracy if done too aggressively. Here, the idea is that when performing the training for your data set, the uppermost layers of the pretrained network are the ones most likely to be significantly changed by the secondary training process. Therefore, it is possible to "freeze" the weights of the lower layers of the pretrained network and only update the weights for the top few layers. Since many architectures have many connections at lower levels than they do at higher levels, this can often result in a significant speedup of training.

Generative models and generative adversarial networks

The type of network we have dealt with so far is aimed at the task of classification. While classification networks are by far the most popular type, they are not the only type used. Several other types of problems can be tackled using machine learning, and one of them is the problem of generative modeling. A *generative model* is one which can be used to "generate" new examples. The basic task is to take in a set of examples and then construct new ones that are unique but similar to the examples the system was trained on. In statistical terms, the goal is to model the distribution from which the training samples were drawn and then generate new samples drawn from the same distribution.

There are several cases in which it might be worthwhile to train such a generative model. First, we might directly want the outputs; that is, generating plausible synthetic examples may be useful in its own right. Second, we may view the generative model as another way of finding potentially useful features, which we could then exploit as a method of unsupervised pretraining. Third, we may be interested in a conditional generative model, which is a fancy way of saying that we are interested in transforming inputs rather than creating them from whole cloth. For instance, taking an image and generating a version in the style of some famous artist. All three of these things are possible using generative models.

In the context of deep neural networks, the most popular generative model is called a *generative adversarial network* (GAN) [11]. GANs work by creating two networks and training them in competition with one another. The first network is called the *generator*, and it is responsible for generating plausible synthetic examples. The second network is called the *discriminator*, and it is responsible for distinguishing between "real" examples (taken from the training set) and "fake" examples (created by the generator). These two networks are effectively playing a zero-sum game in which each player's success must come at the expense of the other. The only

50 CHAPTER 3 Overview of advanced neural network architectures

way for the generator to win is to come up with examples that the discriminator cannot tell apart from real examples; the only way for the discriminator to win is to figure out how to recognize the generator's artificial examples.

Training proceeds by first initializing each network randomly and then alternately updating the first one and then the other. Updating the discriminator network is simple since it is a basic application of backpropagation; since we know which examples are real and which are fake, we have "ground truth" labels. Updating the generator, however, would seem to be more difficult; how do we define an error function for the generator?

The answer is at the heart of what makes GANs such a clever idea: since the output of the generator is used as the input to the discriminator, we can view the combined generator-plus-discriminator as being just one big network. If we take this view, we can then propagate error from the discriminator back all the way down to the bottom of the generator. Of course, when we are training the generator, we want to *minimize* the performance of the discriminator, so we cannot update the weights for the discriminator when we are training the generator. By alternating which network gets updated (always freezing the weights of the other), the overall performance of both players will slowly improve. Every time the generator finds some loophole in the discriminator, the discriminator is forced to fix the loophole. Every time the discriminator discovers flaws in the generated content, the generator must make its output more realistic to fix the flaws.

Thus, over many iterations, the generator slowly converges toward producing synthetic examples that are indistinguishable from the real examples. Of course, in practice, there are several complicating factors that mean the system may not always converge as well as we would like, especially if the distribution we are trying to simulate is highly complex. Nevertheless, GANs are useful in many domains where generating synthetic examples is desirable.

Recurrent neural networks

While convolutional networks perform very well on image and imagelike data, convolution is not always the best choice for all types of problems. One popular alternative for sequential data is a *recurrent* network. The basic idea behind a recurrent network is that for sequential or time-series problems, the "current" input may not be sufficient to successfully solve the problem. In other words, the problem requires some access to past information to solve correctly.

It is possible to address this by simply feeding the network multiple time points concatenated together as a single "input," but this approach has some limitations. In particular, the size of the temporal window is fixed; for example, if we use a window of size 4 (i.e., feed the network the last four timesteps as its input), we cannot capture temporal dependencies longer than four timesteps. Additionally, for every extra timestep of history, we add, the complexity of the network (i.e., the number of nodes and weights) increases dramatically, which often makes the use of long temporal windows impractical.

Reinforcement learning **51**

It is possible to do convolution in the time domain (see, e.g., Ref. [12]), but if longer temporal histories are required, the classic approach is to use a recurrent network. The term "recurrent" here refers to the fact that the hidden nodes have some access to their own previous states; that is, the value of a hidden node at time t is dependent on its inputs at time t as well as its own state at time $t-1$ (and possibly the states of other hidden nodes at time $t-1$).

In the earliest recurrent networks, this was a direct dependence, meaning that the previous states were treated just like any other input to a node; however, this proved to be somewhat inflexible. More recent work typically uses a type of node called a *long short-term memory* (LSTM) [13]. In modern parlance, some use the terms "LSTM" and "recurrent network" interchangeably, though this can sometimes lead to confusion. It is also worth noting that some minor variations on LSTMs may be encountered, but at a high level, they have fairly similar properties (see Ref. [14]). The GRU [15] is probably the most well known of these.

These modern recurrent networks make use of "gates" to allow a context-dependent decision about when to make use of the prior value of a hidden node and when to ignore it. This allows, for example, a sequence delimiter (such as a punctuation mark in text data) to be handled intelligently by the network. In general, LSTMs are significantly more flexible and powerful than their older counterparts, and in the context of modern work, it would be highly unusual to use units without some sort of gate in any type of recurrent network.

As with any other architectural building block, LSTM units are typically arranged in layers, and in modern practice, these layers are often stacked to create deep recurrent networks.

The operation of LSTMs is not quite as well suited to massively parallel GPU implementations as their convolutional counterparts, so training LSTMs is typically slower than training convolutional neural networks (CNNs). However, they are still the right choice for some types of tasks. For example, prior to the introduction of Transformers most modern machine translation systems made use of LSTMs (see Ref. [16] for a popular example from Google).

Reinforcement learning

Most neural network models are trained in a "supervised" fashion, meaning that the system is given a labeled training set consisting of hand-labeled input-output pairs. However, for some types of tasks this is not practical. In particular, for sequential decision problems, it is often the case that we can only accurately assess the performance of the system at certain points. A common example is trying to train a network to play a game like chess or Go; until the game is over, we do not know whether the strategy being employed is a good one or not. This means we cannot train the network to decide what moves to make in the middle of the game using standard supervised learning.

The idea of *reinforcement learning* was developed for situations like this. Inspired by operant conditioning, reinforcement learning works by providing a "reward" signal that tells the system when it is doing well and when it is doing

CHAPTER 3 Overview of advanced neural network architectures

poorly. Over time, the system experiments and, through trial and error, discovers a policy that will maximize its expected reward.

The classic formulations of this algorithm are based on the formalism of a Markov decision process (MDP). An MDP encodes a problem as a discrete set of possible states and actions, along with a transition model that specifies a probability distribution over possible "next" states, given a current state and choice of action. Given a "reward" function that specifies what reward the agent receives for being in any given state, it is possible to use an iterative approximation algorithm to determine the optimal action to take in any state, where "optimal" is defined as the action that leads to the highest expected future reward. It is also common to add a "discount" factor, which allows us to downweight future expected gain based on how distant it is; by tuning this parameter, we can make the agent optimize for more short-term or more long-term rewards.

Unfortunately, this method is limited because the transition model needs to be fully specified as an input; in real-world problems, this is often not the case. If we need to discover the transition probabilities, the problem becomes more difficult. It can still be solved using algorithms like adaptive dynamic programming and Q-learning, but these algorithms tend to converge very slowly. Furthermore, there is an additional difficulty that arises, which is that the agent's own decisions control which state-action sequences it gets observations from. As a result, there is a need to ensure sufficient "exploration" of the state-action space that takes place to avoid converging to poor local optima. Simply taking random actions will result in a high degree of exploration, but in most real-world problems, some areas of the state-action space are more productive than others, so it is inefficient to spend a lot of time exploring choices that we already know lead to bad outcomes. This is known as the "exploration/exploitation dilemma," and it is yet another reason convergence is so slow; if we are too aggressive, we can get premature convergence to a bad solution.

These classic reinforcement learning algorithms are limited in that they tend to work best on small, discrete state spaces, and do not scale well to complex, high-dimensional problems. One of the fundamental issues is that reinforcement learning treats each state in the state space as atomic, meaning that information about each state must be learned separately, even if the states appear to be "similar" to a human. The combinatorics means that the number of training runs needed to get good estimates for large state-action spaces can become intractable.

The solution to this is to use a technique known as *function approximation*, which means that instead of directly learning the expected utility of taking a given action in a given state for each state, we use some sort of regression techniques to learn an approximation. The primary advantage of function approximation is that, in addition to making the combinatorics less bad, it can also allow a means to treat states as nonatomic, meaning we can learn based on similarity between states (i.e., if we have encountered a similar state in the past, we may assume that this state will have a similar expected reward).

It is particularly convenient to use a neural network as the function approximator for reinforcement learning because neural networks can be trained in an online fashion (i.e., one training example at a time), allowing them to be updated after each action. In recent years, deep neural networks have become increasingly popular for this, allowing for better scaling to large and complex problems. Many recent successful solutions to games have made use of reinforcement learning for training, including results on Atari games [17] and Go [18].

The primary downside of reinforcement learning is its lack of computational efficiency; training via reinforcement learning is very slow compared with fully supervised learning. It is typically preferred to use supervised learning where it is an option. However, reinforcement learning opens up the possibility of solving an entire class of problems that would otherwise be intractable, making it a very valuable technique for that class of problems.

Additionally, there are cases in which the two methods can be combined. This typically occurs when we have some training labels for our data, but they do not fully capture the behavior we want. For example, many machine translation systems are trained to optimize the BLEU score [19], even though it is acknowledged to be an imperfect model for human evaluation of translation quality. Training a machine translation system fully using reinforcement learning and human evaluation would require so much time as to be entirely infeasible. However, it is possible to first train the system in a supervised fashion using BLEU scores and then refine the system using reinforcement learning. Since it is only a refinement, much less training data are required, making this more practical. However, it is worth noting that there is no guarantee this will actually improve the final system's performance; some experiments done using Google Translate showed that the benefits of this approach were not as great as had been hoped [16].

Ensembles

A classic machine learning method for boosting the performance of classifiers is to combine several of them together into an *ensemble* classifier. The inspiration is the same as that behind crowdsourcing; in both cases, each individual member may be wrong, but so long as enough of them are right, the "group" gets the correct answer. This relies on the assumption that the members of the ensemble are sufficiently diverse since if they all make the *same* mistakes, there is no benefit. However, there is enough randomness in the standard neural network training algorithms to allow for a reasonable degree of independence, which can be increased if necessary through methods like boosting or bagging.

Even without these methods, ensembles of deep neural networks tend to outperform their members. Thus, a reliable way of increasing the performance of a deep neural network is to simply train four or five different copies of the network and then allow them to "vote" for the correct answer. There are limits to how much performance can be gained this way, but it is a fairly simple means of eking out a few more percentage points of accuracy.

54 CHAPTER 3 Overview of advanced neural network architectures

This is an example of the simplest way to build an ensemble: simply training multiple copies of the *base* classifier and then combining their output. The simplest way of combining classifier output is to allow each classifier to make its own prediction and then choose the plurality prediction as the "final" output. This simple *voting* scheme is easy to implement and easy to understand, but it does not always produce the best possible results. In particular, it gives equal weight to all members of the ensemble, which may not be a good idea. If we have reason to believe that a particular member of the ensemble is likely to be unusually trustworthy (or untrustworthy) for a given example, it would make sense to do *weighted voting*. Here, each classifier still gets to vote, but not all votes count equally. Instead of each classifier contributing +1 to the vote tally for their respective choice, they will contribute an amount proportional to some confidence measure. In the end, the answer with the largest total score is then the one chosen as the ensemble's joint answer. The choice of confidence measure will be problem dependent, but for artificial neural networks, one common choice is to train the networks to produce pseudo-likelihood scores for the various classes as their outputs. The score associated with the "winning" class can then be used as a confidence measure; if the network predicts a class with a value of 0.99, we give it more weight than if the highest output value is only 0.6.

The performance of an ensemble depends on the members making *different* mistakes; therefore, ensemble performance can often be boosted by forcing the member classifiers to be more different from each other, even if that comes at the cost of a small accuracy penalty to each individual member. The simplest example of this is a method called *bagging*, which works by ensuring each member is trained using a slightly different training set. Rather than training each classifier with all of the available training data, each classifier is trained using some fraction of the examples chosen at random. This ensures that different classifiers will have different strengths and weaknesses and helps ensure that the ensemble as a whole does not overfit the training data. The smaller the fraction of the training data given to each classifier, the more diverse they will be, but the lower their performance will also tend to be. At some point, the individual performance degradation overcomes the benefit from increased diversity. Unfortunately, exactly where the sweet spot is located is problem dependent.

One alternative to bagging is a method called *boosting*. Boosting makes the observation that the goal is to ensure that every example is classified correctly by a sufficient number of members and tries to build an ensemble to ensure this is the case. In boosting, the first member classifier is trained normally, but after that each classifier is trained as follows. First, test the performance of the existing ensemble on the training data; for each example, count how many classifiers get it right, and how many get it wrong. Then, weight the examples based on how often they were predicted correctly by the existing ensemble. The more times an example was guessed incorrectly, the more importance weight that is put on that example. Then, a new classifier is trained, factoring in the example weights; this means that the new classifier will be pushed to do well on the examples that other classifiers are getting wrong and will have less incentive to perform well on examples that other classifiers are getting right.

Boosting often performs very well in practice but has a few downsides. In particular, bagging is an algorithm that makes use of data parallelism, meaning all the member classifiers can be trained simultaneously if enough parallel compute power is available. Boosting, on the other hand, is an iterative method; each new classifier needs to know the performance of the previous ones, so they must be trained one at a time. Additionally, boosting works best for classifiers that can easily make use of the importance weights on the examples, which not all possible base classifiers can do.

The only real downside of ensembles is their computational cost; each member of the ensemble must be trained from scratch, since otherwise they will not be sufficiently independent. As a result, producing an ensemble with five members will require five times the total amount of computational resources to train. Since deep neural networks are already expensive to train, this can make the technique impractical. If training a network takes 10 days, most researchers are unwilling to wait another 40 days to build an ensemble. However, the members of the ensemble can be trained in parallel, so if enough computational hardware is available the extra copies can all be trained simultaneously.

Since the gain from forming an ensemble is reasonably consistent, it is common practice for researchers to simply compare single classifiers with each other; whichever wins is likely to win as an ensemble as well, and this saves compute time. When creating production systems, however, it is typically a good idea to make use of ensembles since the training time is a one-time cost and will improve the system's performance.

Finally, it is worth mentioning that many types of ensembles have developed their own names over the years. For instance, a "random forest" is typically a bagging ensemble of decision trees. AdaBoost (Adaptive Boosting) and XGBoost (eXtreme Gradient Boosting) are two common implementations of boosting, also typically using decision trees as their base classifiers.

Genetic algorithms

Genetic algorithms (GAs) are a part of the evolutionary computation (EC), a field that draws inspiration from the biological process of Darwinian evolution. In addition to inspiration, practitioners in this field often use the terminology of biological evolution to describe their algorithms; it is important to understand that these terms are used metaphorically, not literally. Thus, when a term like "genotype" is used, it should be understood as meaning something like "the part of our algorithm analogous to a genotype." As with all computations inspired by biology, the computational models are a massive oversimplification of the actual biological process.

That said, the goal of evolutionary computation is to make use of inspiration from nature to solve difficult computational problems. Genetic algorithms, in particular, are a method for performing a type of search. To train artificial neural networks, GAs are typically used to find good values for the connection weights; in effect, GAs are an alternative to the error backpropagation method. The primary advantage that a GA

56   **CHAPTER 3** Overview of advanced neural network architectures

has over backprop is that the backprop method is a local gradient descent method, making it vulnerable to local minima and bad initializations. GAs operate as a distributed search and, as such, are much more robust against these issues. The primary drawback is computational; GAs are computationally expensive and do not scale as well as backprop to very large numbers of weights. As a result, they are more likely to be applied to smaller networks. Additionally, GAs have a large number of user-tunable hyperparameters, several of which are qualitative, to which they are highly sensitive. As such, the successful deployment of a GA may be more challenging than using backprop. However, for some problems the gains from the distributed search process allow GAs to reach levels of classification performance that are not possible using backprop.

A GA begins with an initial "population" of "individuals," each representing a candidate solution to the problem. Each individual is generated at random, and the population size is a user-tunable parameter. In the case of a neural network, each "individual" will be a complete network. The individual is encoded with a "genome," which is simply a numeric representation of the individual. In some applications, there may be complex encoding functions needed to transform between "genotype" and "phenotype," while in other applications, the two forms might be identical. The key factor is that the genotype encoding must support the functioning of the rest of the genetic algorithm; a poorly chosen encoding can handicap the entire process.

Each individual is evaluated using a "fitness function," which is another term for the thing we are trying to optimize. In the case of a neural net, this is typically just the inverse of the loss function used for backprop (we invert the value because a GA maximizes, where backprop minimizes). In other applications, more complex or interesting notions of "fitness" may be used, including relative measures such as tournament-based selection.

The overall operation of the GA proceeds in rounds, or "generations." In each round, all the individuals are evaluated using the fitness function, and then some of them are selected to "reproduce." The most common selection criterion is fitness proportionate, which simply means the likelihood of selection is proportional to the fraction of the total population's net fitness accounted for by the given individual. Again, the GA literature contains many alternatives, and the choice of a selection strategy is up to the user.

Selected individuals are then used to produce "offspring," or new individuals, using the genetic operators of "mutation" and "crossover." The exact implementation of these operators is defined with respect to the chosen genotype encoding and must be selected on a per-problem basis, but the underlying idea is always the same. Mutation makes a small, random change to the encoding; for a neural network, this might mean adjusting some randomly selected weights with small random perturbations. Crossover operates on two selected individuals and creates one offspring genome by combining the two parent genomes. For a convolutional neural network, this might involve something like taking half the filter kernels at random from one parent and the other half from the other parent. It is important that whatever operators are used must be well suited to the problem; they need to have some reasonable likelihood of producing a child that is better than the parent(s), or the GA will not converge. For this reason, only one or the other is applied in some applications. If there is no useful definition for crossover, it is better to simply rely on mutation.

Once several offspring have been produced, some number of individuals in the "parent" generation are selected for replacement and removed from the population. Again, the choice of how many are replaced at each generation and how they are selected is a user-choosable parameter, but typically, the population size is kept fixed, meaning that for each individual that is removed, a new offspring is added. It is possible to replace the entire population at each generation, known as "full replacement," but partial replacement can yield better results for some types of problems, as it makes for a slower and more conservative search process.

This process of fitness-based selection, reproduction, and replacement is then repeated for many iterations until the GA converges on a solution. Depending on the problem, this may require dozens, hundreds, or even thousands of generations. Typically, the highest-fitness individual discovered by the end of the process is taken as the final "output" of the algorithm.

References

[1] Nair V, Hinton GE. Rectified linear units improve restricted Boltzmann machines. In: Proceedings of international conference on machine learning (ICML); 2010.

[2] He K, Zhang X, Ren S, Sun J. Deep residual learning for image recognition computing research repository. CoRR; 2015.

[3] LeCun Y, L'eon B, Bengio Y, Haffner P. Gradient-based learning applied to document recognition. Proc IEEE 1998;86:2278–324.

[4] Yan LC, Cortes C, Burges C. The MNIST database of handwritten digits., 1999, http://yann.lecun.com/exdb/mnist/.

[5] Krizhevsky A, Sutskever I, Hinton GE. ImageNet classification with deep convolutional neural networks. In: Proceedings of the 25th International Conference on Neural Information Processing Systems—Volume 1 (NIPS'12). Red Hook, NY: Curran Associates, Inc.; 2012. p. 1097–105.

[6] Hinton GE, Salakhutdinov R. Reducing the dimensionality of data with neural networks. Science 2006;313:504–7.

[7] Hinton GE. Training products of experts by minimizing contrastive divergence. Neural Comput 2002;14:1771–800.

[8] Kingma DP, Welling M. Auto-encoding variational bayes. In: The international conference on learning representations (ICLR); 2014.

[9] Ranzato M, Boureau Y, Yan LC. Sparse feature learning for deep belief networks. In: Proceedings of neural information processing systems (NIPS); 2007.

[10] Bengio Y. Deep learning of representations for unsupervised and transfer learning. In: Guyon I, Dror G, Lemaire V, Taylor G, Daniel S, editors. Proceedings of ICML workshop on unsupervised and transfer learning. Proceedings of machine learning research, vol. 27. Bellevue, Washington, USA: PMLR; 2012. p. 17–36.

58 CHAPTER 3 Overview of advanced neural network architectures

[11] Goodfellow I, Pouget-Abadie J, Mehdi M, et al. Generative adversarial nets. In: Ghahramani Z, Welling M, Cortes C, Lawrence ND, Weinberger KQ, editors. Advances in neural information processing systems. Curran Associates; 2014. p. 2672–80.

[12] Gehring J, Auli M, Grangier D, Yarats D, Dauphin YN. Convolutional sequence to sequence learning. In: Proceedings international conference on machine learning ICML'17. JMLR.org; 2017. p. 1243–52.

[13] Hochreiter S, Jürgen S. Long short-term memory. Neural Comput 1997;9:1735–80.

[14] Greff K, Srivastava RK, Koutník J, Steunebrink Bas R, Jürgen S. LSTM: a search space odyssey. CoRR 2015. abs/1503.04069.

[15] Cho K, Merrienboer B, Gulcehre C, et al. Learning phrase representations using RNN encoder-decoder for statistical machine translation. In: Proceedings of conference on empirical methods in natural language processing (EMNLP); 2014. p. 1724–34.

[16] Wu Y, Schuster M, Chen Z, et al. Google's neural machine translation system: bridging the gap between human and machine translation. CoRR 2016. abs/1609.08144.

[17] Mnih V, Kavukcuoglu K, Silver D, et al. Playing Atari with deep reinforcement learning. In: NIPS deep learning workshop; 2013.

[18] Silver D, Huang A, Maddison C, et al. Mastering the game of go with deep neural networks and tree search. Nature 2016;529:484–9.

[19] Papineni K, Roukos S, Ward T, Zhu W-J. BLEU: a method for automatic evaluation of machine translation. In: Proceedings of the 40th Annual Meeting on Association for Computational Linguistics (ACL '02). USA: Association for Computational Linguistics; 2002. p. 311–8. https://doi.org/10.3115/1073083.1073135.

CHAPTER

Complexity in the use of artificial intelligence in anatomic pathology

4

Stanley Cohen

Center for Biophysical Pathology, Rutgers-New Jersey Medical School, Newark, NJ, United States;
Perelman Medical School, University of Pennsylvania, Philadelphia, PA, United States; Kimmel
School of Medicine, Jefferson University, Philadelphia, PA, United States

Introduction

The previous chapters were designed to demystify the machine learning strategies that are at the core of artificial intelligence (AI) by describing its algorithmic building blocks in detail while avoiding mathematical formulation or computer code. As we have seen, the underlying theme of machine learning is classification, which essentially boils down to identifying similarities and differences among things. From this one can progress to understanding relationships among data for prediction and inference. In anatomic pathology, AI can be used to determine if a lesion is cancerous, evaluate the degree of malignancy from subtle morphological differences, predict clinical behavior, and suggest therapeutic strategies.

AI is beginning to be a robust tool at academic centers. For this reason, commercial ventures to exploit it for pathology are proliferating. A major step forward for practical applications has been the approval of whole slide imaging for diagnostic purposes (as digital images are essential for machine learning). However, for anatomic pathology, there are a number of bottlenecks that must be addressed. Some that clearly stand out are (a) the huge databases required for training (true of neural networks in general, and not just in pathology), (b) image complexity, i.e., the large number of features needed to define the samples (instances) within the dataset, and (c) the topological complexity of the data. An example of this complexity is an image dataset, where each sample may contain normal as well as abnormal tissue and abnormal regions may vary in morphology in important ways.

In addition to the above, relationships are important, such as the distribution and location of vasculature, lymphatics, inflammation, mitosis, necrosis, and their spatial relationship to the areas of pathology on the slide. Complicating these factors is the need for the integration of morphological, genetic, and various "-omics" for various predictive analytics. Omic analysis in itself requires the identification of relationships within the data such as clustering and activation patterns, thus making its

Artificial Intelligence in Pathology. https://doi.org/10.1016/B978-0-323-95359-7.00004-2
Copyright © 2025 Elsevier Inc. All rights reserved, including those for text and data mining, AI training, and similar technologies.

integration with morphologic information difficult. Combining omic and morphological data presents a daunting challenge, and translationally relevant work in this area has only just begun.

These kinds of complexity issues are not unique to pathology; a good example involves autonomous driving algorithms, where all of the above come into play. Interestingly, autonomous driving algorithms are developed by data scientists and computer learning experts, with relatively minor input from professional drivers. For the development of AI in pathology, there must be a close interaction between the pathologists (as content providers, content annotators, and ground truth arbiters for training) and the computer scientists who devise the appropriate machine learning algorithms for use. The purpose of this chapter is to expand on previous chapters describing the basics of machine learning and some advanced deep learning strategies for pathologists to bridge that gap, as well as real-world examples of the principles that are involved.

In this chapter, we will first discuss how images can be segmented into regions of interest, using mitosis detection as an example, and then explore various ways of dealing with petascale datasets through weak supervision and possibilities for so-called one-shot learning. Other topics include the choice of an algorithm for a given task and the problem of parameter tuning within that algorithm. This is not designed to be a comprehensive review; subsequent chapters will provide detailed descriptions of state-of-the-art work being done in a number of pioneering laboratories. It will be apparent that the basic strategies for image interpretation can be used across a wide spectrum of other topics of interest to pathologists, including natural language processing and analysis and prediction from genetic arrays, proteomics, transcriptomes, microbiomes, etc.

Life before machine learning

Before proceeding it is important to note that there were many approaches toward the solution of these problems before the widespread use of machine learning. Many of these involve the extraction of important and measurable features from samples and the use of clustering. In image analysis, mathematical operations have been used to distinguish pixel areas on the basis of intensity, color, texture, shape, and geometric relationships. Blobs (connected areas in an image), edges, and boundaries are important to identify in these tasks. One example of a mathematical operation used is the Hough transform, which can be utilized to isolate features of a particular shape within an image. While these techniques could locate nuclei, glands, and other structural elements within the image for the pathologist, they could not generalize from a constellation of these features to an overall classification or diagnosis of the tissue under study. Machine learning facilitated the transition from image analysis to image interpretation.

Multilabel classification

When we think of the use of AI in analyzing images, we tend to think of deep learning structures that can recognize differences between normal and malignant tissues, provide reliable and accurate estimates of degree of malignancy (grading), and provide information regarding prognosis and optimizing therapeutic choice. Often this is based upon overall pattern recognition and classification based on these patterns. However, if there are separate islands of tumor embedded within long stretches of normal stroma or tissue, the segmentation of those tumor fields will allow the program to treat the normal as a kind of background noise. Also, in addition to overall patterns, pathologists look for specific defining features within the image. These might include the presence of mitoses, degree of vascularization, the presence of tumor clusters within lymphatics or small vessels in the tumor, necrosis, and so on. Also important are spatial relationships among these features.

For all these reasons, in pathology, the need for massive datasets is magnified by the need for extensive annotation within each image. This problem is often addressed by decomposing the image into subsets (image patches or tiles) and trying to automate the process of finding regions of interest as much as possible. Dividing an image into subsets obviously increases the number of training instances needed. Moreover, within those regions of interest, there are often specific features we wish to identify as well as identifying and grading the lesion as a whole. We can use traditional segmentation and object detection schemes embedded into deep neural networks or, alternatively, use a single suitably powerful network for all these tasks.

Multilabel classification is based on image segmentation into regions of interest. There are several kinds of image segmentation, as illustrated in Fig. 4.1, which illustrates segregation categories for scenes of cats and dogs as a simple example.

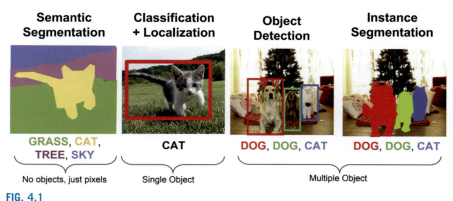

FIG. 4.1

Categories of image segmentation.

Creative Commons, public domain.

62 CHAPTER 4 Artificial intelligence in anatomic pathology

Semantic segregation is separation on the basis of classes within the image; in the illustration, the algorithm has separated classes corresponding to grass, cat, tree, and sky by color without information as to the relative importance of these classes to the human observer (first panel). Object detection involves localizing and classifying the object of interest within the image and either highlighting it for human action or identifying it for further analysis, as shown in the second panel. One can detect multiple classes in the same image (third frame). Finally, there is instance segmentation, in which each of the dogs is individually (and uniquely) identified. In other words, in instance segmentation, we separately identify each instance within a given class, as compared to semantic segmentation, where we identify each class but not the individual elements of that class.

The final goal of this use of machine learning is thus to identify features of interest within the regions of interest and use these in conjunction with overall pattern recognition of the lesion. When dealing with problems involving multilabel classification, a number of machine learning algorithms may be used including multilabel K-nearest neighbor, multilabel decision trees, multilabel support vectors, and multilabel neural networks. A review of multiclassification methods before the advent of deep learning can be found in Ref. [1], with more technical aspects in Ref. [2].

Single object detection

As is the case for AI in general, there is no one algorithm that is optimal for all problems. The use of machine learning for the identification of mitoses provides a good example of this principle and is discussed in detail in Ref. [3]. By way of introduction, it should be noted that detecting mitoses is a difficult problem because of issues relating to interobserver variability that impose constraints on ground truth and because the many papers that have been published make use of widely varying AI approaches. This has been studied in a systematic way. By way of background, prognostic factors related to tumor proliferation have proven to be very important for many tumors. There are a number of histochemical markers for proliferation. Two examples are Ki67 and BRDU. Ki67 is a protein that increases in cells as they prepare to divide. BRDU is a marker that is incorporated into cellular DNA. However, simply counting mitotic figures in tumors by pathologists remains the most widely used form of assessment of proliferation. Although time-consuming and subjective, direct morphological observation can, under ideal conditions, identify mitotic phases. Breast tumors are good examples of cases in which proliferation has prognostic significance. However, even in tumor types where proliferation does not correlate as well with behavior, the presence of morphologically abnormal mitosis can confirm the diagnosis of malignancy.

In 2013, the Assessment of Mitosis Algorithms (AMIDA13) Study was launched (as a competitive challenge, the results were analyzed by Veta et al. and summarized in Ref. [3]). The training data available to all participants consisted of image data accompanied by ground truth mitosis annotations (with due attention to correction for interobserver variability). Two months after the training dataset was released, a testing dataset was made available. This consisted only of image data; annotations were withheld by the challenge organizers to ensure independent evaluation. Results were uploaded to the challenge website for evaluation.

The learning strategies that were compared were computational neural networks, random forests, support vector machines, Bayesian methods, and various boosting strategies. Most of these methods involved a two-step object detection approach in which the first step identified candidate objects that were then classified in the second step as mitoses or nonmitoses. Two of the methods evaluated did not use candidate extraction but instead evaluated every pixel location to obtain a mitosis probability map for each image from which mitoses were detected.

By measures of precision and recall, all of the models described in Ref. [3] gave good performance, with the best being the procedure with multiple alternating convolutional net and max-pooling layers. The F1 score (based on the ratio of the product of precision and recall divided by the sum of precision and recall) of this method was comparable to the interobserver agreement among pathologists. The worst performance, by these measures, made use of random forest classifiers. Interestingly, experiments with combining results from the different methods by majority voting or intersection of the better performing methods showed no performance gain over the best individual method.

Multiple objects

Image segmentation, multilabel classification, and multiple-instance learning (MIL) (see weakly supervised learning section) all have features in common. For example, in the case of histopathologic material, only a small part of the image may be useful for processing and analysis. Thus, a prior step is localization, in which we identify areas of interest. In many applications, this is done by breaking the image up into small contiguous patches and either identifying the "regions of interest" through prior annotation of the training set or by strategies within the algorithm itself. This is similar to the methods used for MIL to be discussed in a subsequent section.

As aforementioned, object detection is based on the use of a classification label for the objects of interest within the sample images rather than simply for the sample as a whole. Various machine learning strategies led to the simultaneous identification and classification of multiple labels within images. As is the case for image analysis in general, neural networks have transformed this field. Typically, one extends a convolutional neural network (CNN) to multilabel classification by transforming it into multiple single-label classification sets. These treat the labels independently and so have difficulty in modeling dependency between labels. Object detection is not limited to neural network analysis. Many of the machine learning strategies can be applied to this task as well. There have been only a few comparative studies of these using standardized datasets. One example has already been described in detail [3]. Bjednordi et al. have compared 32 algorithms for detecting lymph node metastases [4]; 25 were based on deep convolutional networks, while the remainder utilized

CHAPTER 4 Artificial intelligence in anatomic pathology

support vector machines or random forests and statistical and structural texture features. The tasks included both metastasis identification and whole slide image classification. All the algorithms performed well, and several were better than the best performing pathologists in the comparison panel.

Advances in multilabel classification

At this point, it is helpful to compare this image analysis problem to issues that arise in autonomous driving. The algorithms must be able to identify human beings, other cars, road markers, traffic signs, and their specific locations within a scene and thus make use of multilabel classification strategies. As driving is a dynamic process, there needs to be a way of dealing with sequential data, and for this purpose, recurrent neural networks (RNNs) are an obvious choice as they capture temporal relationships. Recently, there have been several advances in the use of visual attention or RNNs combined with CNNs for this purpose. Recall from the previous chapter that whole CNNs are good for dealing with spatially related data, and RNNs are good at handling temporally related data. Previously, RNNs have been used for signal analysis, sequential images (video), and natural language processing (where the temporal sequence of words in a sentence matters). They can also be used for static image analysis. Here, an RNN processes inputs sequentially, dealing with locations within the images one at a time, and incrementally combines information so obtained to create the internal representation.

For example, Wang et al. [5] described a model inspired by the way humans analyze images by continuously moving and their attention from one discriminative object to the next when performing image-labeling tasks. As usual, the CNN extracts the deep feature representations, and the RNN [in this case, a long short-term memory (LSTM) network] iteratively locates class-related regions and predicts label scores for these. LSTM networks are explained in Chapter 3. This model also appears to allow for class redundancies in labeling. For example, for many purposes, labels such as "baby cat" and "kitten" have pretty much the same meaning. Although this study did not involve histopathological materials, more recently, Alom et al. performed similar studies combining residual and recurrent networks [6]. This sophisticated model was applied to blood vessel segmentation, skin cancer segmentation, and lung lesion segmentation. This model yielded state-of-the-art performance in all three tasks.

The possibility of exploiting reinforcement learning for image recognition tasks in pathology is especially interesting. In the case of autonomous driving, reinforcement learning has been applied to dynamic sensory input, but reinforcement learning has only begun to be exploited for tasks involving static image analysis. Mathe et al. [7] have described sequential search strategies that accumulate evidence from a set of image locations utilizing reinforcement learning concepts such as exploration and reward. Basically, traditional object detection is based upon an exhaustive spatial search and is brute force in nature, whereas reinforcement learning involves moving toward a defined goal in a stepwise manner. Chen et al. [8] have combined attentional models with reinforcement learning, where the "reward" is classification correctness.

Interestingly, biological neural networks such as brains utilize reinforcement learning in visual pattern recognition. They also learn to recognize sequences of patterns. Thus, it appears likely that approaches such as these will be fruitful in pathology as well.

Graphical neural networks

Sophisticated implementations such as those above capture relationships among the objects in an image in addition to classifying them. However, there are a number of modern approaches that deal explicitly with such relationships. The majority of these fall into the category of graph neural networks. In mathematical terminology, a graph is a collection of distinct objects (vertices or nodes) and lines (edges) connecting them (Fig. 4.2).

Both neural networks and decision trees are graphs, as are connecting flight maps and highway maps connecting cities. In the case of a neural network, the nodes are the simulated neurons, and the edges show the direction in which weighted information flows toward the next layer of nodes. Edges can be unidirectional or bidirectional. As an example, network analysis, which is a branch of graph theory, is used to understand relationships and patterns within transcriptomes.

If data can be expressed as a network, then it is possible to analyze not only the relationship of a test sample to the classes defined by the training set but also the relationship between the attributes that define that sample. A graphical network

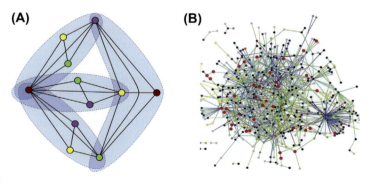

FIG. 4.2

(A) Sample graph showing nodes and edges connecting them, as well as apparent clustering into four subgroups. (B) Actual protein expression network of a treponeme.

(A) Reprinted from Wikipedia Commons, public domain. (B) Reprinted from Wikipedia Commons, with attribution: Hauser R, et al. The syphilis spirochete. PLoS One 2008;3(5):e2292. https://doi.org/10.1371/journal.pone.0002292.

operates at this level. To do so, there must be a way of quantitating the relationships among the nodes. The connections can be described by what is known as an adjacency matrix. The elements of the matrix indicate whether pairs of vertices are adjacent (the element at the intersection of row and column is 1 if there is a connection between them, and otherwise 0). In other words, the adjacency matrix represents the adjacency of vertices. There is another form of representation of a graph that is known as an incidence matrix, which, as its name suggests, represents the incidence of vertices and edges. These matrices are constructed from the graph in visual form. Conversely, it is possible to recreate a pictorial diagram of a graph from its matrices (or one functionally identical or isomorphic to that diagram). As neural networks are based on linear algebra, the graph, as defined by its matrix, is now amenable to computation by the network. The effective implementation of a graphical network is a topic of active investigation. The models are quite complex and beyond the scope of this discussion. However, for the mathematically sophisticated, Zhou et al. [9] have recently published a review of methods and applications involving graphical neural networks.

It is possible to analyze graphical relationships from CNN output without an explicit use of a graphical neural network. For example, newly published work by Jung et al. [10] has demonstrated the power of integrating traditional deep learning with the mathematics of graph theory for the analysis of histopathologic images involving collagen deposition in immunostained images in relation to HIV Infection. A CNN is first used for segmentation of the whole slide image. Parallel processing is applied to convert segmented collagen fibrils into a relational graph, and graph theory is used to extract topological and relational information from the collagenous framework.

In brief, graphical networks are designed to deal with relationships among their nodes, which may represent attributes, attribute sets, or attribute patterns. Conventional CNNs are not very good at this. When dealing with images, there is another area of weakness. A neural network can identify a face from representations that define the overall shape, shape, number, and position of elements within the face (two eyes, one nose, one mouth). A model capable of good generalization will have no trouble with different shapes such as round or oval eyes, broad or narrow noses, and so on. They can also handle translational, rotational, and scaling relationships of the elements. To a CNN, a face in which one eye and the mouth are switched may appear to be similar because they are constructed of similar elements, yet they are quite different to a human observer. The internal data representation of a CNN simply does not take into account important spatial hierarchies between simple and complex objects and thus their meaningful relationships to each other.

Capsule networks

Another newly emerging idea is that of the capsule network, in which the output of each artificial neuron is not a scalar but rather a vector that encodes both a probability of detection of a feature (length) and its state (orientation). Thus, neuronal activities are directly associated not only with feature detection but also with the orientation of that feature. Hinton's group [11] recently described the basic architecture and the process of dynamic routing between capsules.

Capsule networks are relatively new and, in their current form, are difficult to train and not very accurate. Thus, it is still too early to assess their ultimate utility.

Weakly supervised learning

There have been many attempts to achieve unsupervised learning, i.e., to obtain predictive representations from unlabeled data. Although human beings can do this, it has proven very difficult to implement this strategy in a robust way for practical AI. As described in Chapter 2, clustering algorithms such as K-means and principle component analysis are two ways of finding relationships and/or identifying critical features in a database containing unlabeled examples and, in that sense, are unsupervised. Much current research involves employing these and related strategies within the context of deep learning via hybrid strategies. However, for the most part, current efforts in pathology are devoted to weakly supervised rather than unsupervised classification. Weak supervision may be thought of in a simplistic way as the obverse of multilabel classification. In multilabel classification, complexity is introduced by the need to simultaneously identify complex subimages or patterns within each example from a well-defined dataset. In the case of weakly supervised learning, the classifications need not be complex but are based upon an incompletely labeled dataset itself. In both cases, the task is to abstract information that is much more complicated than simply assigning a class label to an unknown sample.

There are three basic types of weak supervision. In incomplete supervision, only a small subset of training data has labels. This is typically the case for image analysis. Huge datasets may be available, but usually, only a small subset of these is annotated. In the case of inexact supervision, some supervision information is given, but it is not as exact as desired. For example, a dataset may consist of breast cancer images, but only the diagnosis is known for the slides. These are coarse labels. Fine-grained information, such as that obtained by annotation by trained pathologists, may only be available for a small subset of these slides. Inaccurate supervision refers to situations in which the label information may not be perfect ground truth. In pathology, this is sometimes due to interobserver variability and sometimes due to the presence of noise, for example, difficulty in distinguishing between a mitotic and a pyknotic nucleus. For a good general survey of these concepts and approaches, see Ref. [12].

With this as background, it is relatively straightforward to understand the concepts behind weakly supervised learning. For incomplete supervision, we can try to make do with the small labeled subset that is available. Recall that a neural network learns by developing increasingly complex (and lower dimensional) representations of the data at each hidden layer within the network. This process is implemented by adjusting the weights of the connections between layers by comparing the output of the network with its input and minimizing a loss or error function.

The starting point involves random weights across the connections, and these are then adjusted via backpropagation. In the previous chapter, several methods for obtaining representational information in the early layers of a neural network using only limited subsets of the data were described. These include autoencoders and generative adversarial networks (GAN). In both of these cases, the initial stages are unsupervised in that the program creates an internal representation by reconstructing an input as its output. This gives rise to internal representations of the data that may then be used as the starting points for a conventional neural network such as a CNN. The initial stages of a GAN or autoencoder thus learn a set of weights that can be used as the starting points of the CNN instead of beginning with a set of random weights. This reduces the size of the training set, and so only a subset of the training set needs to be labeled.

Also, as it was pointed out previously, as early representations may have similar properties across different but similar datasets, it is possible to pretrain a new model by using the early layers of the model previously fully trained on a similar problem and using the weights in those layers as starting points rather than the usual beginning random weights (see Chapter 3). This is known as transfer learning.

In all these cases, a limited set of annotated data is used as for the beginning steps in the training of the algorithm. These are some of the most effective ways of dealing with incomplete supervision. Examples of the use of autoencoders, GANs, and transfer learning, respectively, are found in Refs. [13, 14]. It is also possible to bootstrap learning by using labels assigned to unknown examples by the network with limited labeled data available to it and using these new examples to expand the labeled training set. This is closely related to strategies used to generate synthetic data when there is a class imbalance, as will be discussed in the next section.

For inexact supervision, there are a number of multiple-instance approaches (MIL). In these, the focus is shifted from discrimination of the samples themselves to discrimination of "bags," i.e., breaking up the images into tiles or patches (Fig. 4.3).

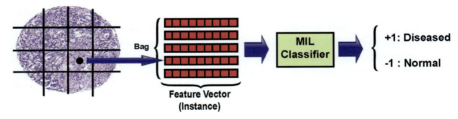

FIG. 4.3

Multiple-instance learning (MIL) for image analysis. The diagnostic image is first split into rectangular patches. A set of features is extracted from each patch, and an instance is formed by the resultant feature vector. Patches belonging to the same diagnostic image are grouped into a bag. The bag is then classified as diseased or healthy by an MIL method.

Reprinted from Kandemirand M, Hamprecht FA, Computer-aided diagnosis from weak supervision: a benchmarking study, Comput Med Imaging Graph 2015;42:44–50.

It should be noted that the MIL approach is not confined to deep learning (neural network) algorithms. It has been applied to support vector machines [15] and in combination with feature clustering [16]. Campanella et al. [17] apply an MIL to a residual neural network in a highly sophisticated manner to the study of prostate, basal cell, and breast cancer at the terabyte scale (discussed in detail in a subsequent chapter).

The above examples all represent semisupervised learning. There are fully unsupervised learning strategies as well. A classic example is the K-means algorithm described in Chapter 2. This and related clustering algorithms require no information about the labels of the sample but instead find clusters in the data based upon geometric closeness in an n-dimensional space. However, to be useful in a diagnostic/predictive setting, the clusters must ultimately be given a label. However, it is not necessary to label every sample in a given cluster individually; one can use the same class label for each element of that cluster. Clustering can be the basis of an unsupervised network for image classification (CUnet) [18] or can be combined with other strategies such as MILs.

For inaccurate supervision, the goal is to both increase the accuracy of ground truth through the use of multiple trained observers and reduce "noise" by statistically based data preprocessing.

Previous chapters have discussed the concept of ensemble learning. In all of the above examples, weakly supervised learning strategies attempt to achieve strong generalizations (performance) from limited amounts of labeled (annotated) data. In contrast, ensemble learning attempts to achieve strong generalizations from the use of multiple learners. While addressing different aspects, these two approaches may be synergistic with each other [19].

Synthetic data

Weakly supervised methods are needed when there are large datasets available but only a limited number of annotated examples within that dataset. There are a number of other situations in which the dataset is well annotated, but there is a class imbalance in that set. For example, there are usually many examples of normal tissue but limited examples of specific kinds of neoplasms of that tissue. Clearly, this creates a situation where there are not enough examples of each class for successful training. In addition, it can lead to a spurious kind of accuracy. As an example, suppose we are studying a neoplasm that has an incidence of 1%. Our training set consists of 10,000 entries, of which 100 are positive and the remainder negative. We can easily generate an algorithm with an accuracy of 99% by simply calling every sample negative. Thus, every AI strategy that generates 99% accuracy is suspect. We can guard against this by calculating metrics such as precision and recall, as described above.

We can expand the number of samples by the use of so-called synthetic data. Synthetic data are created rather than obtained from real-world samples. A typical example is a dataset in which some entries are missing one or more attributes. One can fill in those blanks with approximations based upon a statistical analysis of the collective data attributes in the set. Another method of generating synthetic data is to create a physical model that explains the observations and then reproduce random data using that model because, in many cases, observed features of a complex system can often emerge from simple rules. Patki et al. [20] have described a system that builds machine learning models out of real datasets to create such artificial data, the synthetic data vault. This model captures correlations among the multiple features. In their study, the investigators found no significant performance difference when using synthetic data as compared with using only real-world data.

For the case of image analysis, we can create synthetic data by taking real samples and manipulating them in some controlled way to create randomized new examples that share properties of the real data. There are a number of available programs, such as GANpaint Studio, that allow the creation of realistic images of faces or scenes (available as a free demo from the internet). Such programs have been used to generate so-called "deepfake" images for nefarious purposes, but such algorithms can be used to create synthetic images from real ones for use as supplemental in machine learning protocols. Hou et al. [21] have used GANs to generate histopathology images that model the distribution of real data.

N-shot learning

When performing tasks such as classification using AI, large datasets are used to create internal representations for each class that can be used to generalize to new unlabeled examples. Humans do not learn in this fashion. For example, show an infant a banana a few times, and the child will recognize that banana. It is true that complex patterns require more learning; however, it is well known that one can "train" second-year medical students to distinguish between normal, benign, and malignant examples with much less than 100 samples. In general, humans can classify based on limited examples if they have adequate information about their appearance, properties, and, in some cases, their functionality. This may be a general capability of suitably complex biological brains. Levenson et al. [22] have demonstrated that it is possible to train pigeons to accomplish cancer diagnostic tasks with high degrees of precision in less than 200 trials! An ensemble of 16 pigeons was able to achieve a performance approaching that of a trained CNN.

A "shot" is defined as a single example available for training. In one-shot training, we only have a single example of each class. In N-shot learning, there are N examples. The task is to classify any test image into a class using that constraint. One way of doing this is by the use of "prototypical networks." Unlike the usual deep learning architecture, prototypical networks do not classify the images directly. They learn the mapping of the image into metric space. This means that images of the same class are placed in close proximity to each other, while different classes are placed at further differences. When a new example is presented, the network just checks for the nearest cluster to it. These models are essentially sophisticated forms of the K-means clustering algorithm and are based on CNNs. Models in this category are known as

"Image2Vector" models. More sophisticated models involve neural networks augmented with memory via a neural attention mechanism. These can be complicated to describe, but a good example is seen in Ref. [23]. Although "one-shot learning" should be used to describe the case where $N = 1$, that term is also used more generally when N is very small.

The method for N-shot learning with the greatest potential for one-shot learning currently appears to be the Siamese network. Instead of a model that learns to classify its inputs, the neural network learns to differentiate between two inputs. A typical Siamese network consists of two identical neural networks, each taking one of the two images. Identical means that they have the same configuration with the same parameters and weights. The last layers of the two networks are jointly fed into a contrastive loss function that calculates the degree of similarity between the two images. An example from natural language processing involves scoring an exam question. Here, one input is the question sentence, and the other input is the answer. The output is how relevant the answer is to the question. This is not a simple classification task in that we do not have a database consisting of all possible answers that are judged "correct" and examples that are judged "wrong." Nevertheless, a Siamese network can extract similarity or connection between question and answer.

Very often, the term one-shot learning is used whenever n is a small number. In any event, Siamese Networks have achieved state-of-the-art performance in one-shot image recognition tasks (Koch et al. [24]). In this study, each of the two networks was a standard CNN. The final convolutional layer output was flattened into a single vector, which was then fed into a fully connected layer and then to an additional layer that computes the distance metric between the outputs of each Siamese twin.

There are relatively few examples of one-shot learning involving pathology subject matter. Within the past few months, Yarlagadda et al. [25] demonstrated the use of a one-shot learning model for cervical cancer cell classification in histopathological images.

One-class learning

It is important to distinguish between one-shot learning and one-class learning. Closely related to N-shot learning is the concept of one-class learning. The object of one-class classification is to recognize instances of a class by only using examples of that class. All other classes are lumped together and are known as alien classes. In other words, the difference between one-class and binary classification is due to the absence of training data from a second class. While this approach has received a great deal of attention for simple datasets, it has proven difficult to obtain good results in real-world datasets. Recently, Perera and Patel [26] have achieved impressive results using a CNN model and two novel loss functions (compactness loss and descriptiveness loss) as compared to typical squared difference and/or entropic loss functions used in other deep learning algorithms. They achieved results better than those

CHAPTER 4 Artificial intelligence in anatomic pathology

obtained by competing methods including graphical, K-nearest neighbor, and support vector models. Decision tree-based models were not considered.

One-class learning can be useful for outlier detection, and as it only involves a single class, it is essentially unsupervised. There have been previous approaches to anomaly detection using one-class support vector machines, directly, or hybrid methods combining feature extraction via deep learning with support vector machines [2]. More recently, Chalapathy et al. [27] have described a one-class neural network that uses a novel loss function. Here, the specific features of the hidden layers are constructed for the specific task of anomaly detection. This is in contrast to the generic feature extraction methods used in previous approaches. It is possible to treat cancer as a tissue anomaly. Most such studies have involved classifying tumor transcriptomes as compared to normal (e.g., Ref. [28]). However, it is easy to envision a training protocol in which only normal tissue images are used, and a metastatic focus is then treated as an outlier. As of the date of the preparation of this manuscript, no examples of work in this area were found. However, industrial techniques involving the detection of anomalies in images of surfaces suggest that this is a feasible avenue to explore.

Risk analysis

The purpose of all the methods discussed here is to go beyond simple classification (with grading based on overall morphological appearance) by examining specific and often quantifiable subfeatures of the images mimicking the real-world approach by pathologists. An ultimate goal is to learn how to do this using limited information rather than training sets consisting of thousands of images for each specific classification/diagnostic task. A variety of AI strategies have been explored in this chapter; in most cases, morphologic information is analyzed directly. While strictly not within the purview of this review, it is important to note that a number of studies have appeared that involve a computational approach for learning patient outcomes from digital pathology images by combining traditional deep learning algorithms with a standard risk hazard model. For a number of reasons, including computational ease and the ability to deal with nonlinearity, logarithmic models are used, and in particular, the Cox proportional hazard model. We will discuss two approaches: DeepConvSurv [29] and Survival Convolutional Neural Networks [30]. Both look at image patches from regions of interest within the biopsy specimen. DeepConvSurv is basically a CNN, except that it uses a Cox function as the loss function that is used to train the network. In DeepConvSurv, a large number of semirandomly collected image patches undergo dimensional reduction using K-means clustering. Aggregated clusters are then run through the connected neural network with boosting via Cox's negative log function. Finally, the aggregated patch risk data are equated to patient risk. This, however, is only a loose approximation because patient risk also includes size and location of tumor, all modified by age, sex, and comorbidity, and in modern medicine, genomic and proteomic features.

SCCN is more sophisticated in that it explicitly utilizes genomic data as well as image morphology. It requires annotation of the image dataset, which is then run through a full CNN. The genomic data are inputted directly into the fully connected layers, bypassing the convolutional layers. The network is trained in the usual fashion with a standard loss function, except that the outputs are then entered into a final Cox approximation layer. In effect, SCCN is taking advantage of the universal approximation theorem, namely that a single layer with an appropriate activation factor can approximate any arbitrary function. Here, the function is the Cox partial hazard function. By providing a risk-based heat map overlying histological features in their publication, the authors documented that the algorithm is detecting meaningful risk elements.

Looking ahead, it is likely that deep learning algorithms will become better able to integrate morphological data with the whole spectrum of available patient information, allowing optimal tailoring of therapy to the specific disease as manifested in a specific patient, thus enabling truly personalized precision medicine.

General considerations

In this chapter, the examples chosen for illustration made use of a variety of machine learning strategies, and in some cases, hybrid networks were utilized as well. An important practical question involves the choice of an algorithm to solve a given problem. Unfortunately, there is no correct answer. In this context, it is important to reflect upon the so-called "no free lunch" theorem of Wolpert and Macready [31], which states, "for any algorithm, any elevated performance over one class of problems is offset by performance over another class."

Simply stated, there is no one model that works best for every problem. However, each method may be best aligned with a particular class of problems. This means that it is not possible to generalize that the best performing algorithm in a given study is the best strategy for other machine learning tasks. Thus, it is important to choose the appropriate model for a given problem. Unfortunately, there is no theoretical basis upon which to choose. This must be done empirically through trial and error. The algorithm assessment study [3] provided a good specific example of this general principle.

Regardless of the choice, every model has tunable parameters or hyperparameters. For example, in the case of neural networks, the tunable parameters include the number of nodes in each layer and the number of layers. Backpropagation involves choices of momentum and learning rate. For a CNN, decisions must be made about the size of the convolution matrix. Initial weights must be randomized, but most random numbers are generated by computer starting with a "seed" number. That seed is also a parameter. This list is by no means exhaustive. For random forests, the number of trees, the number of branches, the tree depth, and so on are parameters that must be chosen. For k-means, one must choose the number of clusters, as well as the k random number seeds that start the iterative process of defining clusters. Tunable parameters are a fact of life for the artificial intelligentsia.

The other consideration in selecting a model is the implicit bias of that model. This does not refer to external biases such as those that arise from choices of the elements of the training set or a value-based assessment of the feature or class labels, but rather to the assumptions baked into each model. For example, K-means assumes roughly spherical clusters similar in size to each other (although there are better models that do not have these constraints). Naïve Bayes assumes that the attributes describing the feature vector are independent of each other. Assumptions about data distributions are at the heart of almost every machine learning algorithm. It is important to understand the nature of the dataset in sufficient detail to allow choosing the algorithm whose constraints are least critical for that dataset. Unfortunately, the use of the term "bias" tends to anthropomorphize the AI program and obscures the issues involved.

The detection of implicit bias in a deep learning network is made difficult in that we have no true understanding of how that network is processing its inputs to arrive at the correct outputs. In other words, there are problems with explainability and interpretability. Interpretability is the ability to predict the effect of a change in input or algorithmic (tunable) parameters. Explainability is the ability to understand the basis upon which the algorithm is drawing its conclusion. For example, it might be reassuring to be able to explain to a surgeon the basis upon which the AI came up with a diagnosis of a high-grade malignancy or how it differentiated between two morphologically similar but biologically different tumors. It is interesting that if one questions a reference level, high-trained pathologist as to how he or she arrived at a diagnosis, they will often refer to the "years of experience" that led to the diagnosis. When pressed for more specific criteria, they may make them up, but it is often on an ad hoc basis, to justify the decision already intuitively made. For this reason, the black box nature of neural networks, while disquieting to some, does not bother others of the artificial intelligentsia.

There are ways of gaining some insight as to what is happening behind the scenes so that we can raise the curtain to see the wizard at work. For example, saliency maps create a visualization of the pixels in an image that contributes most to the predictions by the model. By calculating the change in predicted class by applying small adjustments to pixel values, we can measure the relative importance of each pixel to the ultimate output value. This is discussed in Ref. [32]. Other approaches involve trying to determine the activity of neurons in the hidden layers as backpropagation proceeds and, additionally, to obtain visual representations of the increasingly complex outputs of the hidden layers by mapping characteristics of these such as intensity, orientations, color, and shapes.

These approaches may provide some insight as to how the neural network is discriminating among classes but still do not "explain" what is going on in human terms. When an AI makes an unexpected and startling chess move that has no obvious strategic importance to a human observer, and that move initiates a winning sequence, these internal measures provide no clue as to how the program created the move that, in retrospect, turned out to be "brilliant." Thus, they do not create a true learning experience for the human observer. On the other hand, this occasionally happens when human observers watch a human Grand Master at play. Nevertheless, knowledge about the internal decision-making process of a machine learning algorithm may inform the development of better algorithms, so there is something to be said for insisting on AI explainability while accepting our occasional inability to understand human genius.

In spite of all these caveats, studies using different "shallow" learning AI strategies on the same dataset often get similar results. The arrival of deep learning led to a great improvement in machine learning over most, if not all, alternate approaches. When dealing with complex, multidimensional feature sets, neural networks substantially outperform other kinds of machine learning. Even here, when papers dealing with the same dataset but using variations of network models are compared, the improvements claimed by each tend to be incremental. While some are better than others, they all seem to converge around results that achieve greater than 90%–95% accuracy (with similar results for the other evaluation metrics). This, however, may be merely because of the fact that careful attention has been made to optimize the program of choice in each. There are also differences in computational power that must be considered. For this latter reason, performance speed should never be used in comparing one algorithm to another unless they are running on exactly the same platform. Another reason is that often the training sets are simplified and carefully curated versions of data that are found in the wild. This raises the possibility that in less structured situations, greater differences may be observed. Finally, results obtained by a given algorithm in one setting may not always correspond to results using the same algorithm in a different setting. In other words, past performance is not guarantee of future results.

Because of considerations such as these, various ensemble methods have been used, as well as hybrid models involving combining two or more different algorithms sequentially or in parallel. Examples have been presented above for both multilabel detection and weak supervision.

Summary and conclusions

Although we have not discussed the use of AI for clinical laboratory data, natural language processing, or (with one exception) genetic, expression, and protein "-omics," considerations similar to those discussed hold for all these applications. With respect to image recognition and analysis, we are far beyond merely identifying a lesion and predicting outcomes based upon morphological patterns. Our images are complex, and both abnormal structures and the relationship of the normal elements to these must be analyzed. Simultaneously, specific features within the lesion that provide diagnostic or predictive information must be identified and, where possible, quantified. Some of the tools for accomplishing these tasks have been presented in this chapter, and specific references that illustrate their use in real-world settings have been

76 **CHAPTER 4** Artificial intelligence in anatomic pathology

provided. Important caveats have also been noted. The field is growing at an exponential pace; a number of the references chosen did not exist at the beginning of this project. A major consideration for all of these involves the need for huge datasets. As shown, there are a number of approaches to make the generation of training sets more efficient and hopefully to reduce the size of these sets.

To be useful across laboratories, all these approaches are dependent on standardization of each stage of the process, from whole slide imaging to final diagnostic output. It will also require consensus regarding annotation standards. However, given that this is rapidly coming to pass, the future is bright for AI as an augmenter and enhancer of human capability in pathology.

References

[1] Devkar R, Shiravale S. A survey of multi-classification for images. Int J Comput Appl 2017;162:39–42.

[2] Zhang M, Zhou Z. A review on multi-labeling algorithms. IEEE Trans Knowl Data Eng 2014;26:1819.

[3] Veta M, et al. Assessment of algorithms for mitosis detection in breast cancer histopathology images; 2014. arXiv:1411.5825v1.

[4] Bejnordi BE, et al. Diagnostic assessment of deep learning algorithms for detection of lymph node metastases in women with breast cancer. JAMA 2017;2199–210.

[5] Wang Z, et al. Multi-label image recognition by recurrently discovering attentional regions; 2017. arXiv:1711.02816v1 [cs.CV].

[6] Alom MZ, Hasan M, Asari VK. Recurrent residual convolutional neural network based on R2U-Net for medical image segmentation; 2018. arXiv:1802.06955v5.

[7] Mathe AP, Sminchisescu C. Reinforcement learning for visual object detection; 2016. https://doi.org/10.1109/CVPR.2016.316.

[8] Chen T, et al. Recurrent attentional reinforcement learning for multi-label image recognition. In: 32nd AAAI conference on artificial intelligence (AAAI-18); 2018. https://www.aaai.org/ocs/index.php/AAAI/AAAI18/paper/download/16654/16255.

[9] Zhou J, et al. Graph neural networks: a review of methods and applications; 2019. arX1v:1812.08434v3.

[10] Jung H, et al. Integration of deep learning and graph theory for analyzing histopathology whole-slide images. In: IEEE imagery pattern recognition workshop; 2018. https://doi.org/10.1109/AIPR.2018.8707424.

[11] Sabour S, Frosst N, Hinton GE. Dynamic routing between capsules. In: 31st conference on neural information processing systems (NIPS), Long Beach, CA; 2017. p. 1–8.

[12] Zhou Z-H. A brief introduction to weakly supervised learning. Natl Sci Rev 2017;5:44–53. https://doi.org/10.1093/nsr/nwx106.

[13] LeHou, et al. Improving prostate cancer detection with breast histopathology images; 2019. arXiv:1903.05769v1 [cs.LG].

[14] Fischer W, et al. Sparse coding of pathology slides compared to transfer learning with deep neural networks. In: BMC informatics 19, supplement 18; computational approaches for cancer; 2018. p. 109–17. art. 489.

References

[15] Andrews S, et al. Support vector machines for multiple-instance learning. Adv Neural Inf Proces Syst 2002;15:561–8.

[16] Xu Y, et al. Multiple clustered instance learning for histopathology cancer image classification, segmentation and clustering. In: IEEE conference on computer vision and pattern recognition; 2012. https://doi.org/10.1109/CVPR.2012.6247772.

[17] Campanella G, et al. Clinical-grade computational pathology using weakly supervised deep learning on whole slide images. Nat Med 2019;25:1301–9.

[18] Dong L, et al. A compact unsupervised network for image classification. IEEE Trans Multimed 2016. arXiv:11607.01577v1 [cs.CV].

[19] Zou Z-H, et al. When semi-supervised learning meets ensemble learning. vol. 5. Front Range Electrical Engineering; 2010. https://doi.org/10.1007/s11704-009-0000-0.

[20] Patki N, Wedge R, Veeramachaneni K. The synthetic data vault. In: 2016 IEEE 3rd international conference of data science and advanced analytics, vol. 1; 2016. p. 399–420. https://doi.org/10.1109/DSAA.2016.49.

[21] Hou L, et al. Unsupervised histopathology image synthesis; 2017. arXiv:712.05021v1.

[22] Levenson RM, et al. Pigeons as trainable observers of pathology and radiology breast cancer images. PLoS One 2015;10(11). https://doi.org/10.1371/journal.pone.0141357.

[23] Vinyals O, et al. Matching networks for one shot learning; 2017. arX1v:1606.04080v2 [cs.LG].

[24] Koch G, Zemel R, Salakhutdinov R. Siamese neural networks for one—shot image recognition. In: Proceedings of the 32nd international conference on machine learning, Liile, France, vol. 37; 2015.

[25] Yarlagadda DVK, et al. System for one-shot learning of cervical cancer cell classification in histopathology images. In: Presented at 2019 SPIE medical imaging conference, San Diego, CA; 2019. https://doi.org/10.1117/12.2512963.

[26] Perera P, Patel VM. Learning deep features for one-class classification; 2019. p. 1–15. arXiv:1801.05365v2 [cs.CV].

[27] Chalapathy R, Menon AK, Chawla S. Anomaly detection using one-class neural networks; 2018. arX1v:1802.06360v1 [cs.LG].

[28] Quinn TP, et al. Cancer as a tissue anomaly: classifying tumor transcriptomes based only healthy data. Front Genet 2019;10:599. https://doi.org/10.3389/fgene.2019.00599.

[29] Zhu X, et al. WISA: making survival prediction from whole slide histopathological images. In: IEEE conference on computer vision and pattern recognition; 2018. p. 7234–42.

[30] Mobadersany D, et al. Predicting cancer outcomes from histology and genomics using convolutional networks; 2018. www.pnas.org/cgi/doi/10.1o73/pnas.1717139115.

[31] Wolpert DH, Mcready WG. No free lunch theorems for optimization. IEEE Trans Evol Comput 1997;1:67–82.

[32] Simonyan K, Vedaldi A, Zisserman A. Deep inside convolutional networks: visualizing image classification models and saliency maps; 2014. arXiv:1312.6034v2 [cs.CV].

CHAPTER

Dealing with data: Strategies of preprocessing data

5

Stanley Cohen

Center for Biophysical Pathology, Rutgers-New Jersey Medical School, Newark, NJ, United States;
Perelman Medical School, University of Pennsylvania, Philadelphia, PA, United States; Kimmel
School of Medicine, Jefferson University, Philadelphia, PA, United States

Introduction

Most of the applications of artificial intelligence algorithms are dependent upon the use of large labeled datasets (supervised learning). Although there has been great progress in the use of weakly or sparsely supervised learning, for the most part, applications to pathology have not yet exploited these in a systematic way. Similarly, while reinforcement learning makes use of reward strategies rather than loss or cost functions derived from comparison of label to prediction, these have not been typically applied to the topics of interest to pathologists. Recently, however, sophisticated algorithms have been developed using reinforcement learning for image classification [1–3], so this is likely to change in the near future.

Because of the current dependence of these artificial intelligence algorithms on "large data," it becomes important to make sure that data are in an optimal form for input into the chosen algorithm. At the very least, good data are at least as important as a good algorithm, and often they may be more important. In fact, the majority of data science work involves getting data into the format needed for the model to use and in its most compact and informative form. The reason for this is that the corollary of "garbage in, garbage out" is "better data beat fancier algorithms." Real-life data are messy and unruly; it has to be beaten into submission. It is often incomplete or inconsistent and may contain errors. We must make sure that our samples are representative of the various classes we are trying to predict from sample attributes, we must make sure that the samples themselves are relevant, and we must deal with missing attributes among the samples and determine whether outliers are significant or artifacts. Where there is a class imbalance, we must try to correct this either by reducing part of the dataset or constructing synthetic data. These tasks can be broadly categorized as feature engineering and involve feature selection, feature standardization and normalization, feature extraction, and feature correction. These terms are

Artificial Intelligence in Pathology. https://doi.org/10.1016/B978-0-323-95359-7.00005-4
Copyright © 2025 Elsevier Inc. All rights are reserved, including those for text and data mining, AI training, and similar technologies.

CHAPTER 5 Dealing with data: Strategies of preprocessing data

simply descriptions of various aspects of the same process rather than a distinct set of procedures, and they overlap to some extent. Collectively, they fall under the general category of data preprocessing. Other terms used in the literature for this process are data preparation, data cleaning, and data scrubbing. There are both straightforward and commonsense approaches to analyzing the data as well as highly sophisticated mathematical approaches. The former will be discussed first.

Overview of preprocessing

The initial step in working with data is the choice of the most relevant features. This determination depends entirely on the information you are trying to obtain from the data. For example, consider a database for vehicles that includes the following information:

Number of wheels, cargo capacity, height, length, brand, acceleration, roadholding, number of passengers, and type of navigation system. To determine the distribution of trucks versus cars, brand, acceleration, and number of passengers are not as important as height, length, and cargo capacity. For determining the brand, whether the vehicle is a car or a truck may be irrelevant. For identifying a sports car, acceleration and roadholding may provide important information. The presence or absence of a navigation system and what kind it is are not likely to be helpful for any of these potential uses of the data but could be important in another context. In other words, it is important to understand what the data will be used for to determine the appropriate sample attributes to be used.

There is another aspect to preprocessing based upon the so-called "curse of dimensionality."

At first glance, it would seem that the larger the dimensionality of the feature vector (the larger the number of attributes), the easier it would be to distinguish among different classes of the data. However, when we have too many features, observations actually become harder to cluster. While we can think of a feature as a point in some higher dimensional space, it is easy to visualize that as the number of dimensions grows, the further apart each specific observation gets, and no meaningful clusters of the data can be found. At the extreme, each sample will be sufficiently separate from the other so that it becomes a separate class. Effectively, the algorithm memorizes the data, and while it can thus achieve a perfect score on the test data, it has no basis on which to generalize to new data. This is what is meant as the "curse of dimensionality" and is closely related to the problem of overfitting in machine learning.

As the number of features increases, we need to increase the number of samples proportionally to have all combinations of feature values well-represented in the samples, and in the real world, this is often hard to accomplish. This problem is of special importance in neural network-based machine learning models because of the intrinsic need, in most cases, for large amounts of labeled multiattribute data even under optimal conditions.

Feature selection, extraction, and correction

These are best discussed in the context of a specific example related to the sinking of the Titanic based on Kaggle data. Kaggle is an online resource for users to find, publish, explore, and enter competitions and maintains many useful databases for these purposes. In particular, it contains a public domain 60 KB data source (train.csv) of the Titanic manifold for its final voyage [4]. This provides information on a large number of attributes for the passengers, and based on these, the Kaggle challenge is to predict if an arbitrary passenger on the Titanic would survive its sinking. A sample of the data from the manifest is shown in Fig. 5.1.

To summarize, the categories of available information are the following:

Passenger ID	PClass	Name
Sex	Age	SibSp(sibling or spouse)
Ticket	Fare	Parch (parent and child)
Cabin no.	Embarked from	Survival information

If we consider survived and not survived as the 2 classes, then each sample has 11 attributes.

We cannot directly plug this dataset into a machine learning algorithm because of inconsistencies, omissions, and redundancies. Excellent summaries

	PassengerId	Survived	Pclass	Name	Sex	Age	SibSp	Parch	Ticket	Fare	Cabin	Embarked
0	1	0	3	Braund, Mr. Owen Harris	male	22.0	1	0	A/5 21171	7.2500	NaN	S
1	2	1	1	Cumings, Mrs. John Bradley (Florence Briggs Th...	female	38.0	1	0	PC 17599	71.2833	C85	C
2	3	1	3	Heikkinen, Miss. Laina	female	26.0	0	0	STON/O2. 3101282	7.9250	NaN	S
3	4	1	1	Futrelle, Mrs. Jacques Heath (Lily May Peel)	female	35.0	1	0	113803	53.1000	C123	S
4	5	0	3	Allen, Mr. William Henry	male	35.0	0	0	373450	8.0500	NaN	S
5	6	0	3	Moran, Mr. James	male	NaN	0	0	330877	8.4583	NaN	Q
6	7	0	1	McCarthy, Mr. Timothy J	male	54.0	0	0	17463	51.8625	E46	S
7	8	0	3	Palsson, Master. Gosta Leonard	male	2.0	3	1	349909	21.0750	NaN	S
8	9	1	3	Johnson, Mrs. Oscar W (Elisabeth Vilhelmina Berg)	female	27.0	0	2	347742	11.1333	NaN	S

FIG. 5.1

Sample of data obtained from the manifest of Titanic for a final voyage in 1912.

Kaggle Repository.

82   CHAPTER 5 Dealing with data: Strategies of preprocessing data

of the issues involved are found in Refs. [5, 6]. To summarize, 687 of the 891 passengers in the dataset have no cabin information, so it is best to simply eliminate this attribute. In contrast, there are only two empty embarked values, so these samples can simply be dropped. There is no information for 177 passengers. We can create pseudodata for these entries by finding some statistical measures. For example, we can impute an age based on the mean of all those who resided in a given passenger class or some other attribute that appears reasonable. In doing so, it would be helpful to look at the actual distributions involved to see whether they appear to fit a normal distribution. Another approach, where the data may not be amenable to this kind of statistical analysis, is to simply label missing categorical data as "missing" and missing numerical data as "0" just to meet the technical requirement for no missing variables. Using this technique is essentially allowing the algorithm to estimate the optimal value instead of just filling it in with the mean.

The next step is to decide which features are likely to be useful and which are not. While a bit counterintuitive, keeping the entire attribute set may actually harm the predictive accuracy of the algorithm. Passenger ID and passenger name are redundant, so we can eliminate the passenger name column before inputting the data. However, before this step, that information might be useful in uncovering errors. For example, in 1912, if a male and female cabinmate had different surnames, not only were they unlikely to be an unmarried couple, but also they were unlikely to be married to each other. This may simply be a listing error. However, there are some other possibilities, such as a grandparent-child combination or siblings of opposite sex (although this latter might have also been an issue in 1912). This may be useful information in checking the accuracy of SibSp data, for example. One can also cross-check SibSp and Parch data for discrepancies. Age may give clues as well. A 13-year-old is unlikely to have a spouse. An 80-year-old is unlikely to have a parent on board. Erroneous data can be treated as missing data.

In reducing the attribute set, it is important to look for dependencies. For example, fare for a given embarkation point might correlate with passenger class, and if so, one of these might be dropped. Cabin also is likely to correlate with fare and class, but this gives additional information, such as location in relation to deck level and proximity to stairs. These choices involve informed guesses, so different data scientists could wind up with different attribute subsets. For example, in the example used in Ref. [6], the attributes chosen were age, SibSp, Parch, Fare, Sex, Passenger Class, and Embarked.

For the next step, it is important to consider the metric for each. While there are only two possible values for "Sex" (this is 1912) and three possible values for class, there are seven distinct values of SibSp and eight distinct values for Parch in the dataset. There are, however, 281 distinct fares, and it is not likely that fine distinctions would help the model, so it is possible to group the fare values into a smaller number of range categories.

These examples demonstrate that raw data are merely the tip of the iceberg!

Feature transformation, standardization, and normalization

As data are either numeric or coded in numeric form for input to a computer, it may be important for each attribute to fall into a similar numeric range so that one feature does not contribute disproportionally. Normalization usually refers to scaling a variable to have a value between 0 and 1. Standardization, on the other hand, usually means rescaling the database to a mean of 0 and a standard deviation of 1 [7]. Proper scaling can often lead to a boost in algorithmic accuracy even after all its hyperparameters are optimally tuned.

Feature engineering

Feature engineering is the process of using domain knowledge of the data to create features that optimize machine learning algorithms that use those features as input. In other words, feature engineering is about creating new input features from existing ones. This improves computation efficiency and, in some cases, accuracy by reducing the feature set to a more manageable size. It makes use of your domain expertise. The simplest form of feature engineering is to combine features. For example, if two attributes are "weight" and "volume," and absolute size is not necessary, then the single value "density" may be more appropriate. If some of the attributes are not well-represented in the dataset, it may be possible to combine them. In the vehicle example in the previous section, we might combine data on "acceleration" and "road-holding" into a single variable "performance." This may sometimes be helpful for sparse classes as well, for example, the category of non-small cell lung carcinoma, instead of separating these into individual morphologic variants. Finally, categorical or text values may be replaced by binary variables. Brick, wood, vinyl, and stone may be able to be replaced by "brick" and "nonbrick" or 0 and 1, respectively. Many of these techniques lead to dimensional reduction. In the next section, we will describe some formal, mathematically based techniques for data reduction.

The most sophisticated form of feature engineering is the creation of new relevant features rather than merely a combination or replacement. Features useful for face recognition include distances between various facial features, overall size and shape, and so on. All these are quantifiable and become a feature set rather than the raw pixels of the image. Similarly, there are mathematical techniques such as Hough transforms and "blob" analysis that allow the quantification of density contrasts and the detection of edges. These can also be used to create additional features. All these steps involve transforming the data. The use of the data in a machine learning algorithm is merely the final (although admittedly the most sophisticated and mathematically intensive) step.

The power of deep learning systems such as neural networks is that they can implicitly learn to do feature engineering all on their own. The algorithms can automatically create higher-level features from the raw input. The more representation power an algorithm has, the more it is able to derive good features. This is

CHAPTER 5 Dealing with data: Strategies of preprocessing data

approximately based on the number of layers present and the number of artificial neurons in each layer. Creation of an implicit feature set can be done for all forms of highly dimensional data and not just images. Images are a special case because of magnitude, density, and interrelationships of the data they encode, and as has been described in previous chapters, it took the invention of convolutional neural networks to deal with this effectively. However, even in the case of deep learning, feature selection, processing, and human feature engineering are helpful in optimizing algorithmic functions.

Mathematical approaches to dimensional reduction

As described above, there are a number of straightforward approaches to reducing the size of the feature set by taking advantage of dependencies among attributes, eliminating redundant or minimally contributory features, and so on. We can take this a step further by determining which of the features contribute most and which contribute least to classification. This is the basis of a number of approaches to feature dimensional reduction. We will primarily discuss principal component analysis (PCA) as it is one of the most highly used procedures. PCA was first invented at the beginning of the 20th century as a statistical technique and thus precedes the advent of machine learning by half a century. It has become an important tool for data preprocessing across many fields, especially in machine learning. PCA is used to identify those features that provide the most information in regard to distinguishing among examples for the purpose of classification and to remove the others. This is done by transforming the data based upon their contribution to the variance of that data. Another way of thinking about PCA is to note that our data usually contain many variables, some of which may be correlated. This leads to a redundancy in the information in the dataset. In essence, PCA transforms the original variables into linear combinations of these variables that are independent. PCA can be a difficult topic to understand, so we will approach it in two ways: first, a more or less intuitive approach and then a more precise explanation of the underlying mathematical principles.

A good intuitive account of this is found in Ref. [8]. The article begins by noting that variance is a measure of how far a set of numbers is spread out. It describes how much a variable differs from its expected value. If each feature had the same value in all samples, the model using it would have no predictive value because each sample would be identical. Thus, variance is useful as a feature variable in the model. If we define the target as the value to be predicted, then we want features that have nonzero variances with respect to the target. Mathematically, variance is the mean squared difference between the predicted and true values. While variance describes how multiple instances of a single variable vary, covariance is a measure of how much two or more random variables vary together.

As noted above, PCA automates data reduction and does so in an unsupervised manner. It does so on the basis of variance. It begins by finding the direction of

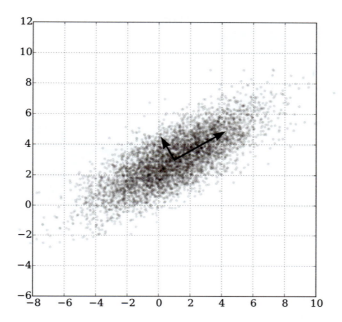

FIG. 5.2

The *longer arrows* show the direction of the first component, and the *shorter arrow* shows the direction of the second arrow. As the data in this example are two-dimensional, there are only two principal components. Each represents an axis along which variance is calculated.

Wikipedia Commons.

maximum variance in the data. This is basically the direction along which the data are most spread out. In analogy with linear regression, we can find a line such that the mean squared difference between the original data points and the nearest point of the line is minimized. This line gives the first principal component. The next principal component will be the line that minimizes loss of information while also being perpendicular to the component already found (Fig. 5.2). This process can be repeated until there are as many components as the original data had dimensions. As will be seen below, these operations represent linear transformations of the data and can be reversed to reconstruct the original data. Thus, both represent equivalent information about the dataset.

At this point, there is no dimensional reduction. However, each component contributes less and less information, as we have started out with the direction in which the data are most spread out. This means that we can discard some of the later components without doing much damage. This will make the component-based dataset dimensionally smaller than the original dataset. The trick is to determine how much bathwater can be discarded while retaining the baby. This is usually done by plotting variance as a function of component number and looking for a cutoff point beyond which there is little variance. Fig. 5.3 provides a good illustration of this. This is a 2D

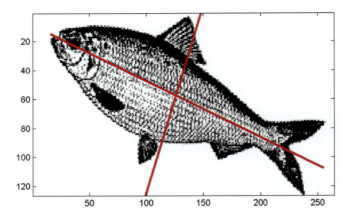

FIG. 5.3

We can easily identify this as a fish without seeing it head-on. The set of data points that define the fish has less variance in the Z direction than in the X and Y directions.

Wikipedia Commons.

sketch of a fish, and it can easily be recognized as a fish. To classify different kinds of fish, we would need enough samples with sufficient variation among the respective attributes to provide enough information to define a generalized set of fish classes, and we can then determine in which class an unknown fish belongs. The interesting thing is that we do not need to see the fish head-on to perform this task. Most fish are not very thick, and therefore there is less of a contribution to variation in that third dimension.

For a more general example, consider an ellipsoid of data points in three-dimensional (3D) space. If the ellipsoid is very flat in one direction, we can recover most of the data structure while ignoring that flat region. If we had samples with **n** attributes rather than just three, then the data are **n**-dimensional, and we can derive principal components along each of the **n** dimensions. We can then construct a "scree plot," which is a simple line segment plot that shows the fraction of total variance in the data with respect to the principal components. Typically, a scree plot shows variance dropping steeply with component number until a point where further changes are minimal. Discarding those later components therefore reduces the dimensionality with only minimal effect of the overall variance in the set of attributes. This is seen in Fig. 5.4.

Here, the amount of variance contributed by a principle component is plotted against component number. The ordinate in Fig. 5.4 is labeled both in terms of variance and "eigenvalues." As shown below, the eigenvalues are calculated from a covariance matrix in PCA, the eigenvalues represent variance, and so the terms can be used interchangeably for graphical analysis. Most scree plots have a point of inflection, and when this occurs, it defines the cutoff for discussing later components. In Fig. 5.4, we would keep the first three components and discard the rest.

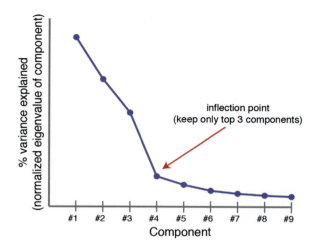

FIG. 5.4

A typical scree plot. The curve has a steep fall-off until the point where changes in eigenvalue are minimal. That point is chosen as the cutoff. By eliminating the principal components that contribute least, we are reducing the dimensionality of the feature vector by that amount while minimizing the effect on the data. In this example, the first three components are kept.

With permission from A. Williams. Everything you did and didn't know about PCA. http://alexhwilliams.info/itsneuronalblog/2016/03/27/pca/.

This is a very oversimplified explanation of PCA. To understand the actual implementation of PCA, a bit of mathematical detail is necessary [9]. What we are actually doing is using linear algebra to transform a feature set into a new set based on variance. We do this by finding linear combinations of variables that best explain the covariance structure of the variables. We implement this task by first constructing a covariance matrix from the raw data matrix (in some applications, a correlation matrix is used; for the purposes of this discussion, we can ignore this distinction). A covariance matrix is a matrix whose element in the i,j position is the covariance between the ith and jth elements of the original vector. The eigenvalues and eigenvectors of this matrix are then calculated. Eigenvectors are vectors that describe a fixed direction under a given linear transformation, and eigenvalues are scaling factors for these vectors. The most intuitive explanation I have come across is that the eigenvalues of a linear mapping provide a measure of the distortion induced by the transformation, and the eigenvectors tell you how the distortion is oriented. In PCA, the matrix is constructed in terms of covariances in the original data, and so the eigenvalues represent variances along the directions (eigenvectors) of the principal components. In other words, the size of the eigenvalues provides information as to how much each contributes to the overall behavior of the original data. The eigenvector associated with the largest eigenvalue is the first principal component, and so on. Thus, the size of the eigenvalue provides information as to

CHAPTER 5 Dealing with data: Strategies of preprocessing data

how much that eigenvalue contributes to the overall behavior of the original data. The final step is to write the feature vector for each data point in terms of the sum of the principal components.

Dimensional reduction comes from discarding the principle components with the smallest eigenvalues. For this reason, a scree plot of variances, such as the one that was shown in Fig. 5.4, is usually shown as a curve in which the eigenvalues are plotted against principal component number rather than of contribution to variance versus component number. These two terms represent the same thing. To put this in concrete terms, for the ellipsoid in a previous example, when we discard the direction of flatness, we do so according to the above calculations. The scree plot would have shown how little information was lost in going from a 3D to a 2D representation.

PCA is a powerful statistical technique that predates machine learning. It has turned out to be useful in virtually all forms of machine learning. Not only does PCA reduce the curse of dimensionality by stripping out less important features in the data, but it also provides a means by which the data can be better visualized in graphical form, as it is usually hard to represent more than three or four dimensions. PCA is not the only technique for doing this. Some others are embedding single value decomposition, linear discriminant analysis, and factor analysis. There are also nonlinear dimensionality reduction methods. These are based on the concept of a manifold. A manifold is a topological space that locally resembles flat Euclidean space near each point. For example, if you think of a linear space as a flat sheet of paper, then a rolled-up sheet of paper is a nonlinear manifold, as each point has a small region near it that is linear. Some nonlinear reduction methods include multidimensional scaling, isometric feature mapping, local linear embedding, and spectral embedding, among others. There are also probabilistic approaches for nonlinear reduction such as t-distributed stochastic neighbor embedding. This technique computes the probability that pairs of data points in the high-dimensional space are related and then finds a low-dimensional embedding that produces a similar distribution. Although beyond the range of this discussion, a good summary can be found in Ref. [9].

Dimensional reduction in deep learning

Neural network-based algorithms provide many ways of performing unsupervised dimensionality reduction. Recall that within the neural network, a representation of the details is developed at increasing levels of complexity. Each of the derived features of the representation is built up from many original features in the dataset and therefore is of lower dimensionality. A convolutional neural network is a good example. The initial feature set may consist of millions of pixels. The convolutional and max pooling layers reduce this to a small number of features that encode the important data in the image. Other ways of generating such representations include autoencoders and generative adversarial networks, as described in Chapter 4. There is an interesting connection between autoencoders and PCA. Autoencoders develop a dimensionally reduced encoding (representation) of the input data by compressing it via convolution and then confirming its accuracy by deconvoluting it and comparing the result to the original input. Although this seems to bear no relationship to PCA, it turns out that if we were to run the data through an autoencoder without internal activation functions, then the mathematics of the resulting network is the same as running it through a PCA algorithm. However, if we include activation functions in the autoencoder, as is typically the case, then the system becomes much more powerful than PCA [10].

Imperfect class separation in the training set

In the above discussion and throughout the book, we have assumed the existence of a labeled training set in which the class of each instance in the dataset is accurately known and distinct from each other. However, there is sometimes a spillover in that one class is contaminated by a small but unknown number of samples from another, either because of imperfect annotation or because of errors in constructing the training set. Fortunately, there are some sophisticated ways of dealing with this. The simplest case is the situation of training a binary classifier to reliably separate a positive and negative class, for example, cancer versus normal. Here, we consider the situation in which the positive set is known, but the negative set is unknown in that it contains both positive and negative examples. In other words, we have to isolate a set of so-called "reliable negatives" for training purposes. Techniques for doing this are known as PU (positive-unlabeled) learning algorithms. The trick is to begin by treating all the negatives as truly negative and train the classifier. Next, using the classifier, score the unknown class and eliminate those for whom the classification is least reliable. Retrain with the positives against that truncated negative set and repeat as necessary until some stopping condition is met [11,12]. Another approach involves bagging the unknown into distinct subsets and evaluating each subset for training [13].

In problems such as this, we assume that the positive set is large enough for effective training. However, often, the available positive set is smaller than the negative set, and ways of dealing with this, such as the creation of synthetic data, have been discussed in previous chapters. It is interesting to note that PU learning itself can be used for classification problems, where there is only a small positive training set available. This involves augmenting the set of labeled positive examples with a set of unlabeled samples that includes both positive and negative examples. In the example of PU learning described above, we assume a sufficiently large set of positive training examples and that the positive samples were drawn from the same distribution as the contaminating positives in the negative set. In this new scenario, in addition to finding a set of reliable negatives, we wish to identify reliable positives hidden within the negative set. An interesting variant of PU using this strategy is known as "learning from probabilistically labeled positives" (LPLP). LPLP has been described for document classification [14] but has obvious applications for image interpretation as well.

90 CHAPTER 5 Dealing with data: Strategies of preprocessing data

Fairness and bias in machine learning

There are two kinds of bias in machine learning: coding or model bias and data bias. Coding bias is important for understanding the validity of a given algorithm for the problem it is addressing, but it is not analogous to human forms of bias. Instead, it refers to assumptions made in constructing the model. For example, in the K-means clustering model, it is assumed that the clusters are of similar size and essentially globular. In naïve Bayes, it is assumed that the feature attributes are independent variables. In neural network backpropagation, it is assumed that local minima can serve as surrogates for the global minimum. Model bias is not a fault; it is basic to the simplifying assumptions that are made in constructing useable models that are computationally efficient while retaining the ability to provide useful and meaningful outputs. Such built-in assumptions are also often necessary simply to allow the problem to be solved within the current state of the art. In any event, it is important to understand the limitations of the algorithm that is being used.

Data bias is much more insidious than coding bias. Most clinicians, in general, and pathologists, in particular, will be users, rather than creators, of machine learning models, so minimizing coding bias may be outside our domain of expertise. However, we will serve as content providers and annotation experts for the training set data. Thus, we are the first line of defense in scrutinizing the data itself for unintended bias.

We can define bias as unfair prejudice in favor or against one thing, person, or group compared with another. For humans, bias can be defined as a particular tendency, trend, inclination, feeling, or opinion, usually preconceived or unreasoned. Bias may be unconscious in that we may have attitudes or stereotypes without conscious awareness of them (implicit bias). As machines lack inclinations or opinions, or as far as we know, awareness, fairness in machine learning can be represented as the relationships between a sensitive attribute (race, gender, nationality, etc.) and the output of the model. In the case of both human and machine, the output is incorrectly classified based upon an attribute that, in point of fact, has no causal relationship to that output. In the case of human bias, the result is often judgmental; in the case of the machine, it can lead to false judgments by humans.

There are many ways that bias can arise in machine learning. Sample bias is where the data used to train the model do not accurately reflect the domain of the problem. For example, the data may be restricted or poorly collated, or where there may be a marked imbalance in the training sets. An example of the former would be a training set in which all cats were placed on red rugs and dogs on brown rugs. An unknown sample of a dog on a red rug might be classified as a cat. Another would be a set of headshots of men and women, where the women all had long hair. An unknown woman with a pixie cut might be classified as a man. Note that these examples do not exhibit prejudice, and they are merely wrong because of "faulty" assumptions about the properties of a population. These are relatively harmless examples. However, imagine a training set for distinguishing Cocker Spaniels from Doberman pinschers, in which the Spaniels were all shown biting their owners while the

Dobermans were shown cuddling. While both breeds can be docile or aggressive, this would paint an unfair picture of the relative temperaments of these two breeds and could lead to an unfair prejudice against Cocker Spaniels.

Set imbalance can contribute to machine bias in a much more insidious and problematic way. For example, in 2015, Google developed an algorithm to identify faces as humans, rhesus monkeys, baboons, gorillas, and other primates. The algorithm classified African Americans as gorillas! This was not neoNazi programming but rather related to a limited set of African Americans in the training set and large numbers of Caucasian examples, as well as large numbers of the various nonhuman primates. Google did not do this intentionally; it worked with the images at hand and did not notice the imbalance in the numbers of people of different races or ethnicities in their database. The neural network had learned skin color as a discriminant between humans and apes! The obvious fix would have been to expand the database, but Google's initial fix was simply to block the algorithm from identifying gorillas altogether.

Data bias can also lead to poor medical care on a population basis. Recently, Obermeyer et al. [15] examined a commercial algorithm used for healthcare decision-making and found that at a given risk score, black patients are considerably sicker than white patients. Remedying this disparity would increase the percentage of black patients receiving additional help from 17.7% to 56.5%. The bias was not because of judgmental issues but rather because of the underlying data structure. The algorithm used healthcare costs as a surrogate variable for degree of illness, on the apparently reasonable notion that the sicker you are, the more you spend on medical care. However, this equivalence rests on the assumption of equal access to healthcare. However, unequal access to healthcare and/or socioeconomic factors means that we spend less money caring for black patients. In this case, the fix is simpler: look at more appropriate measures of illness (or better still, having identified the societal problem, fix it).

These results should be contrasted with another study published by the same author in 2003 [16]. In this study, a large bank of fictitious job resumes was created. He also used fictitious names such as Emily and Greg versus Lakisha and Jamal and assigned them randomly to these resumes. The resumes were then sent to employers who posted job advertisements in the media of the day. White-sounding names received 50% more callbacks for interviews than black-sounding names. Applicants living in better neighborhoods received more callbacks, but this parameter was unaffected by perceived race. The amount of discrimination was uniform across occupations and industries. These results suggested that racial discrimination was still a prominent feature of the labor market in 2015.

It is instructive to note the differences in the causes of bias in these two examples. In the latter, which is much more pernicious, it is presumably based on unfair stereotyping. In the former, it is because of an error in the choice of attribute (cost vs. actual illness) rather than an unfair generalization. Nevertheless, both can lead to significant negative consequences.

CHAPTER 5 Dealing with data: Strategies of preprocessing data

Issues relating to understanding the features used for analysis are not confined to machine learning but are also problematic in traditional statistical analysis. In Ref. [17], Smith describes a study that, using statistical analysis, found that patients with sepsis with high pH levels are more likely to be readmitted to the hospital after discharge than patients with low pH. This implied that having a low pH is a good thing. This was surprising as it is well-known that acidosis in sepsis is associated with more severe disease. Analysis of the data revealed that the data included patients who had died during their hospital stay. The patients least likely to be readmitted are the ones discharged to the mortuary!

It can be very difficult to evaluate fairness in machine learning models. We initially defined fairness in machine learning in terms of the relationships between a sensitive attribute (race, gender, nationality, etc.) and the output of the model. However, it is obvious from the above examples that one cannot evaluate fairness in this context without considering the data generation strategies for the model and the causal relationships embedded within the data. For example, differences in disease prevalence based on genetic makeup may not reflect unfair bias in the data. On the other hand, differences in disease prevalence because of environmental or socioeconomic factors may reflect an underlying unfair bias. Recently, scientists at Google's DeepMind have used a technique known as causal Bayesian network analysis to explore these issues in an attempt to build fairer learning systems [18,19].

Summary

Good performance of a machine learning model requires a large set of features to allow classification and prediction from the data. There is an optimal point for this. Increasing the attribute number beyond that point can degrade performance. This can usually be overcome by increasing the complexity of the network in terms of both layer width and layer depth. However, the downside here is that the network may lose its ability to generalize from the known examples to novel ones. In addition, irrelevant features may deteriorate performance by the network focusing on these rather than the ones that are truly important. For all these reasons, there must be a balance between algorithmic power and the structure of the data. A data scientist must pay as much attention to data preprocessing as to the choice of a machine learning model. There are also ethical considerations with respect to data such as restricted or poorly constructed datasets leading to implicit biases on the part of the model as well as privacy issues. These are very important and have come under increasing study as artificial intelligence is becoming more and more ubiquitous in society. For all these reasons, understanding the data as well as the algorithms is essential to address these issues properly.

References

[1] Chen T, et al. Recurrent attentional reinforcement learning for multi-label image recognition. arXiv 2017. 1712.07465c1 [cs.CV].

[2] Wang Z, Sarcar S. Outline objects using deep reinforcement learning. arXiv 2018. 1804.04603v2 [cs.CV].

References **93**

[3] Mathe S, Pirinem A, Sminchisescu C. Reinforcement learning for visual object detection. IEEE Conf Comput Vis 2016. https://doi.org/10.1109/CVPR.2016.316.

[4] https://www.kaggle.com/hesh97/titanicdataset-traincsv.

[5] https://towardsdatascience.com/a-beginners-guide-to-kaggle-s-titanic-problem-3193cb56f6ca.

[6] https://medium.com/@andrewadelson/testing-different-models-for-the-titanic-dataset-be5f725b7ec0.

[7] Raschka S. Feature normalization and scaling and the effect of standardization for machine learning algorithms, https://sebastianraschka.com/Articles/2014_about_feature_scaling.html.

[8] Yiu T. Understanding principal component analysis, https://towardsdatascience.com/understanding-pca-fae3e243731d.

[9] Raj JT. A beginner's guide to dimensionality reduction in machine learning, https://towardsdatascience.com/dimensionality-reduction-for-machine-learning-80a46c2ebb7e.

[10] Glassner A. Chapter 25: Autoencoders. In: Deep learning: from basics to practice; 2018. p. 1414.

[11] Liu B, Dai Y, Li X, et al. Building text classifiers using positive and unlabeled examples. In: Third IEEE international conference on data mining; 2003. p. 179–86.

[12] Fusilier DH, Montes-y-Gómez M, Rosso P, Guzmán CR. Detecting positive and negative deceptive opinions using PU-learning. Inf Process Manage 2015;51:433–43. https://doi.org/10.1016/j.ipm.2014.11.001.

[13] Mordelet F, Vert J-P. A bagging SVM to learn from positive and unlabeled examples. Pattern Recogn Lett 2014;37:201–9. https://doi.org/10.1016/j.patrec.2013.06.010.

[14] Li X-L, Liu B, Ng S-K. Dealing with data: strategies of preprocessing data. In: Kpketal JN, editor. Machine learning: ECML 2007: 18th European conference on machine learning. Springer-Verlag; 2007. p. 201–13.

[15] Obermayer Z, et al. Dissecting racial bias in an algorithm used to manage the health of populations. Science 2019;366:447–53. https://doi.org/10.1126/science.aax2342.

[16] Bertrand M, Mullainathan S. Are Emily and Greg more employable than Lakisha and Jamal? A field experiment on labor market discrimination. In: National bureau of economic labor studies program; 2003. NBER Research Working Paper No. (9873).

[17] Smith G, Cordes J. Using bad data. In: Smith G, Cordes J, editors. The nine pitfalls of data science. Oxford University Press; 2019.

[18] Chiappa S, Isaac WS. A causal Bayesian networks viewpoint on fairness. arXiv 2019. 1907.06430v1 [stat.ML].

[19] Chiappa S, Gilliam PS. Path-specific counterfactual fairness. arXiv 2018. 1802.018139v1 [stat.ML].

CHAPTER

Artificial intelligence in pathology: Easing the burden of annotation

6

Benjamin R. Mitchell[a], Marion C. Cohen[b], and Stanley Cohen[c]

[a]*Department of Computer Science, Swarthmore College, Swarthmore, PA, United States,*
[b]*CSAI, PA, United States,* [c]*Center for Biophysical Pathology, Rutgers-New Jersey Medical School, Newark, NJ, United States*

Introduction

"Artificial Intelligence"(AI) is becoming a pervasive set of tools that promise to create as powerful a paradigm shift as that which occurred in the industrial revolution, and its emerging role in medicine and science is no exception. Strictly speaking, artificial intelligence is neither. It represents an extension and enhancement of human capability in the same sense as a bicycle, which is an improvement over running. It often involves the improvement of computer performance through experience, which is learning rather than intelligence per se. However, current machine learning requires vast amounts of labeled, human-annotated information for training. This often presents a barrier to adoption, especially in pathology, where the datasets often include mega and gigapixel images. This is not how biological brains work; it represents a significant hurdle to overcome. New models that better mimic how the brain learns are being developed, but even before this inflection point, there are many strategies for coping with our data addiction. In this chapter we will discuss such strategies.

Artificial intelligence 101

Although deep learning neural network (NN) models have achieved dominance in recent years, less sophisticated forms of machine learning remain useful in pathology. Those models are based on geometric (clustering and nearest neighbor), probabilistic (Bayesian), or stratification (random forest) strategies, among others. One of the most powerful varieties of the former is the support vector machine (SVM), which involves finding a decision boundary between different classes in the data (reviewed in Ref. [1]).

All of these models utilize training sets consisting of known, labeled examples. Each sample is defined in terms of a finite set of attributes (features), and its attribute set is known as a feature vector. The feature vector can be represented as a point in

Artificial Intelligence in Pathology. https://doi.org/10.1016/B978-0-323-95359-7.00006-6
Copyright © 2025 Elsevier Inc. All rights are reserved, including those for text and data mining, AI training, and similar technologies.

n-dimensional space, where n is the number of attributes needed to classify the sample. For example, we can distinguish the classes of caterpillars and ladybugs by using two attributes, length and width, so the feature vector is 2-dimensional. In this case, the decision boundary is a line. For an n-dimensional feature vector, the boundary is a hyperplane, and it has $(n - 1)$ dimensions.

Neural networks [2,3] also make use of data in the form of feature vectors, but do not do so by explicitly calculating hyperplanes. Unlike an SVM, the operation of a neural network (NN) is (very loosely) based on an analogy with a biological neuron that receives multiple inputs from other neurons. If the sum of those inputs exceeds a threshold value, the neuron sends a signal to the next neuron. In a basic NN, the nodes (neurons) are arranged in layers, with every neuron in one layer being directly connected to all the neurons in the next. Each input is given a weight, and initially these are randomly assigned. The weights are modified (via a process known as "error back-propagation" [4] based on a training set of known examples until the model becomes good at predicting the class of an unknown sample that it has never seen before. By virtue of the multiple interconnections from one layer to the next, each layer provides an increasingly complex representation of the feature vector.

Although all these approaches to machine learning remain important, in recent years NN have come to dominate the field of artificial intelligence because of their power and flexibility. NN are often referred to as examples of "deep learning" because of their multi-layer architecture. Although it seems that NN represent a new evolution of machine learning, its roots had their beginning in the mid-twentieth century at approximately the same time as the simpler kinds of machine learning were being used. It was the limitations of early NN that delayed their adoption until innovative research and advances in computer hardware allowed NN to evolve to overcome those limitations.

For megapixel images (as well as many other kinds of large data arrays), the basic NN must be modified so as to allow the early layers of the NN to find representations of the data that are much smaller than the original number of input features, and the NN architecture that does this best for images is known as a convolutional neural net (CNN) [5]. Most, but not all, modern AI is based on deep learning. A detailed explanation of the basics of machine learning and examples of their use in pathology, with emphasis on deep learning, can be found in Ref. [1].

The human in the loop: Harvesting usable data

As indicated above, in all these machine learning algorithms, the computer learns best from a set of labeled (annotated) examples (samples) that fall into several classes, and based on this training, it can then make predictions about unknown samples it has not previously seen. Since we are teaching the computer by example, this is known as supervised learning. Note that as with all AI, this is 'learning' in the sense of improving performance through experience and is not intelligent behavior as we understand it in humans.

Because they are much more powerful, NN have largely, but not completely, replaced the other forms of machine learning in fields such as pathology. However, since most neural nets are very complex and real-world data have large numbers of attributes, they require vast amounts of known data for training. Large datasets for pathology are very hard to come by other than in some major medical centers and require a great deal of human input for proper labeling (annotation). There is also the problem of class imbalance, as for example in tumors, where there is often a smaller set of samples of specific tumor types than there are of corresponding normal tissue. Additionally, NN trained on material from one center may not perform as well on data from another center. Another problem is that all these algorithms are one-trick ponies; it is hard to train a model to handle multiple tasks, including even related ones. The computer basically suffers from "savant syndrome": it can do one task incredibly well but does poorly on everything else.

These limitations represent a major challenge for moving forward in using AI in pathology. A hint as to how this can be overcome can be found in the observation that computer learning, as described above, has very little to do with the learning that biological brains are capable of. For example, it doesn't take more than a few bananas to teach a toddler what a banana is, or even the difference between a guitar and a radio, even though they both make music. Pigeons can discriminate between cancer and normal tissue based on as few as 200 examples [6] by rewarding correct responses via reinforcement learning.

The brain uses what is known as Hebbian learning [7], in which changes in synaptic strength depend only on activity across the <u>local</u> connections to and from a neuron (colloquially, neurons that fire together wire together). Thus, biological learning is essentially local. Nonlocality is an emergent process that arises globally through interconnections involving many such local networks distributed throughout the brain. In contrast, current models of AI are intrinsically nonlocal, in that the weight updates depend on <u>all</u> the neural layers, working backward from output to input. It is likely that models that process data like a brain may reduce the amount of training data to that required for a human or even a pigeon. Unfortunately, such models are not yet tractable for large-scale problems, such as those routinely encountered in pathology. Until such new paradigms are fully developed, one can modify current NNs in various ways by relaxing dependence on strictly supervised learning. The purpose of this review is to provide an overview of some of these approaches.

Reducing the need for annotated data

It would be nice for a computer to learn in a completely unsupervised manner without annotated training input from a human being. However, even human beings don't learn in this fashion. While there are algorithms for unsupervised learning, the usual approach is to combine some unsupervised steps with some supervised ones (semi-supervised learning). For sparsely supervised learning, only a small number of labeled, annotated examples are available. For weakly supervised learning, there may be only partial annotation; for example, we may only know if a sample has cancer or not, but nothing about the distribution or extent of the cancer within the tissue sample, or local differences in its morphology. An alternative approach called "transfer learning" is to take advantage of previously trained models from other domains similar to the one that is being studied. We will deal mainly with sparsely supervised learning; weakly supervised learning utilizes similar techniques in conjunction with segmentation algorithms. In all cases, semi-supervised learning begins with an unsupervised strategy.

Overview of unsupervised pretraining

The basic idea behind the use of unsupervised pretraining is that the "unsupervised" step is essentially finding a feature-based complex representation of the data suitable for re-constructing the original inputs without knowing their class labels. These representations are then used as starting points for subsequent supervised training. So long as a similar set of features is used for classification, pretraining the network in an unsupervised setting can allow the supervised learning stage to achieve good convergence even in cases where the amount of labeled data would otherwise be too small to allow it.

It should be noted that unsupervised pretraining is not a magic bullet; in particular, it can reduce the amount of labeled data required to a point, but there will still be a point at which there is simply too little labeled data to teach a model that generalizes well. Additionally, it is nearly always the case that unsupervised pretraining will be outperformed by fully supervised training using the same amount of data (i.e., if you have labels for all your data, you are better off simply doing supervised training from the beginning). That said, this is still a potentially useful method, particularly in cases where acquiring large amounts of labeled data is considerably more expensive than acquiring large amounts of unlabeled data.

Transfer learning

One of the simplest ways of dealing with data that has a small number of labeled examples is to piggyback them onto a model that has been trained on other datasets that are similar but not related [8].

Recall that the NN builds up a series of more complex representations of the data from layer to layer. The intuition behind transfer learning is that if two problems involve similar kinds of data, then the low-level features of the earlier representations are likely to be similar. If this is true, instead of beginning with random weightings of the inputs in the early layers of the new NN, we can use the weightings that were obtained in training the previous model. The nice thing about transfer learning is that one can start with an NN trained by someone else. There are a number of pretrained networks available in publicly accessible deep learning toolkits, and pretrained NN are becoming increasingly available for pathologists. The more closely the datasets of the pretrained and the to-be-trained models match, the more effective this strategy will be. For example, if one were training an NN to distinguish benign from malignant prostate neoplasms, it would be better to start with a known model trained on breast tissue for this purpose, rather than a model that has been trained to distinguish roses from orchids.

The actual transfer learning process involves taking the network that is already trained and removing the fully connected higher level(s) from the network while keeping the weightings of the earlier layers frozen. The upper layer(s) are replaced with a new fully connected upper level with randomly initialized weights, as the resulting hybrid is trained using the new data set. The primary advantage is that the training can succeed with a much smaller labeled training set than would be required to train the same network "from scratch" (i.e., starting with randomly initialized weights). Transfer learning is thus an important part of the digital pathologist's armamentarium.

Unsupervised pretraining via clustering

Geometric cluster-based algorithms can be used as purely unsupervised learning for the early stages of a more complex model. The simplest such model is the K-means algorithm (KM) [9]. KM attempts to find clustering in unlabeled data. One starts by choosing a value of K (the number of clusters) based on the number of classes to be distinguished. Next, one makes a random guess as to where the central points of each cluster are so as to create K pseudo-centers. These should be reasonably distant from each other but need to have no relationship to the actual clusters and can be chosen randomly (unless there is some prior knowledge about the data structure). Next, assign each point of the dataset to its nearest pseudo-center. We now form clusters, each comprising all the data points associated with its pseudo-center. Next, update the location of each cluster's pseudo-center so that it is now in the center of all its members. Do this repeatedly until there are no more changes to cluster membership. There are now K clusters, each consisting of a group mainly, if not exclusively, of one class. Now, it is only necessary to label a small number of samples from each cluster and assign those labels to every member of that cluster. This creates a large dataset without having to manually annotate every sample of that dataset. There is now sufficient training data for an NN. Clustering is unsupervised; the resulting NN is sparsely supervised.

A more sophisticated use of clustering is a method known as N-shot or few-shot learning [10]. A "shot" is defined as a single example available for training. In N-shot training there are only a few examples of each class (for one or a small number of examples, it is often called one-shot learning). For images, the task is to classify any test image to a class using that constraint. One way of doing this is by the use of "prototypical networks" [11]. Unlike the usual deep learning architecture, prototypical networks do not classify the images directly. Instead, they learn how to map the image into metric space. This means that images of the same class are placed in close proximity to each other, while different classes are placed at further differences. This, like K-means, is a geometric model. However, instead of using a basic K-means as a preprocessing step, the whole structure is built into a convolutional neural network. Algorithms in this category are often known as "Image2Vector" models. Even more sophisticated models involve NNs augmented with a neural attention mechanism; a good example of a disease-related application of few-shot learning to image classification is seen in Ref. [12].

One class learning

It is important to distinguish between one-shot learning and one-class learning. The object of one-class classification, also known as anomaly detection, is to recognize instances of a class by only using examples of that class [13]. All other classes are lumped together and are known as alien classes. In other words, the difference between one class and binary classification is due to the absence of training data from a second class. While this approach has received a great deal of attention for simple datasets, it has proven difficult to obtain good results in real-world datasets such as those encountered in pathology. However, this is probably more like human learning than the large dataset-dependent multi-class models, so it remains a topic worthy of investigation. Recently, Perera and Patel [14] and Chalapathy et al. [15] have described one-class learning for anomaly (outlier) detection. More directly related to pathology, this strategy has been used to distinguish tumor transcriptomes as compared to normal in this manner [16]. Such approaches could, in principle, detect foci of cancer in otherwise normal tissue by considering those foci as anomalies; in other words, based on a training protocol in which normal tissue images are used and a metastatic focus is then treated as an anomalous outlier. As of the date of the preparation of this manuscript, no examples of work in this area have been found. However, industrial techniques involving the detection of anomalies in images of surfaces suggest that this is a feasible avenue to explore.

A similar strategy involves the use of Siamese networks [17,18]. Instead of a model that learns to classify its inputs like the usual NN, the neural network learns to differentiate between two inputs. A typical Siamese network consists of two identical neural networks, each taking one of the two images. Identical means that they have the same configuration with the same parameters and weights. The last layers of the two networks are jointly fed into a contrastive loss function that calculates the degree of similarity between the two input images. An example from natural language processing involving scoring an exam question may make this approach more intuitive. Here, one input is the question sentence, and the other input is the answer. The output is how relevant the answer is to the question. This is not a simple supervised learning task, in that we do not have a database consisting of all possible answers that are judged "correct" and examples that are judged "wrong." Nevertheless, a Siamese network can extract similarity or a connection between question and answer. In Ref. [18] this approach was used for image classification in breast cancer.

In addition to detecting similarities and differences in pairs such as cancer versus noncancer, Siamese networks can also be used to analyze patterns within image patches from whole slide images, and thus may be useful for segmentation. For example, Gildenblat and Klaiman [19] have used a sophisticated model for image retrieval from patch networks using a Siamese network consisting of two branches of a modified residual neural network (ResNet-50) pretrained on the ImageNet dataset for retrieving areas of tumor from scanned images.

Unsupervised pretraining via autoencoders

There are other strategies for discovering useful representations without class labels that are based on generative models. A classic example of this is to first train one or more layers of a neural network as an autoencoder [20]. The autoencoder is trained simply to re-construct its input; that is, the desired output is merely a copy of the input. If the network has enough hidden nodes, this task becomes trivial, but if the network is restricted in capacity, then it is forced to learn a low-dimensional representation of the input data. This can be thought of as analogous to a lossy compression scheme; the "compressed" data is encoded by the activations of the hidden nodes.

Once the unsupervised model has been trained, it is then used as the basis for a supervised model, typically by simply adding one or two extra fully connected layers to the top. At that point, the entire network is trained using a standard supervised approach. This results in more efficient training of an NN and may in some cases reduce the number of annotated examples needed for training.

If an even more abstract and compressed representation is desired, we can use the output of one autoencoder layer as the input for another. The result is called a stacked autoencoder. Other variations of autoencoders, such as variational autoencoders and sparse autoencoders, are also available. Autoencoders are not only useful in classification but they can also be used for tasks such as stain normalization as a preprocessing step for both human and computer image interpretation [21].

Unsupervised pretraining via generative adversarial networks

The type of network we have dealt with so far is aimed at the task of classification or similarity discrimination. While these networks are by far the most popular type, they are not the only type used. There are several other types of problems that can be tackled using machine learning, and one of them involves generative modeling. A generative model is one which can be used to generate new examples. The basic task is to take in a set of examples and then construct new examples that are unique but are similar to the examples the system was trained on. In statistical terms, the goal is to model the distribution from which the training samples were drawn and then generate new samples drawn from the same distribution.

There are several cases in which it might be worthwhile to train such a generative model. First, we might directly want the outputs; that is, the generation of plausible synthetic examples may be useful in its own right. Second, we may view the generative model as another way of finding potentially useful features, which we could then exploit as a method of unsupervised pretraining (similar to the use of autoencoders). Third, we may be interested in a conditional generative model, which is a fancy way of saying that we are interested in transforming inputs, rather than creating them from whole cloth (for instance, taking an image and generating a version in the style of some famous artist). All three of these things are possible using generative models.

In the context of deep neural networks, a powerful generative model is the generative adversarial network (GAN) [22]. GANs work by creating two networks and training them to compete with one another. The first network is called the generator, and it is responsible for generating plausible synthetic (fake) examples. The second network is called the discriminator, and it is responsible for distinguishing between real examples (taken from the training set), and fake examples (created by the generator). These two networks are effectively playing a zero-sum game in which each player's success must come at the expense of the other. The way for the discriminator to win is to figure out how to recognize the generator's artificial examples, and the way for the generator to win is to make better fakes. After many training iterations, the fakes become indistinguishable from the real thing. At that point, the generator will have created a compact internal representation of the original. We can then replace the discriminator with a traditional neural net and proceed with training. This is analogous to using early internal representations from another model as the starting point for the new model, as was the case for transfer learning. In both cases, we are starting supervised learning with informed guesses about the weights of the earlier layers, rather than random values, so we can get away with less labeled examples. Another use would be to tweak the representations created by the generator so as to add a bit of variance to them and then use them as additional labeled examples for training. This is one of the ways of generating synthetic data. By expanding the labeled dataset in this manner, we can start with a smaller set of labeled data than what would otherwise be required.

GANs can do much more than this, but a deeper dive is beyond the scope of this chapter. However, Tschuchnig et al. have recently published an excellent review of the use of GANs in digital pathology [23].

Reinforcement learning (RL)

To reiterate, most neural network models are trained in a supervised fashion, meaning that the system is given a labeled training set consisting of hand-labeled input-output pairs. However, for some types of tasks, this is not practical. In particular, for sequential decision problems, it is often the case that we can only accurately assess the performance of the system at certain points. A common example is trying to train a network to play a game like Chess or Go; until the game is over, we don't know whether the strategy being employed is a good one or not. This means we cannot train the network to decide what moves to make in the middle of the game using standard supervised learning.

The idea of reinforcement learning was developed for situations like this. Inspired by operant conditioning, reinforcement learning works by providing a "reward" signal that tells the system when it is doing well and when it is doing poorly. Over time, the system experiments and through trial and error discovers a policy that will maximize its expected reward.

The classic formulations of this algorithm are based on the formalism of a Markov decision process (MDP). An MDP encodes a problem as a discrete set of possible states and actions, along with a transition model that specifies a probability distribution over possible next states given a current state and choice of action by an "Agent," Given a "reward" function that specifies what reward the agent receives for being in any given state, it is possible to use an iterative approximation algorithm to determine the optimal action to take in any state, where "optimal" is defined as the action that leads to the highest expected future reward. It is also common to add a "discount" factor, which allows us to down-weight future expected gain based on how distant it is; by tuning this parameter, we can make the agent optimize for either more short-term or more long-term reward. It is important to stress that the agent is not given a set of defined paths as a training set; in this sense, reinforcement learning is a kind of unsupervised learning.

Recently, RL has been employed to solve classification problems. Here, instead of a series of actions leading to a goal, such as avoiding obstacles, the agent performs a classification action at one sample; the classification action is evaluated and returns a reward to the agent [24,25]. This turns out to be useful in cases where there are imbalanced data distributions, which is a situation that is common in pathology where there are usually many more examples of normal than diseased tissue samples. Lin et al. [25] do so by simply making the reward from the minority class sample larger than the reward from the majority class sample. They have shown that this approach outperforms more traditional approaches dealing with imbalance. The primary downside of reinforcement learning is its lack of computational efficiency; training via reinforcement learning is very slow compared to fully supervised learning. Furat et al. [26] recently described "PixelRL," an algorithm that combines RL with a fully convolutional network. "Hybrid" algorithms such as these hold great promise for future advances in artificial intelligence.

Self-supervised learning

All approaches described above suffer from a major limitation of AI as it exists today. They are basically trained to do tasks that don't require active thinking. They are not designed for the task that humans excel at, namely reasoning. The challenge is to create deep learning systems that can learn and plan complex action sequences and decompose tasks into subtasks. In other words, deep learning systems are good at providing end-to-end solutions but bad at breaking them down into specific interpretable and modifiable steps that are the building blocks of reasoning about complex tasks, and lever previously acquired knowledge. There are some baby steps in this direction, such as capsule networks [27] and compositional learning [28]. LeCun, who is the father of convolutional networks, has taken what are arguably the first systematic steps toward the development of self-supervising machines through an approach that he defines as variable energy-based models [29]. The approach is to learn an energy function that takes low values on the data manifold and higher values everywhere else through mappings based on hidden variables. To quote: "Energy-Based Models (EBMs) capture dependencies between variables by associating a scalar energy to each configuration of the variables. Inference consists in clamping the value of observed variables and finding configurations of the remaining variables that minimize the energy. Learning consists in finding an energy function in which observed configurations of the variables are given lower energies than unobserved ones. The EBM approach provides a common theoretical framework for many learning models, including traditional discriminative and generative approaches."

EBM allows for the initial stages of training to be accomplished by the use of large amounts of unlabeled data for initial training and looking for relevant associations. For example, in analyzing a video stream, an EBM can predict a future frame from the current one (as well as a past frame from the current one). It can also infer missing parts of a sample. EBMs can be used for classification tasks. In a typical neural architecture, the image is fed to a convolutional neural network. The dependencies at the granular level are captured and given some probability scores. However, in the case of EBMs, classification is based solely on energy values. This approach significantly reduces the number of labeled, annotated samples necessary.

Zero shot learning

It is even possible to combine transfer learning and one-shot learning with semantic information so that a model trained to identify one class can identify a new class that it has never seen before. This is known as zero-shot learning and is an exceedingly complex approach that is under active investigation. Current versions, known as CLIP (contrastive language-image pretraining), are based on strategies similar to those developed for modern large language models. An overview of zero-shot learning can be found in Ref. [30,31].

Drowning in data: Quantum computing to the rescue

This review has been focused on methods of dealing with situations where annotated data is limited. There is another potential problem. What happens when annotated data becomes overwhelming? It has been estimated that humans collectively produce about 2.5 exabytes of data per day. We are at the limits of the data processing power of traditional computers, and data keeps growing faster than improvements in their circuitry. It is therefore now possible to pose problems that because of scale cannot be solved in timescales that are reasonable by human standards. Fortunately, computer architectures that take advantage of quantum effects such as super-positioning and entanglement have been developed that allow for a massive parallelism that can enable the solution of such problems in seconds, rather than many years to accomplish. These algorithms were originally simulated on conventional computing devices to determine feasibility but have now been realized in actual quantum devices. There are a number of ways to create detectable quantum states. One example is a device manufactured by D-wave systems that is based on superconducting electrical loops. These are wired together to allow them to interact magnetically. There are many other quantum circuit topologies that have also been implemented. In addition to D-wave, IBM, Microsoft, Google, and others have functioning quantum computers. Some of these are available on the cloud for general access by the public.

Although the mechanisms of quantum computation are beyond the scope of this discussion, it suffices to say that a major difference between traditional and quantum computing lies in the computational building blocks used. While a traditional computer uses "bits" that can take on either the values of zero or one, the "qubit" is neither; it exists in an indeterminate state that is neither zero nor one until it is measured. Also, since particles are also waves, qubits are associated with a phase value. Thus, a qubit can be said to exist in a large number of possible states (superposition) until it is measured, at which point it collapses to a single value. The quantum algorithm is a strategy that, for a given input, causes an array of qubits to collapse into the correct configuration (solution).

Artificial intelligence, for the most part, requires annotated data, which is a small subset of the total amount of existing data. However, for images alone, there are a number of centers with millions of annotated digitized slides, and in shortly, there will be another 50 million from but one single additional source. Abstracting meaningful patterns from data of this magnitude will require quantum computation, leading to the hybrid field of quantum artificial information (QAI). One big advantage of quantum computing applied to AI is due to the number of dimensions it can process. In the time that a classical artificial neuron can process an input of N dimensions, a quantum perceptron can process 2^N dimensions [32].

An early application of QC, Grover's algorithm, was designed to identify a unique sample in a large, unsorted database. A classical computer can do this kind of sorting problem for N samples in N steps, but Grover's algorithm requires only the square root of N. Grover's algorithm has an obvious analogy with the problem of pattern recognition or image classification, and recently, it has been applied to that task as well. There are currently analogs of classical perceptrons and Hopfield networks, and quantum learning has been applied to unsupervised learning and reinforcement learning as well. The integration of QC with neural networks is described in detail in Ref. [33] and other examples of machine learning. Ultimately, QAI will be able to integrate very different data sets (such as transcriptomes and morphology), which is difficult without human intervention at present.

Interestingly, the human brain, with its massive parallelism and ability to process multiple computational tasks simultaneously, seems to function a bit like a quantum computer; however, it does so at room temperature rather than the ultra-low temperatures that are currently required to retain coherent states. Quantum effects may also play a role in other biological and biochemical processes such as photosynthesis (reviewed in Ref. [34]).

Summary and overview

Rather than provide an overview of current research articles relating to artificial intelligence in the pathology literature, this review has focused on the strategies for dealing with the huge datasets that we routinely encounter. Much of this data may be either unlabeled, or in the situations where segmentation is needed, incompletely annotated. Although not explicitly stated, the discussion has been mainly image-focused (pun intended), although the strategies, in broad outline, hold for other tasks such as natural language processing, risk prediction, and so on. For simplicity, but without loss of generality, the discussion has been based upon simple CNNs, but in the real world, more advanced architectures, such as residual networks, recurrent neural networks, "deep belief systems," and so on have been used successfully. The purpose of this review is to outline and discuss various strategies for reducing the number of annotated samples necessary for training artificial intelligence. Even as we improve in this area, it is highly unlikely that AI will supplant human pathologists for many reasons. One such reason is related to the differences between human errors and machine errors. Utilizing the approaches described here and in the more specific examples of the subsequent reviews in this issue, it seems that humans and computers make approximately the same number and kinds of mistakes. What is interesting is that they make different mistakes. For this reason, a partnership between human and machine will likely be better than either one alone. A small step in this direction is seen in reports of interactions between man and machine for collaborative annotation based on scribbling algorithms [35]. Perhaps the most sophisticated approach to AI-assisted annotation is the "Tissuewand" algorithm [36]. This kind of interaction between man and machine has been described as a "centaur" model [37]. With this collaboration, AI will evolve from being a pathologist's digital assistant to being a (nontenured) pathologist's partner and colleague, rather than a replacement for the pathologist.

References

[1] Cohen S. The basics of machine learning: strategies and techniques. In: Cohen S, editor. Artificial intelligence and deep learning in pathology. Elsevier Press; 2021. p. 13–4.
[2] McCulloch WS, Pitts W. A logical calculus of the ideas immanent in nervous activity. Bull Math Biophys 1943;5:115–33.

[3] Rosenblatt F. The perceptron: a probabilistic model for information storage and organization in the brain. Psychol Rev 1958;65:385–408. https://doi.org/10.1037/h0042519.

[4] Rumelhart DE, Hinton GE, Williams RJ. Learning representations by back-propagating errors. Nature 1986;323:533–6. https://doi.org/10.1038/323533a0.

[5] LeCun Y, Kavukcuoglu K, Farabet C. Convolutional networks and applications in vision. In: Circuits and systems (ISCAC), Proceedings of 2010 IEEE international symposium; 2010. p. 253–6.

[6] Levenson RM, et al. Pigeons as trainable observers of pathology and radiology breast cancer images. PLoS ONE 2015;10(11). https://doi.org/10.1371/journal.pone.0141357.

[7] Hebb DO. The organization of behavior. Psychology Press; 1949. ISBN: 978-0805843002S.

[8] Pan SJ, Yang Q. A survey on transfer learning. IEEE Trans Knowl Data Eng 2010;22 (10):1345–59. https://doi.org/10.1109/TKDE.2009.191.

[9] Wu J. Cluster analysis and K-means clustering: an introduction. Springer; 2012. p. 1–16.

[10] Sankesara H. N-shot learning: learning more with less data., 2019, https://blog.floydhub.com/n-shot-learning/.

[11] Snell J, Swersky K, Zemel RS. Prototypical networks for few shot learning. arXiv 2017. preprint arXiv:1703.05175,v2 [cs.LG].

[12] Chen X, Fan Z, et al. Molecular subgrouping of medulloblastoma based on few-shot learning of multitasking using conventional MR images: a retrospective multicenter study. Neuro-oncol Adv 2020;2:1–11. https://doi.org/10.1093/noajnl/vdaa079.

[13] Chalapathy R, Menon RA, Chawla S. Anomaly detection using one-class neural networks. arXiv 2019. 1802.06360v2.

[14] Perrera P, Patel VM. Learning deep features for one-class classification. IEEE Trans Image Process 2019;28:5450–63. arX1v:1801.054652v2.

[15] Sokolov A, Paull EO, Stuart JM. One class detection of cell states in tumor subtypes. Pac Symp Biocomput 2016;2016:405–16. https://doi.org/10.1142/9789814749411_0037.

[16] Quinn TC, Nguyen N, Lee SC, Venkatesh S. Cancer as a tissue anomaly: classifying transcriptomes based only on healthy data. Front Genet 2019. https://doi.org/10.3389/fgene.2019.00599.

[17] Koch G, et al. Siamese networks for one-shot image recognition. In: Proc. 32 international conference on machine learning vol Vol. 3, 7 Liile, France wo15.alakhutdinov; 2015.

[18] Cano F, Cruz-Roa A. An exploratory study of one-shot learning using Siamese convolutional neural network for histopathology image classification in breast cancer from few data examples. In: Proc. SPIE 11330, 15th international symp. on medical information processing and analysis, 113300A; 2020. https://doi.org/10.1117/12.2546488.

[19] Gildenblat J, Klaiman E. Self-supervised similarity learning for digital pathology. arXiv 2019. 1905.08139v3.

[20] Hinton GE, Salakhutdinov R. Reducing the dimensionality of data with neural networks. Science 2006;313:504–7.

[21] Janowczyk A, Basavanhally B, Maadabhuahi A. Stain normalization using sparse auto-encoders (StaNoSA): application to digital pathology. Comput Med Imaging Graph 2019;57:50–61.

[22] Goodfellow I, Pouget-Abadie J, Mehdi M, et al. Generative adversarial nets. In: Ghahramani Z, Welling M, Cortes C, Lawrence ND, Weinberger KQ, editors. Advances in neural information processing systems 27. Curran Associates; 2014. p. 2672–80.

[23] Tschuchnig ME, et al. GANs in digital path: survey of trends & future potential. arXic 2020. 2004.14936v2 [eess.IV].

[24] Wiering MA, Hasselt HV, et al. Reinforcement learning algorithms for solving classification problems. In: 2011 IEEE symposium on adaptive dynamic programming and reinforcement learning (ADPRL); 2011. https://doi.org/10.1109/ADPRL.2011.5967372.

[25] Lin E, Chen Q, Qi X. Deep reinforcement learning for imbalanced classification. arXiv 2019. 1901.01379v1 [cs.LG].

[26] Furuta R, Inoue N, Yamasaki T. PixelRL: fully convolutional network with reinforcement learning for image processing. IEEE Trans Multimed 2020;22:1704–19. https://doi.org/10.1109/TMM.2019.2960636.

[27] Sabour S, Frosst N, Hinto GF. Dynamic routing between capsules. arXiv 2017. 710.09829v2 [cs.CV].

[28] Tokmakov P, Wang YX, Hebert M. Learning compositional representations for few shot recognition. arX1v 2019. 1812.09213v3 [cs.Cv].

[29] Lecun Y, Chopra S, et al. Energy-based models. In: Bousquet G, et al., editors. Predicting structured data. MIT Press; 2007. p. 191–246.

[30] Pourpanah F, et al. A review of generalized zero-shot learning methods. IEEE Trans Pattern Anal Mach Intell 2022;45(4):4051–70. https://doi.org/10.1109/TPAMI.2022.3191696.

[31] Arie LG. CLIP: Creating image classifiers without data., 2023, https://datascience.com/clip-creating-image-classifiers-without-data-b21c72b741fa.

[32] Acampora G. Editorial. Quantum Mach Intell 2019;1:1–3.

[33] Wittek P. Quantum machine learning. Pattern recognition and neural networks. In: Quantum Machine Learning. Elsevier; 2014. p. 63–72.

[34] Cao J, Cogdell RJ, Coker DF, et al. Quantum biology revisited. Sci Adv 2020;6 (14). https://doi.org/10.1126/sciadv.aaz4888.

[35] Can YB, Chaitanya K, et al. Learning to segment medical images with scribble-supervision alone. In: International workshop on deep learning in medical image analysis and multimodal learning for clinical decision support; 2018. p. 236–44. arXiv: 1807.04668 [cs.CV].

[36] Lindvall M, Sanner A, et al. Tissuewand, a rapid histopathology annotation tool. J Pathol Inform 2020;11:1–27.

[37] Case N. How to become a centaur. J Des Sci 2018. https://doi.org/10.21428/61b2215c.

CHAPTER

Digital pathology as a platform for primary diagnosis and augmentation via deep learning

7

Anil V. Parwani and Zaibo Li

Department of Pathology, The Ohio State University Wexner Medical Center, Columbus, OH, United States

Introduction

The field of pathology is entering an exciting time with the more widespread use of digital imaging in pathology. Pathologists have been using images for a long time, but only in the last 20 years there has been an increased effort to digitize the workflow [1,2]. The electronic communication of digitized images, digital pathology, and the ability to transfer a microscopic image between one pathologist and another physician or member of the health care team has applications today in the national as well as global pathology community to be used for primary diagnosis, education, peer-review, intraoperative consultation, second opinion consultations, and quality reviews [1,3–15].

The most exciting aspect of digital pathology is the development and widespread use of whole slide imaging (WSI) technology [6,16]. WSI allows the scanning of entire glass slides, with an outputting of an image file that is a digitized reproduction of the glass slide with images that boast diagnostic quality standards [2,5,17,18]. The College of American Pathologists has published recommendations for the validation of WSI scanners for clinical use [16,19–21]. In recent years, the US Food and Drug Administration (FDA) has approved two WSI vendors to market devices for primary diagnosis, which have been major milestones in the journey toward adopting and implementing digital pathology systems [22,23]. A new set of regulatory pathways is now potentially going to be implemented as tools like deep learning and artificial intelligence start their way from research labs and clinical labs [14,24–30].

WSI scanners and the infrastructure surrounding the use and storage of digital images have advanced significantly since the description of the first one in 1997 by Ferreira et al. [31,32]. In the last decade, innovations and significant technical advancements in WSI have contributed to the ability of clinical laboratories to image large numbers of slides automatically, rapidly, and at high resolution [6]. The ability to routinely digitize entire slides not only allows pathologists to better apply computer

Artificial Intelligence in Pathology. https://doi.org/10.1016/B978-0-323-95359-7.00007-8
Copyright © 2025 Elsevier Inc. All rights are reserved, including those for text and data mining, AI training, and similar technologies.

power and network connectivity to the study of histomorphology but has significantly improved the quality and accessibility of biospecimen repositories [33], particularly large image datasets that can be used for research in the areas of deep learning and artificial intelligence [34–40]. The tools have also paved the way toward creating remote networks of digital pathology scanning centers to enable rapid sharing of clinical cases [41]. The latter inventions and innovations in the field of artificial intelligence are paving the way to augment the clinical diagnostic workflow and primary diagnosis in pathology [28,42]. This chapter will focus on providing an overview of digital pathology and particularly whole slide imaging as a unique platform for clinical applications such as primary diagnosis and the potential to augment this diagnosis via deep learning and artificial intelligence [11,20,26,43,44].

Digital imaging in pathology

Imaging plays a critical role in modern pathology practice and is a driving technology to enhance workflows within the clinical laboratory. Pathology images are unique in the healthcare environment and present significant challenges in labeling, storage, and management. Imaging workstations can be thought of as PCs with image-specific software and the appropriate hardware peripherals needed to handle digital images [32,45–47].

With improved computing power and infrastructure increasingly supporting more powerful computers, more robust networking, and cheaper storage, this is a perfect time when pathologists are able to review large images rapidly and easily as they can now manage text. Many of the laboratory information system (LIS) vendors are now able to provide integration with WSI systems [18]. In recent years, we have been witnessing a vast improvement in the capture, storage, and retrieval of images handled by the LIS [48]. Digital images continue to be used for teaching, quality assurance studies, and even consultation. The basic ability to capture the digital image faithfully must be validated before further applications can be developed [49–51].

Imaging software comes in many flavors and serves many functions. Software for image processing allows the adjustment of brightness, contrast, hue, and resolution and is therefore extremely useful for providing pathologic images for academic use [52]. Many laboratories have implemented image acquisition and management systems that acquire images during all stages of the specimen workflow (requisition scanning, gross and microscopic examination) and store them in either customized home-grown systems or commercial systems [52–54]. These images are linked to the surgical accession numbers, which in turn are linked to the electronic medical record and the laboratory information systems (LIS) [49]. the most critical aspect of this workflow is the ability to integrate with the LIS and have images readily available for patient care [46,51].

Several laboratories have utilized image-embedded reports to improve the care of patients by providing additional information, which can also include annotations, tables, graphs, and diagrams. Such reports can, in addition, provide the clinician with added insight into the practice of pathology, facilitating greater understanding between disciplines. Image-embedded reports also serve as a powerful marketing tool for expanding the outreach capabilities of the laboratory [55,56]. The advent of artificial intelligence tools will pave the way in the future to automate many of the tasks of "choosing" the right image based on clinical data and other relevant information. There are also significant advances occurring in parallel in the field of molecular pathology, which, together with digital pathology and artificial intelligence, will provide the next generation of powerful and advanced diagnostic tools [53].

Digital images in pathology originate from several sources including autopsy, gross room, cameras mounted on microscopes, or scanned using scanners. There are numerous situations where digital imaging in pathology has provided useful solutions to pathologists including their deployment in pathology reports, conferences, presentations, quality assurance studies, and creation of teaching sets of rare and archival material. These images may be published as a collection of archived images on the Internet or local intranet for teaching. Digital images are also acquired for legal situations [54,57,58].

Telepathology

Telepathology technologies allow light microscopic examination and diagnosis from a distance. These systems may be combined with deep learning algorithms to assist the pathologist to screen a remote case for consultation or extract features from these images for assisted diagnosis. Telepathology use cases include primary diagnosis, teleconsultation, intraoperative consultation, telecytology, telemicrobiology, quality assurance, and tele-education [59–61]. There are three general approaches to telepathology. These include dynamic, static, and hybrid systems [32,62–67].

Dynamic systems are remote-controlled microscopes that provide the pathologist with a "live" view of the distant microscopic image while allowing them to control the stage, focus, and change the magnification remotely. An advantage of the dynamic systems is that they give the remote pathologist control and flexibility in viewing any region of the entire slide and do so at any power [68]. These systems tend to be more expensive with proprietary software requirements [69–71].

The second method of telepathology is "store-and-forward" telepathology, in which digital images are first captured via cameras or acquired via WSI systems for transmission to the consultant. These images can be captured in standard file formats and can be viewed via a remote viewer. In the future, this mode of telepathology may be used to send images to the cloud, where image analysis may be performed, or deep learning algorithms may be run to prescreen for rare event detection or for diagnosis or grading of cancer or for identifying microorganisms [69,72–74].

The third method of telepathology is the use of hybrid systems, which can combine features of both "store and forward" and dynamic systems. Hybrid telepathology systems can provide a "live" view using robotic-controlled microscopy with high-resolution still image capture and retrieval. This dual functionality makes them valuable. Some of the hybrid systems in the market today have capabilities for WSI as well [2].

Future developments in telepathology are likely to focus on three areas. The first is the adoption of standards that will help with interoperability. The second area will be to improve the integration with the LIS and electronic medical records to facilitate image viewing, reporting, and billing capabilities. The third area is the development of WSI for use in clinical practice. The latter will open up the area of applying deep learning and artificial intelligence (AI) algorithms for remote locations [2,4,35,39,69].

Whole slide imaging (WSI)

Much like the evolution of efficiency and effectiveness in radiology, the pressure on pathologists to reduce time to make diagnoses and develop more efficient workflows is trending toward digitalization, and WSI is turning out to be a technology to enable this digital workflow [16,75–77]. WSI scanning provides the ability to digitize an entire glass slide, which can be viewed as a digital file, providing capability similar to a radiology workflow including the use of picture archiving communication system (PACS) [78]. In other words, WSI technology allows the glass slides to be treated like digital files and shared at multiple locations simultaneously, providing means to collaborate, make a diagnosis remotely, and potentially apply deep learning algorithms either prior to viewing the image or after sign-out to check for errors and diagnostic quality. Robust infrastructure and the reduction in costs for network, storage, and computing power have made the use of WSI even more feasible. There are large implications for the practice of pathology with the potential for major efficiency and quality gains [6,11,79,80].

A WSI scanner is basically a microscope that is under full robotic control, attached to highly specialized cameras which contain high-performance photo sensors. There are two main approaches to creating WSI files including the use of tile-based imaging, in which a square photo sensor is used to capture multiple "tiles" adjacent to each other, or a line scan-based imaging, in which an oblong photo sensor is used to continually capture strips of image data as it sweeps through the slide. Once the slides are digitized and converted to millions of pixels, the massive pixel pipelines will allow pathologists to remotely view the images and share them [6,81].

In addition, and more notably, these images can now be standardized and analyzed using deep learning (DL) and AI tools to look for specific features such as mitotic figures, infectious agents such as acid-fast bacilli, or even cancer grade. With the ease of availability of cloud computing, powerful processors, and robust infrastructure today, it is possible to create a pixel pipeline-based workflow which allows for creating AI-based prognostic or diagnostic algorithms [26,82–84].

WSI scanners have changed through several generations since the first commercially available scanner in 2001. In the last decade, there have been several vendors who have been focusing on producing automated, high-speed WSI scanners with an emphasis on scanning speed, image quality, and high-throughput capabilities. A modern WSI scanner today can run in batch mode and can capture and compress an image of a slide with a 1.5×1.5 cm tissue section in under a minute with spatial sampling as low as $0.25\,\mu m/$pixel. These scanners have now also implemented cutting-edge optics, illumination, and sensors which allow for high-speed image capture [31,32].

WSI scanners of the future still have many chances to improve on many aspects of the mechanics and performance such as speed, point spread function, scan to view timing latency, and other methods of capture such as Z-stacking, immunofluorescence, and phase contrast. These enhancements should result in substantial improvements in speed, throughput, and resolution in the future. Z-stacking is an especially important consideration for organizations and/or laboratories that will be scanning cytology specimens and/or blood smears.

Many recent studies have demonstrated the utility of WSI in teaching and assessment [7,85,86]. The high-speed WSI robot industry is becoming highly diverse, with a wide range of optics, detectors, slide handling devices, and software, resulting in an increasing range of capabilities and costs. The quality of the images produced is generally of diagnostic quality, and with the viewing software, it is possible to have annotations and other information presented with the image [8,9,87,88].

The quality of focusing is limited by multiple optical and mechanical constraints, notably the numerical aperture (NA) of the objective and movement resolution on the vertical (z-) axis. Newer devices are implementing nontraditional optics, illumination, and sensors designed specifically for very high-speed image capture, resulting in significant improvements in speed, throughput, and resolution. Several scanners come with automatic algorithms to determine the optimal focus throughout the slide, but other scanners only have the capacity to image one level of the z-axis at a time.

The ability of a WSI system to create sharp, well-focused digital images in true color is determined by a combination of several components of the system's optical setup, including platform stability, optics, camera, adapter, and condenser setup. In general, the resolution feature represents an objective comparison measurement of the capability of an optical-based system to represent fine details in an image, usually represented as $\mu m/$pixel. Resolution correlates to the combined numerical aperture (NA) of the objective lens and condenser. Higher NA values result in higher image resolutions [6,89–91].

Increasing resolution also presents compromises to the overall scanner performance, including increased scanning time and decreased depth of field of focus, resulting in more frequent focus adjustments. The increased operational and algorithmic performance needs, coupled with lower sensitivity to histological preparation artifacts, may result in higher rescan rates. However, higher resolution is critical for interpretation of digital images of certain tissue types/applications such as transplant kidney biopsies [36]. For example, a scanning resolution approaching $0.15\,\mu m/$pixel enables the pathologist to determine the presence of clinically important, subtle details within the digital side that are typically visible while utilizing a traditional optical microscope [89].

The main component of the optical microscope is the objective lens. High NA objectives are typically used in optical microscopes provided within a WSI system

[92]. NA affects image resolution and scanning conducted with higher NA objectives creates sharper images. Lenses are typically nonproprietary to the WSI manufacturer. Vendors are beginning to incorporate multiple objective lens configurations within a single device. The majority of WSI systems are supplied with a $20\times$ objective lens for scanning purposes, as the $20\times$ objective lens configuration provides the best match of field size and resolution [93]. However, interpretation of specific pathology tissue types, such as neuropathology or transplantation pathology, may require higher resolutions that are obtained using higher magnification lenses and configurations [93].

The scanning resolution offered by currently marketed WSI systems using a bright field light source with a $20\times$ magnification objective configuration ranging from 0.22 to 0.5 µm/pixel, and when using a $40\times$ magnification objective configuration, is about 0.25 µm/pixel.

The balance between image resolution and magnification of the resulting digital image displayed on a computer monitor is an important point to consider when selecting and configuring a WSI system. It is difficult to match a WSI image to an image obtained using a traditional microscope. The design of a bright field optical system is composed of multiple stages or planes from where light is collected and focused on the specimen through the objective and connected tube lens/camera adaptor to reach the camera chip sensor and ultimately to the human eye via monitor display. To understand and optimize the resulting image quality on a monitor display, it is important to relate the resolution of the various components of this system [94,95].

Innovations in illumination are being utilized by WSI vendors; the traditional bright field halogen lighting, although still currently used in many scanners, is being replaced with newer LED-based modules in order to provide improved color consistency. New techniques in illumination also include LED "strobing" or flash illumination instead of constant illumination, as this minimizes motion artifact, enabling faster acquisition [92]. Automated multichannel fluorescent whole slide imaging is now readily available from some vendors [96,97].

Whole slide image viewers

All WSI systems are provided with software that supports viewing of WSI image files (digital slides) at the scanner control computer workstation. Imaging viewing software allows viewing of digital slides on a computer screen in a manner similar to the viewing of glass slides using a traditional optical microscope [98]. Most vendors provide feature-rich image viewing software solutions, offering many functions for image navigation, panning and zooming movements, and image manipulation/editing with an enhanced ability to add annotations and microscopic image measurements [6,78].

Image viewer software, unique to each vendor, is often installed on the local user's computer and may require administrative rights or the equivalent for deployment. However, in order to support the frequent need for remote image viewing, vendors often provide a web-based version of the viewing software. The latter usually offers slightly less functionality in terms of image modification, annotation, and measurement but allows for off-site viewing using protocols built off standard web technologies to facilitate ease of deployment.

The image viewer provided with each WSI system typically enables the viewing of WSI image files saved using a proprietary image format unique to their WSI system and does not support viewing of images created by other WSI scanner systems. However, several companies offer image viewers that currently support viewing of differently formatted files. The native, vendor-provided image viewing software and/or applications and their coupling to the corresponding vendor-specific pathology workflow software may offer several advantages over "generic" or viewing systems [78].

The main advantage of vendor-unique viewers includes access to specific vendor-encoded features such as user comments and/or annotations. Other advantages may include access to data while using third-party solutions, such as viewing multichannel fluorescence data, as limited access to such solutions is available while using "generic" viewers. To support comparison of different slides, multiple slide viewing and navigation options are provided by various vendor-specific image-viewers [58,98,99]. This feature is used to compare different biomarkers for a single case.

Digital slide viewers allow the pathologist to see a virtual slide on a screen in a similar fashion as they would looking through a traditional microscope. With these viewers, pathologists are able to view virtual slides at different magnifications. Furthermore, some of these viewers provide additional functionality like viewing of virtual slides at different magnifications in two windows or tools to annotate. Some viewers may also be linked to laboratory information systems, leading to a true digital workflow experience for the pathologist [100,101].

Whole slide image data and workflow management

Image and data management databases and tools are typically available as supporting software programs included with the majority of WSI systems intended for clinical or research use [50,102]. Nearly all WSI systems offer a choice of visualization and workflow/data management packages that include support for WSI file formats. Such software packages allow pathologists and histology technicians to view, manage, and analyze the digital slides and the associated clinical data [103].

Image and information management is an important benefit of creating a digital workflow for enabling the pathologist to access the relevant patient information in a digital cockpit or workstation [11,14,18,104]. By providing rapid access to this information, time is saved, and the quality of the information is enabled. Once this information and images are managed in the workstation, the second important aspect of the value pyramid can become available, which is the ability of "image sharing" [6]. This image sharing can be used for access to experts for consultation, showing cases to a colleague for clinical and treatment decisions, for quality assurance, or for "workload balancing". As these applications of image sharing are rapidly used, the value of digital pathology workflow will make an enormous impact on the care of cancer patients. These benefits are not nearly as powerful as the emerging third value of digital pathology, which is the field of image analysis and computer-aided diagnosis. Advances in the area of image analysis will provide the pathologists with powerful tools or "helper" applications to enable the pathologist to make manual and labor-intensive tasks more

CHAPTER 7 Digital pathology as a platform for primary diagnosis

automated including detecting "rare events" such as quantifying biomarkers, finding tumor metastasis within a lymph node, accurately quantitating the volume of tumor in specimen, grading tumors as a part of histological assessment, and screening tissue for cancer detection [6,14,44,105].

Vendors provide a variety of anatomic pathology workflow solutions, from simple image management (e.g., tagging images with basic metadata for searching) to advanced integration with the Laboratory Information System (LIS). Specific evaluation of these features should be based on the particular user or institutional requirements. However, the ability to integrate vendor-provided web-based viewers into third-party applications is becoming commonplace. This eliminates the need to conform a customer workflow to specific vendor functions and gives total control of any needed customization, provided the customer can provide the necessary IT and software development expertise.

The vast majority of high-throughput scanners commercially available today possess automated slide matching/distribution/tracking capabilities. Image management software can support anatomic pathology workflow by supporting the automation process, matching slides to cases, and permitting distribution of cases to the appropriate pathologist. Such systems can also improve case tracking abilities and provide immediate access to prior cases. Many WSI vendors provide dedicated workflow support solutions. Interoperability between the APLIS and the WSS can automate data exchange between the systems and lead to an improved anatomical pathology workflow. Interoperability challenges remain with respect to integrating multiple vendor WSI systems into a unified laboratory workflow solution. This is mainly due to the various advanced imaging features available across WSI platforms and the optimized way in which this data is encoded per vendor. For example, many WSI systems allow scanning and encoding of multiple focal planes that can later be revisited "digitally" upon digital slide viewing [103].

In pathology, international standardization initiatives such as Digital Imaging and Communications (DICOM) have helped create an imaging standard so that pathology images too can be incorporated into PACS [103,106–108]. The DICOM supplements 122 and 145 provide flexible object information definitions dedicated respectively to pathology specimen description and WSI acquisition, storage, and display [107]. According to these recommendations, large multiresolution, multiplanar WSI can be stored by mapping subregions from each layer into a DICOM series. With such a tiled organization images are stored in squares or rectangular tiles and subsequently arranged in two-dimensional arrays. Multiple resolutions are stored in this manner, with the highest resolution forming the base of a pyramid [109].

Selection criteria for a whole slide scanner

Today, much progress has occurred in the industry, and there are multiple WSI scanners on the market in the USA and globally. These devices have all of the following basic features: automated tissue identification, autofocus, autoscanning, automated image compression, and automated barcode reading. Some of the key features that need to be considered when evaluating a new WSI scanner are listed in Table 7.1.

Evolution of whole slide imaging systems **117**

Table 7.1 Key features to consider when evaluating a new whole slide scanner.

- **Physical/General**
  - Scanner dimensions
  - Scanner weight
  - Environmental tolerance
  - Service support

- **Glass Slide Handling**
  - Slide size
  - Slide capacity
  - Slide handling robotics

- **Image Acquisition**
  - Objective lens
  - Bright-field illumination source
  - Scanning resolution
  - Camera features

- **Image Digitization**
  - Scanning method
  - Scanning speed and time
  - Tissue detection method
  - Scanning region size
  - Color calibration
  - Z-stacking scanning capabilities
  - Fluorescence scanning capabilities

- **Workflow**
  - Laboratory Information system integration
  - Bar code support

Evolution of whole slide imaging systems

Early generation WSI systems were mostly built on a robotic compound microscope frame and often utilized the "stop-start" imaging methods to produce an initial low-resolution overview or thumbnail image of the glass slide for tissue detection [76,78,110]. Once the tissue was located, each field of view was individually focused and captured, repeating this process until all present tissue areas were captured. The resultant image was then represented in full color (typically 24 bits per pixel) with a dimension of many thousands of pixels in both width and height. Image fields were captured with an overlap to provide context for a "stitching" process used to join the images together at a later step in the process. Due to their large data sizes, these images were represented as a pyramid-like structure organized into multiple layers, each layer containing a different resolution. This stratification of data by resolution allowed for more efficient viewing via specialized vendor-supplied software [76].

Second-generation WSI systems minimized image acquisition time by utilizing alternate camera technologies such as line scanning [6]. These techniques allowed slide movement at a near-constant velocity or speed, thereby eliminating start/stop delays, as well as minimizing the amount of algorithmic image "stitching" required to create the final digital slide image. Focus detection algorithms were also improved to allow for interpolated focus adjustment during the scanning process. Expanded imaging methods enabled capture and digital revisiting of multiple focal planes. Camera speed improvements led to additional optical configuration options, including higher NA objectives (higher resolution imaging) and oil immersion modalities. In addition, more fail-safe mechanisms of slide loading mechanisms were developed, and WSI system slide-loading capacity was increased [41].

The current generation of scanners have slide loaders with slide capacities of up to 1000 slides per run, with the added ability to support various nonstandard histology glass sizes such as whole mounts (e.g., $2'' \times 3''$ slides). Lastly, the availability of more economical data storage options in recent years has enhanced workflow, making archival and remote web-based image viewing possible for routine daily workloads [6]. The newer scanners not only have faster processing and camera performance (measured by frames per second) but have also implemented enhancements to the continuous scanning techniques used by second-generation scanners by adopting illumination methods such as high-intensity light-emitting diode (LED) strobes to eliminate delay in image capture due to motion artifact. Focus mapping, also known as predictive autofocusing, has remained a viable technique, as it gives a priori information about focus position, but multiple tissue scans are required, thus decreasing scanning efficiency. An alternative technique, termed "frequency"-based sampling, saves time by incorporating the concept of focus maps with real-time continuous focus adjustments, whereby the system performs an autofocus for every number (N) of imaging fields scanned. Nonetheless, the optimal focus is difficult to maintain at higher NAs without increasing autofocus frequency (i.e., reducing N), requiring increased time for each complete scan performed at higher magnification. Newer WSI systems also offer advanced imaging functionality such as an epi-fluorescence (fluorescence) capability in addition to oil-immersion microscopy, thereby providing additional flexibility regarding potential uses of the WSI to support the anatomic pathology lab operations and needs. Over time, image capture time has decreased, image quality has improved, and the overall utility of the WSI has moved steadily toward a recognized increase in core value to both clinical and research imaging operations [6,29,41,111,112].

Infrastructure requirements and checklist for rolling out high-throughput whole slide imaging workflow solution

Step 1: Deploy scanners and initiate validation of anatomic pathology slides [113–115].

1. Depending on the volume of slides to be scanned, the first step is to begin installing the desired number of scanners in a designated digital scan lab facility, ensuring that each scanning appliance is afforded a 10 Gbit link to the

WSI server cluster, via a private, unshared channel. These facilities should ideally be in the histology lab or in close proximity to the histology lab [113].

2. The WSI scanner needs to be installed along with the associated server in a secure, clinical-grade data center with networking to the scan facilities and collective scanning appliances, using multiple dedicated 10 Gbit channels.

3. The next step is to install, in a sequenced fashion, the cohort of designated pathology workstations in various locations, with each having an independent 1 Gbit channel back to the server. This will enable the remote viewing of the WSI images by the pathologists in their work areas.

4. The next milestone is to establish connectivity between respective integration engines, such that patient demographics, case data, and other clinically relevant data, as deemed necessary in support of primary case sign-out, can be populated into the Digital Pathology Cockpit. This is accomplished by a bidirectional HL7 interface engine interface between scanners and the laboratory information.

5. This unified bidirectional HL7 Interface will provide support for ADT and demographic updates, accession logs, specimen/block/slide logs, and barcode IDs, internal histology orders, case history, slide archive inventory, and case narrative texts [116].

6. Finally, the pathologist workstation or cockpit will be ready for routine intramural case sign-out capability. Moreover, this capability serves as the foundational work needed for subsequent optimization in support of full-case-volume sign-out upon the availability of use of digital pathology solutions for primary case sign-out, utilizing FDA-approved devices.

7. The laboratory information system serves as a single source of truth (SSOT) for preserving the overall hospital-level referential integrity of data [115].

8. All pathologist sign-out workstations/cockpits have to be passed with certification protocols, and all digital pathology workstation locations need to have the availability of full bandwidth under maximal concurrency conditions [116].

9. Routine pathology slides are scanned prior to sending them out to pathologists, and images will be available in the pathology cockpit for digital sign-out [18].

10. After a digital sign-out is achieved, the next step is to validate and implement AI tools such as automated quantification of prognostic and predictive biomarker assays, automated lymph node metastasis detection, screening tissue for cancer detection, etc.

WSI and primary diagnosis

The first WSI scanner was approved by the FDA in 2017 and the second one in 2019, paving the way for using WSI as a primary diagnostic tool [117,118]. Prior to this, most of the financial gain from implementing WSI system was for remote consultations, frozen sections [119], education [120], and research. Utilizing WSI for primary diagnostics is helping labs with increased productivity and cost mitigation by increasing patient safety by providing pathology images for easy sharing for expert consultation, which in turn will lead to a reduction in over-calls and under-calls in pathology diagnosis. One of the impacts of improved access to images and sharing images is also in the rapid inpatient review of cases and the potential to decrease the number of hospital days of care and facilitate earlier discharge of patients where delay in pathology diagnosis may be resulting in discharge decisions. In a study published by the Journal of Pathology Informatics, a group from the University of Pittsburgh studied the positive effects of implementing digital pathology [121]. Workflow improvement benefits as a result of employing digital pathology included the support of lab automation (e.g., bar coding and tracking of assets, bidirectional interfaces with the scanners) and the potential to increase an individual pathologist's productivity by at least 13% due to improved organization and racking of surgical pathology cases [121]. Other possible areas for financial gains will be to increase the number of pathology consults coming to a facility, and indirectly, by recognition of these new technologies creating a band recognition, more patients will come to a facility for surgery and treatment [121]. An example of digital workflow for primary diagnosis and workstation setup is illustrated in Fig. 7.1.

The most significant benefit of WSI technology is to allow pathologists to sign out remotely. In 2020, the coronavirus disease 19 (COVID-19) pandemic emerged as the most significant global health crisis of our time [122]. In the United States, the Centers for Medicare & Medicaid Services (CMS) responded to the pandemic by issuing a waiver that allowed pathologists to work remotely in order to address new safety and practice restrictions. During the pandemic, laboratories and hospitals have seen a significant increase in the demand for digital pathology and remote pathology services [123,124]. A digital pathology workflow provides increased flexibility for pathologists, who can work from home, from office, or from another dedicated space and also reduces in-person face-to-face interaction during the pandemic. A digital workflow also improves efficiency by making digital images available prior to the receipt of glass slides (especially important for centralized histology lab); providing rapid retrieval of prior cases; directly accessing all digital images by pathologists, trainees, and other lab staff; allowing efficient sharing of digital images with colleagues; sending either the case digital image link or using the built-in collaboration function for intradepartmental consultation, etc. [125]. WSI technology has several other advantages over conventional microscopy, such as the availability of annotation and measurement tools allowing precise microscopic measurements of tumor size, depth of invasion, and distance to surgical margins, improvement of office ergonomics, and the ability to make use of computer-aided diagnostic tools such as image algorithms and 3D reconstruction used to convert poorly oriented specimens into well-oriented ones. WSI has also created new business models in pathology [126]. One such example is the virtual Immunohistochemistry service provided by large national laboratories. After the remote reference laboratory performs technical staining and slide scanning services, the referring pathologist is provided with full access to these immunostained slides for their interpretation or referral to a teleconsultant. This has allowed some pathology practices to

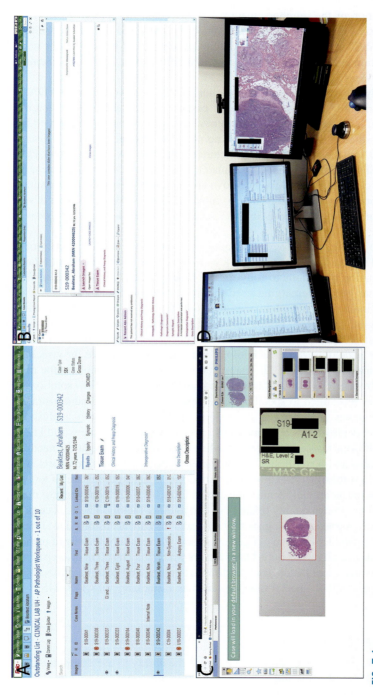

FIG. 7.1

A digital workflow for primary diagnosis and workstation setup. (A) Pathologist's worklist in laboratory information system (LIS). The cases with "eye" icon have digital images ready for view. (B) A direct link to digital images is incorporated into the case result in LIS. (C) Digital images can be viewed in image managing system (IMS) with image tools. (D) A standard workstation set up with three monitors: the left one for literature/email view, the middle one for LIS, and the right one (FDA approved) for viewing digital slides.

re-capture a portion of the reimbursement for professional interpretation services. Finally, some issues that still need to be satisfactorily addressed to truly accommodate the use of WSI for making primary diagnoses are related to regulations (e.g., FDA and validation guidelines), malpractice and liability issues, and reimbursement for technical services [118,127–129].

More recently, Digital Pathology Association (DPA) has worked with FDA to downgrade the risk of WSI scanners. Under certain circumstances manufacturers can opt to submit a de novo application for a moderate or low-risk device which has not yet been classified and for which there is no predicate device. The DPA's discussions with the FDA on the notion that manufacturers submit de novo applications. FDA indicated the WSI for primary diagnosis is, rather than a PMA, a candidate for de novo applications as a class II device. Once a de novo application receives marketing authorization, other companies will then be able to submit 510(k) for WSI. The first device was approved for primary diagnosis in 2017, and the second device was approved in 2019.

The practice of histopathology evaluation by benchtop optical microscopes remains the gold standard for pathology review. However, as the workflow of related specimen information transitions to fully digital systems, including the use of WSI scanners, there has been an emerging need to incorporate the traditional glass slide histopathology findings as digital images to move toward a completely slideless and seamless digital workflow [11,50,87,130,131]. This need has been increasingly met over the past decade by the development of robotic telepathology systems and evolved to whole slide digitization technology and associated software for visualization and data management [114,132]. The selection of systems available in the marketplace ranges from offering broad-based Anatomic Pathology workflow solutions to those with feature sets suited to specific areas of pathology. Hospital administrators and C-suite executives are taking a closer look at the cost-to-benefit analysis of WSI systems, and deployments continue to increase steadily around the world [121].

Some of the lessons learned from early "validation" studies highlight the need for diagnostic standards and better design of study conditions for a direct comparison of light and WSI modalities [9,132–140]. Some important considerations are to look at intraobserver discrepancies and identify the causes [106,141,142]. Many times, these discrepancies are not image quality related but due to other factors. These discrepancies may be because subject pathologists sign out cases they are accustomed to signing out, while general pathologists sign out general cases, whereas subspecialized pathologists sign out subspecialty cases only. Other factors which may result in discrepancies may be due to preanalytical factors such as different cutting and staining protocols as well as using scanners with special functions such as Z-stacking [118,143–145].

WSI and image analysis

One of the major advantages of digitizing slides is the ability to use computer algorithms to conduct reproductive and accurate measurements (size, cell counting, mitosis counting, etc.), quantify biomarkers, detect lymph node metastasis, screen tissue for cancer detection, predict disease molecular changes, and predict disease prognosis and survival [146–151]. Many WSI vendors have engaged in cooperation with companies developing computer-assisted diagnosis (CAD) and providing quantitative image analysis by providing a selection of solutions from specific, defined assays (e.g., HER2/neu) to rule-based image segmentation [152,153]. These application packages can interoperate with WSI formats, either via vendor-supplied interfaces or via provided programming interfaces. Several vendors, but not all, include image analysis software packages with a WSI system [154,155].

FDA-approved quantitative algorithms for diagnostic use are available from a limited number of vendors. Approved tests include breast cancer biomarkers—progesterone receptor (PR), estrogen receptor (ER), human epidermal growth factor receptor 2 (HER2/neu), the cell proliferation index marker Ki67, and p53 [149,156–158].

WSI and deep learning

As more laboratories adopt digital pathology and digital images are incorporated into the diagnostic pathology workflow, there will be better case management and resulting cases because of overall improvement in the information management surrounding a patient case [99,159–167]. AI tools can now be applied prior to sign-out of the case to help collate and integrate the information that is needed for case review, but ultimately, the pathologist will review the case and make a diagnosis [14,79,151,168–170] [171]. The use of DL and AI tools combined with digital pathology images can extend the value of digital pathology far beyond what is possible today and quantified above. This is the true value of digital pathology, and the AI tools that have been developed and are being developed today will help transform diagnostic pathology [14,172].

As anatomic pathologists implement digital pathology, they will now have the capability to start using smart deep learning tools in daily practice [28,171]. This is easy to illustrate by considering a specific area in surgical pathology, prostate cancer diagnosis, and grading. Adenocarcinoma of the prostate is one of the most common cancers diagnosed in men worldwide, second only to lung cancer. The increase in diagnosis stems from the increase in screening for prostate cancer, and it has led to unnecessary testing and overtreatment [173,174]. Prostate cancer has an excellent 5-year survival rate of greater than 98% [159–167]. For this reason, it is important to determine if treatment will have an added benefit to the patient or simply decrease his quality of life. Current standards utilize Gleason scoring to determine prostate cancer prognosis and treatment benefit [174]. Nagpal et al. have published a study on the development and validation of a deep learning algorithm for improving Gleason scoring of prostate cancer [175]. This study aimed to address the issue of grading variability, improve prognostication, and optimize patient management. A two-stage deep learning system (DLS) was developed to perform Gleason scoring and quantitation on prostatectomy specimens [175].

Twenty-nine pathologists reviewed the validation dataset as a comparison to the performance of the DLS. The mean accuracy of the 29 pathologists was 0.61 compared to an accuracy of 0.70 for the DLS ($p=0.002$). Gleason grade decision thresholds were also investigated with the DLS achieving area under the receiver operating characteristic curves between 0.95 and 0.96 at each threshold. At a Gleason grade ≥ 4, the largest difference was seen, and the DLS showed greater sensitivity and specificity than 9 out of 10 pathologists [175]. Due to its clinical significance, Gleason pattern quantitation was compared among the groups: DLS had 4–6% lower mean absolute error than the average pathologist, and the DLS predicted the same pattern as the pathologist 97% of the time. Lastly, the DLS was able to predict a more gradual transition from well to poorly differentiated by utilizing "fine-grained Gleason patterns," such as 3.3 or 3.7, instead of the traditional methods of Gleason scoring [175].

Bulten et al. reported on a deep learning method for the automated Gleason grading of prostate biopsies. The purpose of this latter study was to create a deep learning system (DLS) that was fully automated for detecting cancer and performing Gleason grading without the need for training with manual pixel-level annotations [176]. This DLS attained a quadratic kappa agreement of 0.918 with the reference set. Misclassification occurred at the decision boundaries of grade groups 2 and 3 as well as groups 4 and 5. Determining benign vs malignant biopsies resulted in an area under the receiver operating characteristic curve of 0.990 with a specificity of 82%. In a subset of 100 cases assessed by an external panel of 15 pathologists, the median inter-rater agreement kappa value was 0.819, compared to the system kappa value of 0.854 [176]. Additional analyses showed that the DLS performs better than pathologists with less than 15 years of experience ($p=0.036$) and is not significantly different from those with greater than 15 years of experience ($p=0.955$). No significant difference was found between the DLS and the panel of pathologists for detection of benign and malignant specimens ($p=0.704$). Lastly, the DLS was applied to an external data set of 641 cores that were annotated by two pathologists independently, and quadratic kappa values were 0.711 and 0.639. These values fall within the range of inter-observer variability but are lower than those achieved by the reference set [176].

Both studies above showed that deep learning systems can achieve pathologist-level performance [175,176]. This is important in the field of pathology, with increasing demand and not enough physicians going into pathology to keep up. The first study developed a DLS that was more accurate than several general pathologists in the quantitation of Gleason patterns. Additionally, it created a system that could discriminate these patterns at a finer level. The second study created a DLS using an automated training system that did not require extensive additional pathologist work hours and thus saved valuable resources. It used a straightforward and easily interpretable system that is similar to the methods pathologists currently use in clinical practice, which allows simple communication between the DLS and clinicians. Both studies created systems that can not only detect the presence of a tumor but accurately grade the aggressiveness of the tumor utilizing Gleason scoring [175].

There are important limitations to both studies that would need to be addressed before the implementation of these tools into clinical practice [175,176]. The first study employed a time-consuming and expensive method to train the DLS with 112 million pathologist-annotated image patches requiring almost 900 pathologist hours. The purpose of creating the DLS becomes obsolete if pathologists have to spend exuberant amounts of time training them. This study used prostatectomy specimens to determine the Gleason grade. This means the grade obtained can only inform postoperative treatment decisions instead of whether a prostatectomy was recommended to begin with [175]. The second study solved this issue by using biopsy specimens to determine Gleason grade but only obtained data from a single center. This caused the DLS performance to be lower for the external test set [176]. Both studies utilized digital slide review, which is not universally accepted by all pathologists. They also did not show that either DLS could stage or review the slides for unusual pathology, other tumor types, or foreign tissue that could be present in prostate biopsies.

Overall, there have been promising advances in DLS for the automated Gleason grading of prostate cancer. Both studies discussed above were conducted at approximately the same time in slightly different manners. They used different training methods for each DLS as well as different types of specimens. It appears that semi-automatic labeling for training the DLS of the second study paired with the multicenter data retrieval of the first study could produce even more generalizable results. The ability to further classify regions into "fine-grained Gleason patterns" could increase classification accuracy and agreement within the DLS. Utilizing the Gleason grade of biopsy specimens instead of prostatectomy specimens is essential in providing insight for determining the next steps in clinical care for patients with prostate adenocarcinoma. Other investigators have also developed and built algorithms for the diagnosis and detection as well as grading of prostate cancer [177–180], and it is expected that similar methods and tools will continue to flourish as digitization of pathology workflow continues to increase with increasing adoption of these tools.

Conclusions

Significant technological gains have led to the adoption of innovative digital imaging solutions in pathology. WSI, which refers to scanning of conventional glass slides to produce digital slides, is the most recent imaging modality being employed by pathology departments worldwide. WSI continues to gain traction among pathologists for diagnostic, educational, and research purposes. The quality of the images produced is of diagnostic quality, and with the viewing software, it is possible to have annotations and clinical metadata presented with the image, resulting in a virtual microscope with all the clinical information needed to sign-out the case [181]. Many recent studies have demonstrated the utility of WSI in teaching, quality assurance assessment, frozen consultation, telepathology, image analysis, and research use

[172,182,183]. Some of the advantages of using digital pathology, particularly WSI, has been the ability to is to help address several problems with conventional light microscopy, including eliminating the inefficiencies of "snapshot" imaging or taking images of regions of interest for meetings and conferences. Digital pathology has now paved the way to facilitate the sharing of cases/slides across distances. Now we can have access to experts, irrespective of patient location. Digital workflow also contributes to our ability to share cases with multiple colleagues simultaneously. We can start creating more smart worklists, particularly aiming in the prioritization of caseloads and workload balancing. It is now possible to reallocate cases across a network, which in turn enables the experts to do expert work and generalists to do general pathology. There is access to archived images, which in turn reduces and may eliminate the physical, more expensive movement of glass slides.

Applications of artificial intelligence and machine learning techniques such as deep neural networks may be trained to not only recognize specific patterns on a whole slide image of an H&E slide but, in addition, AI tools may also help in the interpretation of features in the tissue that are predictive and/or prognostic. With deep learning software deployment, pathologists will gain knowledge in the three areas of pathology workflow: information management, image sharing, and image analysis [14,26,184,185].

References

[1] Evans AJ, et al. Implementation of whole slide imaging for clinical purposes: issues to consider from the perspective of early adopters. Arch Pathol Lab Med 2017;141 (7):944–59.

[2] Farris AB, et al. Whole slide imaging for analytical anatomic pathology and telepathology: practical applications today, promises, and perils. Arch Pathol Lab Med 2017;141 (4):542–50.

[3] Bauer TW, Slaw RJ. Validating whole-slide imaging for consultation diagnoses in surgical pathology. Arch Pathol Lab Med 2014;138(11):1459–65.

[4] Dangott B, Parwani A. Whole slide imaging for teleconsultation and clinical use. J Pathol Inform 2010;1.

[5] Cornish TC, Swapp RE, Kaplan KJ. Whole-slide imaging: routine pathologic diagnosis. Adv Anat Pathol 2012;19(3):152–9.

[6] Zarella MD, et al. A practical guide to whole slide imaging: a white paper from the digital pathology association. Arch Pathol Lab Med 2019;143(2):222–34.

[7] Pantanowitz L, et al. Whole slide imaging for educational purposes. J Pathol Inform 2012;3:46.

[8] Mukhopadhyay S, et al. Whole slide imaging versus microscopy for primary diagnosis in surgical pathology: a multicenter blinded randomized noninferiority study of 1992 cases (pivotal study). Am J Surg Pathol 2018;42(1):39–52.

[9] Amin S, Mori T, Itoh T. A validation study of whole slide imaging for primary diagnosis of lymphoma. Pathol Int 2019;69(6):341–9.

[10] Cohen S, Furie MB. Artificial intelligence and pathobiology join forces in the American Journal of Pathology. Am J Pathol 2019;189(1):4–5.

References   **127**

[11] Hanna MG, et al. Whole slide imaging equivalency and efficiency study: experience at a large academic center. Mod Pathol 2019;32(7):916–28.

[12] Malarkey DE, et al. Utilizing whole slide images for pathology peer review and working groups. Toxicol Pathol 2015;43(8):1149–57.

[13] Romero Lauro G, et al. Digital pathology consultations-a new era in digital imaging, challenges and practical applications. J Digit Imaging 2013;26(4):668–77.

[14] Parwani AV. Next generation diagnostic pathology: use of digital pathology and artificial intelligence tools to augment a pathological diagnosis. Diagn Pathol 2019;14(1):138.

[15] Farahani N, Parwani AV, Pantanowitz L. Whole slide imaging in pathology: advantages, limitations, and emerging perspectives. Pathol Lab Med Int 2015;7:23–33.

[16] Amin W, Srintrapun SJ, Parwani AV. Automated whole slide imaging. Expert Opin Med Diagn 2008;2(10):1173–81.

[17] Pantanowitz L, et al. Review of the current state of whole slide imaging in pathology. J Pathol Inform 2011;2:36.

[18] Park Y, et al. Whole slide scanner and anatomic pathology laboratory information system integration with HL7 interface to support digital pathology sign-out workflow. Mod Pathol 2019;32.

[19] Pantanowitz L, et al. Validating whole slide imaging for diagnostic purposes in pathology: guideline from the College of American Pathologists Pathology and Laboratory Quality Center. Arch Pathol Lab Med 2013;137(12):1710–22.

[20] Parwani AV, et al. Regulatory barriers surrounding the use of whole slide imaging in the United States of America. J Pathol Inform 2014;5(1):38.

[21] Saco A, et al. Validation of whole-slide imaging in the primary diagnosis of liver biopsies in a University Hospital. Dig Liver Dis 2017;49(11):1240–6.

[22] Evans AJ, et al. US Food and Drug Administration approval of whole slide imaging for primary diagnosis: a key milestone is reached and new questions are raised. Arch Pathol Lab Med 2018;142(11):1383–7.

[23] Boyce BF. Whole slide imaging: uses and limitations for surgical pathology and teaching. Biotech Histochem 2015;90(5):321–30.

[24] Allen TC. Regulating artificial intelligence for a successful pathology future. Arch Pathol Lab Med 2019;143(10):1175–9.

[25] Chang HY, et al. Artificial intelligence in pathology. J Pathol Transl Med 2019;53(1):1–12.

[26] Colling R, et al. Artificial intelligence in digital pathology: a roadmap to routine use in clinical practice. J Pathol 2019;249(2):143–50.

[27] Harmon SA, et al. Artificial intelligence at the intersection of pathology and radiology in prostate cancer. Diagn Interv Radiol 2019;25(3):183–8.

[28] Niazi MKK, Parwani AV, Gurcan MN. Digital pathology and artificial intelligence. Lancet Oncol 2019;20(5):e253–61.

[29] Acs B, Hartman J. Next generation pathology: artificial intelligence enhances histopathology practice. J Pathol 2020;250(1):7–8.

[30] Abels E, Pantanowitz L. Current state of the regulatory trajectory for whole slide imaging devices in the USA. J Pathol Inform 2017;8:23.

[31] Ferreira R, et al. The virtual microscope. Proc AMIA Annu Fall Symp 1997;449–53.

[32] Park S, et al. The history of pathology informatics: a global perspective. J Pathol Inform 2013;4:7.

[33] Amin W, et al. Use of whole slide imaging for tissue microarrays: the cooperative prostate cancer tissue resource model. Mod Pathol 2009;22:379a.

[34] Niazi MKK, et al. Advancing clinicopathologic diagnosis of high-risk neuroblastoma using computerized image analysis and proteomic profiling. Pediatr Dev Pathol 2016.

[35] LeCun Y, Bengio Y, Hinton G. Deep learning. Nature 2015;521(7553):436–44.

[36] Barisoni L, et al. Digital pathology imaging as a novel platform for standardization and globalization of quantitative nephropathology. Clin Kidney J 2017;10(2):176–87.

[37] Niazi MKK, Arole TETV, Parwani BAV, Lee C, Gurcan MN. Automated T1 bladder risk stratification based on depth of lamina propria invasion from H&E tissue biopsies: a deep learning approach. SPIE Med Imaging 2018;10581. https:/doi.org/10.1117/12.2294552.

[38] Wong STC. Is pathology prepared for the adoption of artificial intelligence? Cancer Cytopathol 2018;126(6):373–5.

[39] Wang S, et al. Artificial intelligence in lung cancer pathology image analysis. Cancers (Basel) 2019;11(11).

[40] Reyes C, et al. Intra-observer reproducibility of whole slide imaging for the primary diagnosis of breast needle biopsies. J Pathol Inform 2014;5(1):5.

[41] Saylor C, et al. Evaluation of whole slide imaging systems prior to establishing a model digital pathology network for the air force medical service: methods and selection criteria. Mod Pathol 2013;26:382a.

[42] Tizhoosh HR, Pantanowitz L. Artificial intelligence and digital pathology: challenges and opportunities. J Pathol Inform 2018;9:38.

[43] Aeffner F, et al. Whole-slide imaging: the future is here. Vet Pathol 2018;55(4):488–9.

[44] Salto-Tellez M, Maxwell P, Hamilton P. Artificial intelligence-the third revolution in pathology. Histopathology 2019;74(3):372–6.

[45] Aeffner F, et al. Digital microscopy, image analysis, and virtual slide repository. ILAR J 2018;59(1):66–79.

[46] Furness PN. The use of digital images in pathology. J Pathol 1997;183(3):253–63.

[47] Yagi Y, Gilbertson JR. Digital imaging in pathology: the case for standardization. J Telemed Telecare 2005;11(3):109–16.

[48] Leong FJ, Leong AS. Digital imaging applications in anatomic pathology. Adv Anat Pathol 2003;10(2):88–95.

[49] Park S, Pantanowitz L, Parwani AV. Digital imaging in pathology. Clin Lab Med 2012;32(4):557–84.

[50] Park S, et al. Workflow organization in pathology. Clin Lab Med 2012;32(4):601–22.

[51] Park SL, et al. Anatomic pathology laboratory information systems: a review. Adv Anat Pathol 2012;19(2):81–96.

[52] Marchevsky AM, et al. Storage and distribution of pathology digital images using integrated web-based viewing systems. Arch Pathol Lab Med 2002;126(5):533–9.

[53] Brachtel E, Yagi Y. Digital imaging in pathology – current applications and challenges. J Biophotonics 2012;5(4):327–35.

[54] Amin M, et al. Integration of digital gross pathology images for enterprise-wide access. J Pathol Inform 2012;3:10.

[55] Cross S. Websites review-pathology images. Histopathology 2001;38(4):376.

[56] Mai KT, et al. Creating digital images of pathology specimens by using a flatbed scanner. Histopathology 2001;39(3):323–5.

[57] Fung KM, et al. Whole slide images and digital media in pathology education, testing, and practice: the Oklahoma experience. Anal Cell Pathol (Amst) 2012;35(1):37–40.

[58] Camparo P, et al. Utility of whole slide imaging and virtual microscopy in prostate pathology. APMIS 2012;120(4):298–304.

[59] Monaco SE, et al. A "virtual slide box" using whole slide imaging for reproductive pathology education for medical students. Lab Invest 2011;91:132a.

[60] Park S, et al. Whole slide image teaching sets to support graduate medical and pathology education. Histopathology 2012;61:44.

[61] Cucoranu IC, Parwani AV, Pantanowitz L. Digital whole slide imaging in cytology. Arch Pathol Lab Med 2014;138(3):300.

[62] Weinstein RS. Telepathology: practicing pathology in two places at once. Clin Lab Manage Rev 1992;6(2):171–3. discussion 174–175.

[63] Evans AJ, Kiehl TR, Croul S. Frequently asked questions concerning the use of whole-slide imaging telepathology for neuropathology frozen sections. Semin Diagn Pathol 2010;27(3):160–6.

[64] Dietz RL, Hartman DJ, Pantanowitz L. Systematic review of the use of telepathology during intraoperative consultation. Am J Clin Pathol 2020;153(2):198–209.

[65] Chong T, et al. The California telepathology service: UCLA's experience in deploying a regional digital pathology subspecialty consultation network. J Pathol Inform 2019;10:31.

[66] Laurent-Bellue A, et al. Telepathology for intra-operative frozen sections. Ann Pathol 2019;39(2):113–8.

[67] Girolami I, et al. Diagnostic concordance between whole slide imaging and conventional light microscopy in cytopathology: a systematic review. Cancer Cytopathol 2020;128 (1):17–28.

[68] Weinstein RS, Holcomb MJ, Krupinski EA. Invention and early history of telepathology (1985–2000). J Pathol Inform 2019;10:1.

[69] Weinstein RS, et al. Overview of telepathology, virtual microscopy, and whole slide imaging: prospects for the future. Hum Pathol 2009;40(8):1057–69.

[70] Weinstein RS, et al. Telepathology overview: from concept to implementation. Hum Pathol 2001;32(12):1283–99.

[71] Fritz P, et al. Experience with telepathology in combination with diagnostic assistance systems in countries with restricted resources. J Telemed Telecare 2019. 1357633X19840475.

[72] Oberholzer M, et al. Modern telepathology: a distributed system with open standards. Curr Probl Dermatol 2003;32:102–14.

[73] Wang S, et al. Computational staining of pathology images to study the tumor microenvironment in lung cancer. Cancer Res 2020.

[74] Niazi MKK, Parwani AV, Gurcan MN. Computer-assisted bladder cancer grading: α-shapes for color space decomposition. In: SPIE medical imaging. International Society for Optics and Photonics; 2016.

[75] Henricks WH. Evaluation of whole slide imaging for routine surgical pathology: looking through a broader scope. J Pathol Inform 2012;3:39.

[76] Ghaznavi F, et al. Digital imaging in pathology: whole-slide imaging and beyond. Annu Rev Pathol 2013;8:331–59.

[77] Amin M, et al. Implementation of a pathology teaching website with integrated whole slide imaging for greater compatibility with tablets/smartphones. Lab Invest 2013;93:375a–6a.

[78] Ho J, et al. Use of whole slide imaging in surgical pathology quality assurance: design and pilot validation studies. Hum Pathol 2006;37(3):322–31.

[79] Feldman MD. Beyond morphology: whole slide imaging, computer-aided detection, and other techniques. Arch Pathol Lab Med 2008;132(5):758–63.

[80] Velez N, Jukic D, Ho J. Evaluation of 2 whole-slide imaging applications in dermatopathology. Hum Pathol 2008;39(9):1341–9.

[81] Wilbur DC, et al. Whole-slide imaging digital pathology as a platform for teleconsultation: a pilot study using paired subspecialist correlations. Arch Pathol Lab Med 2009;133 (12):1949–53.

[82] Xing F, Yang L. Robust nucleus/cell detection and segmentation in digital pathology and microscopy images: a comprehensive review. IEEE Rev Biomed Eng 2016;9:234–63.

[83] Gurcan MN, et al. Histopathological image analysis: a review. IEEE Rev Biomed Eng 2009;2:147–71.

[84] Cruz-Roa A, et al. High-throughput adaptive sampling for whole-slide histopathology image analysis (HASHI) via convolutional neural networks: application to invasive breast cancer detection. PloS One 2018;13(5), e0196828.

[85] Foster K. Medical education in the digital age: digital whole slide imaging as an e-learning tool. J Pathol Inform 2010;1.

[86] Dziegielewski M, Velan GM, Kumar RK. Teaching pathology using "hotspotted" digital images. Med Educ 2003;37(11):1047–8.

[87] McClintock DS, Lee RE, Gilbertson JR. Using computerized workflow simulations to assess the feasibility of whole slide imaging full adoption in a high-volume histology laboratory. Anal Cell Pathol (Amst) 2012;35(1):57–64.

[88] Gilbertson JR, Yagi Y. Clinical slide digitization – whole slide imaging in clinical practice. In: Gu OE, editor. Virtual microscopy and virtual slides in teaching, diagnosis and research. Boca Raton: Taylor and Francis; 2005.

[89] Sellaro TL, et al. Relationship between magnification and resolution in digital pathology systems. J Pathol Inform 2013;4:21.

[90] Shrestha P, Hulsken B. Color accuracy and reproducibility in whole slide imaging scanners. J Med Imaging (Bellingham) 2014;1(2), 027501.

[91] Yagi Y. Color standardization and optimization in whole slide imaging. Diagn Pathol 2011;6(Suppl 1):S15.

[92] Varga VS, Molnar B, Virag T. Automated high throughput whole slide imaging using area sensors, flash light illumination and solid state light engine. Stud Health Technol Inform 2012;179:187–202.

[93] Rojo MG, et al. Critical comparison of 31 commercially available digital slide systems in pathology. Int J Surg Pathol 2006;14(4):285–305.

[94] Rojo MG, Bueno G. Analysis of the impact of high-resolution monitors in digital pathology. J Pathol Inform 2015;6:57.

[95] D'Haene N, et al. Comparison study of five different display modalities for whole slide images in surgical pathology and cytopathology in Europe. In: Medical imaging 2013: digital pathology; 2013. p. 8676.

[96] Varga VS, et al. Automated multichannel fluorescent whole slide imaging and its application for cytometry. Cytometry A 2009;75(12):1020–30.

[97] Rojo MG, Daniel C, Schrader T. Standardization efforts of digital pathology in Europe. Anal Cell Pathol (Amst) 2012;35(1):19–23.

[98] Ameisen D, et al. Towards better digital pathology workflows: programming libraries for high-speed sharpness assessment of whole slide images. Diagn Pathol 2014;9 (Suppl 1):S3.

References   **131**

[99] Al-Janabi S, et al. Whole slide images for primary diagnostics of gastrointestinal tract pathology: a feasibility study. Hum Pathol 2012;43(5):702–7.

[100] Lebre R, et al. Collaborative framework for a whole-slide image viewer. In: 2019 IEEE 32nd international symposium on computer-based medical systems (CBMS); 2019. p. 221–4.

[101] Ameisen D, et al. Towards better digital pathology workflows: programming libraries for high-speed sharpness assessment of whole slide images. Diagn Pathol 2014;9.

[102] Thorstenson S, Molin J, Lundstrom C. Implementation of large-scale routine diagnostics using whole slide imaging in Sweden: digital pathology experiences 2006-2013. J Pathol Inform 2014;5(1):14.

[103] Marques Godinho T, et al. An efficient architecture to support digital pathology in standard medical imaging repositories. J Biomed Inform 2017;71:190–7.

[104] Girolami I, et al. The landscape of digital pathology in transplantation: from the beginning to the virtual E-slide. J Pathol Inform 2019;10:21.

[105] Yeh FC, et al. Automated grading of renal cell carcinoma using whole slide imaging. J Pathol Inform 2014;5(1):23.

[106] Campbell WS, et al. Whole slide imaging diagnostic concordance with light microscopy for breast needle biopsies. Hum Pathol 2014;45(8):1713–21.

[107] Jodogne S, et al. Open implementation of DICOM for whole-slide microscopic imaging. In: Proceedings of the 12th international joint conference on computer vision, imaging and computer graphics theory and applications (Visigrapp 2017), Vol. 6; 2017. p. 81–7.

[108] Clunie D, et al. Digital imaging and communications in medicine whole slide imaging connectathon at digital pathology association pathology visions 2017. J Pathol Inform 2018;9:6.

[109] Lajara N, et al. Optimum web viewer application for DICOM whole slide image visualization in anatomical pathology. Comput Methods Programs Biomed 2019;179.

[110] Khvatkov V, Harris E. Pathology: the last digital frontier of biomedical imaging. MLO Med Lab Obs 2013;45(11):28–9.

[111] Déniz O, et al. Multi-stained whole slide image alignment in digital pathology. In: SPIE medical imaging. International Society for Optics and Photonics; 2015.

[112] Schonmeyer R, et al. Automated whole slide analysis of differently stained and co-registered tissue sections. In: Bildverarbeitung Fur Die Medizin 2015: Algorithmen - Systeme - Anwendungen; 2015. p. 407–12.

[113] Hartman DJ, et al. Enterprise implementation of digital pathology: feasibility, challenges, and opportunities. J Digit Imaging 2017;30(5):555–60.

[114] Ho J, et al. Needs and workflow assessment prior to implementation of a digital pathology infrastructure for the US air force medical service. J Pathol Inform 2013;4:32.

[115] Lloyd M, et al. How to acquire over 500,000 whole slides images a year: creating a massive novel data modality to accelerate cancer research. Mod Pathol 2018;31:592–3.

[116] Guo H, et al. Digital pathology and anatomic pathology laboratory information system integration to support digital pathology sign-out. J Pathol Inform 2016;7:23.

[117] Boyce BF. An update on the validation of whole slide imaging systems following FDA approval of a system for a routine pathology diagnostic service in the United States. Biotech Histochem 2017;92(6):381–9.

[118] Evans AJ, et al. US Food and Drug Administration approval of whole slide imaging for primary diagnosis a key milestone is reached and new questions are raised. Arch Pathol Lab Med 2018;142(11):1383–7.

[119] Pradhan D, et al. Evaluation of panoramic digital images using Panoptiq for frozen section diagnosis. J Pathol Inform 2016;7:26.

[120] Dangott B, Parwani AV. Implementation of a whole slide image database for pathology residency education. Mod Pathol 2010;23:124a.

[121] Ho J, et al. Can digital pathology result in cost savings? A financial projection for digital pathology implementation at a large integrated health care organization. J Pathol Inform 2014;5(1):33.

[122] Mohanty SK, et al. Severe acute respiratory syndrome coronavirus-2 (SARS-CoV-2) and coronavirus disease 19 (COVID-19)–anatomic pathology perspective on current knowledge. Diagn Pathol 2020;15(1):1–17.

[123] Williams BJ, et al. The future of pathology: what can we learn from the COVID-19 pandemic? J Pathol Inform 2020;11.

[124] Cimadamore A, et al. Digital pathology and COVID-19 and future crises: pathologists can safely diagnose cases from home using a consumer monitor and a mini PC. J Clin Pathol 2020;73(11):695–6.

[125] Lujan GM, et al. Digital pathology initiatives and experience of a large academic institution during the coronavirus disease 2019 (COVID-19) pandemic. Arch Pathol Lab Med 2021;145(9):1051–61.

[126] Lloyd M, et al. Sectioning automation to improve quality and decrease costs for a high-throughput slide scanning facility. Lab Invest 2019;99.

[127] Wack K, et al. A multisite validation of whole slide imaging for primary diagnosis using standardized data collection and analysis. J Pathol Inform 2016;7:49.

[128] Evans JL, et al. A scalable, cloud-based, unsupervised deep learning system for identification, extraction, and summarization of potentially imperceptible patterns in whole-slide images of breast cancer tissue. Cancer Res 2019;79(13).

[129] Rojo MG, Castro AM, Goncalves L. COST action "EuroTelepath": digital pathology integration in electronic health record, including primary care centres. Diagn Pathol 2011;6(Suppl 1):S6.

[130] Natan M. Achieving simultaneous improvements in workflow and multiplexing in whole-slide tissue imaging. Cancer Res 2018;78(13).

[131] Sarwar S, et al. Physician perspectives on integration of artificial intelligence into diagnostic pathology. NPJ Digit Med 2019;2:28.

[132] Menter T, et al. Intraoperative frozen section consultation by remote whole-slide imaging analysis -validation and comparison to robotic remote microscopy. J Clin Pathol 2019.

[133] Jukic DM, et al. Clinical examination and validation of primary diagnosis in anatomic pathology using whole slide digital images. Arch Pathol Lab Med 2011;135(3):372–8.

[134] Thrall MJ. Validation of two whole slide imaging scanners based on the draft guidelines of the College of American Pathologists. Lab Invest 2013;93:487a.

[135] Bongaerts O, et al. Conventional microscopical versus digital whole-slide imaging-based diagnosis of thin-layer cervical specimens: a validation study. J Pathol Inform 2018;9:29.

[136] Lee JJ, et al. Validation of digital pathology for primary histopathological diagnosis of routine inflammatory dermatopathology cases. Am J Dermatopathol 2018;40(1):17–23.

[137] Bonsembiante F, et al. Diagnostic validation of a whole-slide imaging scanner in cytological samples: diagnostic accuracy and comparison with light microscopy. Vet Pathol 2019;56(3):429–34.

[138] Tabata K, et al. Whole-slide imaging at primary pathological diagnosis: validation of whole-slide imaging-based primary pathological diagnosis at twelve Japanese academic institutes. Pathol Int 2017;67(11):547–54.

[139] Evans AJ, Asa SL. Whole slide imaging telepathology (WSITP) for primary diagnosis in surgical pathology: a comprehensive validation study at university health network (UHN). Lab Invest 2012;92:391a.

[140] Pekmezei M, et al. Concordance between whole-slide imaging and light microscopy for surgical neuropathology. Lab Invest 2014;94:509a.

[141] Ordi J, et al. Validation of whole slide imaging in the primary diagnosis of gynaecological pathology in a university hospital. J Clin Pathol 2015;68(1):33–9.

[142] Baidoshvili A, et al. Validation of a whole-slide image-based teleconsultation network. Histopathology 2018;73(5):777–83.

[143] Saco A, et al. Validation of whole-slide imaging in the primary diagnosis of liver biopsies in a University Hospital. Dig Liver Dis 2017;49(11):1240–6.

[144] Sturm B, et al. Validation of diagnosing melanocytic lesions on whole slide images-does z-stack scanning improve diagnostic accuracy? Virchows Arch 2017;471:S15.

[145] Eze O, et al. Validation of multiple high-throughput whole slide imaging (WSI) systems for primary diagnoses and clinical applications at a large academic institution. Lab Invest 2019;99.

[146] Robertson S, et al. Digital image analysis in breast pathology-from image processing techniques to artificial intelligence. Transl Res 2018;194:19–35.

[147] Dimitriou N, Arandjelovic O, Caie PD. Deep learning for whole slide image analysis: an overview. Front Med 2019;6.

[148] Aeffner F, et al. Introduction to digital image analysis in whole-slide imaging: a white paper from the digital pathology association. J Pathol Inform 2019;10:9.

[149] Lara H, et al. Quantitative image analysis for tissue biomarker use: a white paper from the digital pathology association. Appl Immunohistochem Mol Morphol 2021;29(7):479–93.

[150] Alam MR, et al. Recent applications of artificial intelligence from histopathologic image-based prediction of microsatellite instability in solid cancers: a systematic review. Cancers (Basel) 2022;14(11).

[151] Pantanowitz L, et al. An artificial intelligence algorithm for prostate cancer diagnosis in whole slide images of core needle biopsies: a blinded clinical validation and deployment study. Lancet Digit Health 2020;2(8):e407–16.

[152] Kong H, Gurcan M, Belkacem-Boussaid K. Partitioning histopathological images: an integrated framework for supervised color-texture segmentation and cell splitting. IEEE Trans Med Imaging 2011;30(9):1661–77.

[153] Voelker F, Potts S. Whole-slide image analysis with smart tissue finding on common toxicological problems. Toxicol Pathol 2009;37(1):146.

[154] Lin HJ, et al. Fast ScanNet: fast and dense analysis of multi-gigapixel whole-slide images for cancer metastasis detection. IEEE Trans Med Imaging 2019;38(8):1948–58.

[155] Niazi MKK, Downs-Kelly E, Gurcan MN. Hot spot detection for breast cancer in Ki-67 stained slides: image dependent filtering approach. In: SPIE medical imaging. International Society for Optics and Photonics; 2014.

[156] Strack M, Sandusky G, Rushing D. Digital whole slide quantitative image analysis of TOPO II and P53 in sarcomas evaluating both primary and lung metastatic tumors. Cancer Res 2016;76.

[157] Shafi S, et al. Integrating and validating automated digital imaging analysis of estrogen receptor immunohistochemistry in a fully digital workflow for clinical use. J Pathol Inform 2022;13, 100122.

[158] Hartage R, et al. A validation study of human epidermal growth factor receptor 2 immunohistochemistry digital imaging analysis and its correlation with human epidermal growth factor receptor 2 fluorescence in situ hybridization results in breast carcinoma. J Pathol Inform 2020;11(1):2.

[159] Romo-Bucheli D, et al. Automated tubule nuclei quantification and correlation with oncotype DX risk categories in ER plus breast cancer whole slide images. In: Medical imaging 2016: digital pathology; 2016. p. 9791.

[160] Shin H-C, et al. Deep convolutional neural networks for computer-aided detection: CNN architectures, dataset characteristics and transfer learning. IEEE Trans Med Imaging 2016;35(5):1285–98.

[161] Sirinukunwattana K, et al. Locality sensitive deep learning for detection and classification of nuclei in routine colon cancer histology images. IEEE Trans Med Imaging 2016;35(5):1196–206.

[162] Bejnordi BE, et al. Context-aware stacked convolutional neural networks for classification of breast carcinomas in whole-slide histopathology images. J Med Imaging 2017;4(4).

[163] Chen M, et al. Automated segmentation of the choroid in EDI-OCT images with retinal pathology using convolution neural networks. Fetal Infant Ophthalmic Med Image Anal 2017;2017(10554):177–84.

[164] Cruz-Roa A, et al. Accurate and reproducible invasive breast cancer detection in whole-slide images: a deep learning approach for quantifying tumor extent. Sci Rep 2017;7.

[165] Jamaluddin MF, Fauzi MFA, Abas FS. Tumor detection and whole slide classification of H&E lymph node images using convolutional neural network. In: 2017 IEEE International Conference on Signal and Image Processing Applications (Icsipa); 2017. p. 90–5.

[166] Korbar B, et al. Looking under the Hood: deep neural network visualization to interpret whole-slide image analysis outcomes for colorectal polyps. In: 2017 IEEE conference on computer vision and pattern recognition workshops (Cvprw); 2017. p. 821–7.

[167] Nirschl J, et al. Deep learning classifier to predict cardiac failure from whole Slide H&E Images. Mod Pathol 2017;30:532a–3a.

[168] Cruz-Roa A, et al. High-throughput adaptive sampling for whole-slide histopathology image analysis (HASHI) via convolutional neural networks: application to invasive breast cancer detection. PloS One 2018;13(5).

[169] Gurcan MN, et al. Image analysis for neuroblastoma classification: segmentation of cell nuclei. In: 28th annual international conference of the IEEE engineering in medicine and biology society. IEEE; 2006.

[170] Di Cataldo S, et al. Automated segmentation of tissue images for computerized IHC analysis. Comput Methods Programs Biomed 2010;100(1):1–15.

[171] Chen CM, et al. A computer-aided diagnosis system for differentiation and delineation of malignant regions on whole-slide prostate histopathology image using spatial statistics and multidimensional DenseNet. Med Phys 2020.

[172] Volynskaya Z, Evans AJ, Asa SL. Clinical applications of whole-slide imaging in anatomic pathology. Adv Anat Pathol 2017;24(4):215–21.

[173] Epstein JI. Is there enough support for a new prostate grading system factoring in intraductal carcinoma and cribriform cancer? Eur Urol 2019.

[174] Gandhi JS, et al. Reporting practices and resource utilization in the era of intraductal carcinoma of the prostate: a survey of genitourinary subspecialists. Am J Surg Pathol 2019.

[175] Nagpal K, et al. Development and validation of a deep learning algorithm for improving Gleason scoring of prostate cancer. NPJ Digit Med 2019;2:48.

[176] Bulten W, et al. Automated deep-learning system for Gleason grading of prostate cancer using biopsies: a diagnostic study. Lancet Oncol 2020.

[177] Zhou NY, Gao Y. Optimized color decomposition of localized whole slide images and convolutional neural network for intermediate prostate cancer classification. In: Medical imaging 2017: digital pathology; 2017. p. 10140.

[178] Campanella G, et al. Clinical-grade computational pathology using weakly supervised deep learning on whole slide images. Nat Med 2019;25(8):1301.

[179] Huang CH, Racoceanu D. Automated high-grade prostate cancer detection and ranking on whole slide images. In: Medical imaging 2017: Digital pathology; 2017. p. 10140.

[180] Kyriazis AD, et al. An end-to-end system for automatic characterization of Iba1 immunopositive microglia in whole slide imaging. Neuroinformatics 2019;17(3):373–89.

[181] Hanna MG, et al. Whole slide imaging equivalency and efficiency study: experience at a large academic center. Mod Pathol 2019;32(7):916–28.

[182] Parimi V, et al. Validation of whole frozen section slide image diagnosis in surgical pathology. Mod Pathol 2016;29:399a–400a.

[183] Fertig RM, et al. Whole Slide Imaging. Am J Dermatopathol 2018;40(12):938–9.

[184] Ibrahim A, et al. Artificial intelligence in digital breast pathology: techniques and applications. Breast 2019;49:267–73.

[185] Stevenson CH, Hong SC, Ogbuehi KC. Development of an artificial intelligence system to classify pathology and clinical features on retinal fundus images. Clin Exp Ophthalmol 2019;47(4):484–9.

CHAPTER

Artificial intelligence model development, deployment, and regulatory challenges in anatomic pathology

Jerome Y. Cheng[a], Jacob T. Abel[b], Ulysses G.J. Balis[a], Liron Pantanowitz[c], and David S. McClintock[d]

[a]Department of Pathology, University of Michigan, Ann Arbor, MI, United States, [b]Department of Laboratory Medicine and Pathology, University of Washington, Seattle, WA, United States, [c]Department of Pathology, University of Pittsburgh, Pittsburgh, PA, United States, [d]Department of Laboratory Medicine and Pathology, Mayo Clinic, Rochester, MN, United States

Introduction

Recently, there has been much hype regarding the potential for machine learning (ML) and artificial intelligence (AI)-based systems to replace or supplant physicians in many medical specialties, including anatomic pathology (AP) [1,2]. In consonance with this trend, several recent studies and reports in the general media have been published, claiming that ML models are able to surpass human performance in various scenarios [1,3–5]. In spite of these stated capabilities for higher predictive accuracy, it's unlikely such AI-based tools will wholly replace physicians anytime in the near future. Moreover, AI-based solutions can frequently encounter failure modes that often exhibit unpredictable and/or incorrect results when confronted with newly encountered data or patterns [6,7]. In such circumstances, ML algorithms are said to operate in *divergent* or *extrapolatory* modes, where it is well recognized that spurious results are possible or even likely [6]. Because of this intrinsic limitation, in the immediate short term AI tools will see their greatest use and offer maximal benefit by providing supplemental assistance to providers [8–10]. If medical specialties are willing to embrace this new approach, repetitive tasks currently carried out by physicians, and specifically pathologists, may be suitable candidates for AI-based assistance and automation.

Indeed, amassed evidence to date demonstrates that a competent medical generalist, in cooperation with a properly developed AI tool, is able to surpass the performance of a medical specialist practicing without an AI-based solution [11–13]. This is congruent with the findings of a survey involving 487 pathologists from 54 countries, where over 71% felt AI tools could increase their diagnostic efficiency, although the majority of those queried thought the diagnostic decision-making process should remain predominantly a human task [14]. Caution is also necessary with the implementation of machine-based assistance in clinical settings, as it has been shown that pathologists' diagnostic decisions are prone to be influenced by AI, introducing novel sources of bias [12]. A limited number of AI-based prediction tools have actually made their way into the market and have proven effective in practical clinical situations [1,15]. However, while many of the studies highlight the ability of these narrow AI algorithms to excel at a particular task, most have neglected the broader complexity of real-world conditions. In real-world use, these tools would likely be exposed to a much wider array of heterogeneous cases and data patterns never seen before, leading such AI tools to either fail or exhibit lower accuracies than initially reported. One well-known example of this phenomenon includes a previously described sub-par IBM Watson implementation [16].

In another example, an AI-based diabetic retinopathy detection system was initially reported to have a diagnostic sensitivity and specificity of over 90% under controlled conditions; however, it faced several challenges in real-world settings wherein 21% of retinal images uploaded into the system were rejected due to quality issues, such as poor lighting conditions [17].

Building a machine or deep learning model entails a lengthy process, from weeks to months, and at times years, of development. It begins with the identification of a worthy problem for which AI could be helpful (i.e., what is the problem you are trying to solve?), followed by data collection and aggregation, data transformation, and ultimately model training [10,18,19]. If successful, subsequent deployment of a reliably performing algorithm in a clinical setting (clinical laboratory or otherwise) should undergo rigorous validation studies [20,21]. Commercial AI products also have to obtain market approval, e.g., CE marking (Conformité Européenne, i.e., European Conformity) or United States Food and Drug Administration (FDA) approval or clearance, before they can be legally marketed for clinical use [22,23]. The ultimate test is whether the final product can then be successfully integrated into pathologists' workflows. An early exemplar in this regard is the FDA-cleared computer-assisted automated Pap smear screening devices in cytopathology practice, which have evolved over time from early field of view image capture and analysis devices to novel whole slide imaging devices capable of capturing multiple z-planes per scan with automatic AI analysis [24]. At the time of this writing, only one company has received FDA approval for an AI-based surgical pathology algorithm for clinical use (Paige) [25], two companies have received FDA Breakthrough Device designation [26] for their histopathology AI software platforms (Paige, Ibex) [27,28], and many companies have received CE marking for AI-based cancer detection algorithms (e.g., Mindpeak, Lunit, Paige, Ibex, Roche, Aiforia) [29–34]. While there has been progress made in the AI clinical space for anatomic pathology, there are still many more AI tools to be developed, each with its own hurdles to conquer. The primary aim of this chapter is to therefore address the challenges in AI development, deployment, and regulation that need to be overcome prior to artificial intelligence gaining widespread adoption in anatomic pathology. Please note that any

Development challenges **139**

open-source or commercial products mentioned in this chapter are for informational purposes only; the authors do not promote any product over another.

Development challenges
Problem identification

Pathologist buy-in is critical for the successful adoption and integration of AI-based algorithms into practice. Therefore, such applications must have both clinical and practical utility, aiming to fill gaps or unmet needs without disrupting clinical workflow. Examples include rare-event identification, tumor percentage calculation, mitosis detection, immunohistochemistry scoring, and other tasks found to be monotonous, tedious, or prone to higher interobserver variability [35–37]. One notable example is Ki-67 index scoring in neoplasms, where hundreds to thousands of tumor cells are counted on tissue sections in a repetitive task more suitable for computational automation. The current manual process is unnecessarily time-consuming, compelling some pathologists to resort to an improvised workaround, at the expense of accuracy and reproducibility, of estimating the Ki-67 index via light microscopy through a process referred to as "eyeballing." [38] This example clearly shows where AI tools can be developed with pathologist input to improve both diagnostic quality and pathologist efficiency. However, if AI tools are merely developed for their novelty or intellectual appeal, they have a low likelihood of being utilized for routine clinical practice. Thus, AI startups need to be cautious of the "shiny object syndrome" and avoid the trap of always searching for the elusive killer app. Instead, these companies should focus on tools integral to pathologists' work, such as ensuring consistent whole slide image quality, recognizing that some of the low-hanging fruit in this regard may be relatively mundane but nonetheless important tasks [39–41].

Dataset curation and annotation

Determining one's dataset to use for AI algorithm generation does not follow a one-size-fits-all approach and by itself can be a complicated task. For example, convolutional neural networks (CNNs) typically require large training sets, from hundreds to thousands of slides, in order to achieve high model accuracy, significant performance gains, and greater generalizability [42,43]. Conversely, in certain use cases involving transfer learning (see below), small datasets consisting of less than 100 digital slides may suffice [44]. For rare diseases, only an extremely limited number of slides may be available, leading some to simulate larger datasets through data augmentation techniques [45]. Therefore, the actual number of slides needed for any given AI task may vary greatly depending on the problem at hand.

Adding more categories to an AI classification task will also necessarily increase the number of slides/images necessary for algorithm training (i.e., classifying two types of cancer will need less training samples than building a model for classifying five types of cancer). Publicly available datasets can be helpful in supplementing locally curated image repositories; however, partially due to confidentiality, copyright, and budgetary issues, there are not many such datasets available in pathology [46]. The Cancer Genome Atlas (TCGA) is one publicly available dataset offering a substantial number of cases with digital slides and molecular metadata [43,47]. Unfortunately, TGCA has limited cases from many diagnostic subsets, with these often insufficient in number to train clinical-grade histopathology AI models. Another helpful, albeit limited, source of datasets is in the form of public challenges offered for developing deep learning algorithms [48].

A related, consistent issue facing many AI tool development groups involves the digitization of curated glass slide collections, the capture of their associated metadata, and licensing access to the aggregated datasets. As an example, an academic center may have a large collection of cases/slides identified for a given condition; however, this collection may not have been made fully digital. Further complicating matters is the issue of scanning historical cases that do not have barcode labels, easily exportable metadata from the current or legacy laboratory information system (LIS), or even worse, paper-based (and, if old enough, handwritten) records only. For this reason, most slides readily available for training machine learning algorithms come from recent cases (typically from the 2010s and later) for which associated metadata can be easily aggregated. Once the slides have been scanned and the data aggregated, data use and/or licensing agreements must then be developed that govern access to the data, how it can be used, and how intellectual property generated from the data will be shared. Currently, there are efforts underway by some medical centers to digitize their historical data [49], ultimately with the goal of creating large-scale data sets to be shared for research collaborations or for licensing/partnerships with commercial entities [50].

After data collection and curation, the next task in deep learning algorithm development is labeling the data (e.g., malignant vs. benign, necrosis vs. fibrosis, etc.). Annotation types vary based on the deep learning algorithm type as follows:

(i) image classification: images + image label
(ii) image segmentation: image + pixel-level labeling (ground truth map)
(iii) object detection and classification: bounding box coordinates + class of image within the bounding box
(iv) weakly supervised methods: slide-level classifications (vs. labeling each image tile)

In general, the expert-driven annotation process is very time-consuming and laborious, especially considering the large number of images needed and the significant person-hours required for review and annotation. For this reason, multiple instance learning, where all tiles belonging to the same slide are associated with the same slide-level classification, can be helpful; in one published study involving thousands of whole slide images and multiple instance learning, it was still possible to attain very high accuracy scores [42]. Of note, the accuracy of trained models may be affected by the inherent variability associated with annotation performance, particularly if the task is hard (e.g., PD-L1 staining of tumor cells vs. macrophages). This expert-driven annotation bottleneck has proven to be expensive for manufacturers; however, AI-assisted annotation techniques, in addition to annotation-free deep learning techniques, have now emerged as potential solutions by reducing the amount of manual effort needed to label image datasets [51,52].

Crowdsourcing is another proven and viable method of labeling sizable datasets, for example, through quantity-based payments, gamification, and marketing elements [53]. ImageNet, arguably the most popular publicly available image dataset, leveraged crowdsourcing through Amazon's Mechanical Turk to label over 14 million images from 21,000+ classes [54]. However, crowdsourcing should be used with caution given the inherent noise introduced by such diverse annotation methods and the variable expertise of those involved. For those who prefer to annotate images themselves, there are many commercial solutions (e.g., Visiopharm, Indica Labs, Supervisely), stand-alone open-source applications (e.g., QuPath), or web-based tools (e.g., Markit, Piximi Annotator), available for use [55–60]. However, users should note that these tools are not one size fits all and may come with their own difficulties, such as steep learning curves, imperfect user interface design, or high cost.

Model development and training

A significant amount of trial and error is needed in order to develop a model with superior performance metrics. A deep learning system's model design can have a major bearing on bias and therefore can also affect the success of an algorithm's predictions [61]. The following points highlight the most common steps involved in model development and training:

1. Splitting the data into training, testing, and holdout datasets [62]
2. Selecting a machine learning framework (e.g., PyTorch, TensorFlow)
3. Choosing the best machine learning technique (e.g., CNN for images) or building your own model
4. Selecting the learning method (e.g., transfer learning)
5. Evaluating the model through performance metrics (e.g., AUC, F1-score, accuracy)
6. Fine-tuning hyperparameters

For training, a large enough dataset comprising digital slides of the diagnoses of interest may not always be available. In these circumstances, one can use transfer learning, a machine learning technique that leverages knowledge gained from one model and applies it to a different model, in many cases maintaining high accuracy [63]. Models pretrained on public datasets, such as the ImageNet dataset consisting of everyday objects (e.g., trees, pens, animals, etc.), are a popular option for transfer learning and can be effective in making predictions on the new subject matter [54,63,64]. Within pathology, one can surmise that transfer learning will become more effective once models trained on large histopathological datasets become

142 **CHAPTER 8** AI challenges in AP

publicly available. In this scenario, having a large dataset covering all the variances found in diagnoses of interest should lead to model outcomes with higher accuracy versus transfer learning algorithms trained on smaller datasets. However, subsequent clinical validation still requires datasets independent from the one utilized for analytical (technical) validation of the algorithm. Currently, deep-learning-based algorithms are more convenient to develop in nonmedical fields given that pretrained models are readily available and can be used immediately with little modification.

Hardware and cost

AI training is a time-consuming process when large datasets and complex models are involved [42,43]. While using advanced hardware with multiple graphics processing units (GPUs) can speed up training, this can become prohibitively expensive. When building a high-end deep learning workstation, it is beneficial to choose faster central processing units (CPUs), increased system and GPU random access memory (RAM), and readily accessible and adequate storage space. Pre-2020, sub-$1000 deep learning workstations were available for those who wanted to train simpler AI models using smaller datasets, or one could add a lower cost GPU to a personal computer for computationally less demanding projects. However, the Sars-CoV-2 (COVID-19) global pandemic, computer chip shortages, increased cryptocurrency mining, and increased personal and business computing demands have significantly made acquiring and building your own GPU-based deep learning workstation both costly and difficult [65,66], raising the bar for those who wish to begin exploring AI tools.

Deployment challenges

There are multiple deployment challenges one must overcome prior to implementing a pathology AI system within a clinical work environment. To start, defined business use cases and pathologists' assurances to use the AI system are fundamental requirements to be met before spending significant time, effort, and money on AI-associated software, hardware, and implementation. Next, the transition from glass to digital workflows is an absolute requirement for performing AI in AP laboratories, with departments needing to acquire and deploy digital pathology (DP) systems, including whole slide imaging/slide scanning devices and image management systems. Data management and storage capabilities will need to be increased (or acquired), in addition to planning for any extra full-time equivalents (FTEs) needed to accommodate changes in both the histology laboratory and pathologist's workflows. These systems will need to integrate with locally generated pathology laboratory data hosted on established enterprise information technology (IT) infrastructures, either through on-premises or cloud-based computing solutions.

Finally, one must both acknowledge the clinical laboratories and pathologists' relative inexperience with AI and understand the range of issues every department will face prior to adoption. One example of failed adoption of a promising healthcare AI technology that failed to take these factors into account involved an academic medical center spending approximately $62 million over 5 years on IBM Watson to assist clinical decision-making. Regrettably, IBM Watson was never used by the medical center for a single clinical decision on actual patients, the implementation being afflicted by procurement delays, unexpected costs, and learning problems from data ingest issues due to the unstructured nature of medical narrative data [67]. Fortunately, in the digital pathology AI space there are promising solutions today, including multiple self-service AI platforms for pathologists that do not require coding or prior programming experience to create AI algorithms with reproducible results (e.g., Aiforia Create, DeePathology™ STUDIO, Visiopharm Oncotopix Discovery) [68–70].

Pathologist buy-in and transitioning to a digital workflow

First and foremost, one must gain pathologist buy-in for major changes to what is, admittedly, a century-old workflow. Given that change doesn't happen overnight, change management issues for pathologist end-users should be anticipated separately from the technical hurdles. To start, institutional leadership must commit to the adoption of both DP and AI within AP, demonstrating to pathologists how the long-term benefits of digital workflows outweigh their risks. Historically, resistance from pathologists to adopt DP and AI has been attributed to a lack of trust, technophobia, uncertain liability, and concerns that AI may 1 day replace physicians [2]. The fact that many AI tools are a "black box" to the user further compounds clinician reluctance to adopt AI. Instead, "explainable AI" solutions now exist for AP, offering a promising alternative where ML-based methods are created to be more easily understood by humans and providing pathologists with clear explanations for how predictions are generated [71].

Fundamentally, the transition to digital pathology must occur before AI deployment is possible in AP. Implementing digital pathology has a number of proven use cases, including primary and secondary clinical diagnosis, telepathology, convenient slide sharing, robust research data set creation, and pathology education/teaching [72,73]. Digital images can be acquired in multiple ways, from simple cameras on microscopes to complex whole slide imaging (WSI) systems and, most recently, augmented reality microscopes where digital images are overlaid on glass slide fields of view [74,75]. With recent high-throughput WSI systems, faster and cheaper bulk data storage, increased network bandwidth, and improved information system interoperability, digital pathology workflows are now a feasible option in pathology departments [76]. However, even with these advances, only a handful of pathology practices currently sign out most of their cases digitally [77,78]. Overall, there has been much resistance globally to go fully digital [79], drawing parallels to clinicians resisting the change from paper charts to electronic health records or radiologists from film to digital scans [80]. Recently, the changes brought on by the COVID-19 global pandemic

144 CHAPTER 8 AI challenges in AP

have thrust remote digital pathology use into the forefront of pathology practices, with increased DP adoption being one of the few silver linings to come out of the pandemic [81,82]. The increased demand for remote work, combined now with the hype and ensuing desire to leverage AI within medicine, has bolstered the business case for implementing digital pathology for many clinical laboratories and pathology practices.

IT infrastructure: Cloud computing vs. on-premises solutions

The decision whether to go with a cloud-based or on-premises AI solution depends on several factors, namely one's preferred workflow, how often the tools will get used, software and hardware costs, and the inclination of one's IT cybersecurity risk group to allow the use of cloud-based solutions. When considering a cloud-based AI solution, cost is one of the primary issues, with vendors often charging for GPU computational use on a usage basis [83] in addition to all required storage costs. Further, the potential for a data breach and/or inappropriate use of patient data increases with each external data handoff—a solid business associate's agreement, full end-to-end encryption, and clear data use agreements are essential for cloud-based systems in order to maintain one's HIPAA obligations [84]. However, even with these potential barriers, the ability to outsource the AI algorithm development and expertise, in addition to the external management and support of all hardware and software, may make these solutions the preferred option. To date, multiple healthcare institutions have announced cloud computing partnerships with major IT companies, such as Google and Amazon, to drive digital transformation efforts [85–87].

Conversely, purchasing and implementing an on-premises solution may be optimal if the laboratory has access to a robust IT infrastructure and plans to perform frequent and computationally intensive AI calculations. Moreover, some AI algorithms can function adequately on low-cost hardware, allowing for a smaller initial deployment with incremental adoption as AI needs increase. Being able to fully control the data pipeline and custom configure an AI solution to one's clinical workflow is an added advantage, especially considering that one size rarely fits all. From a cybersecurity perspective, there is fundamentally decreased risk with on-premises solutions because these, by default, restrict data from leaving the institution. Overall, the decision to go with a cloud or on-premises solution comes down to cost, security, and departmental preferences [88].

Lack of pathologist's experience with AI

Most pathology departments have never used AI for surgical pathology in their typical case sign-out routine [10,43], leading to a number of concerns when one begins implementing machine learning algorithms within clinical workflows. The following are common questions/issues encountered:

What is the right evidence standard for AI to be embedded in practice?

There is no clear evidence standard for using AI in pathology at this time. It remains to be determined whether AI in routine clinical practice will be based upon algorithm performance, regulatory approval, validation with large clinical trials, or published evidence in peer-reviewed scientific literature. In general, there is consensus that any ML algorithms designed for clinical use should undergo formal clinical validation prior to implementation for patient care. Further, efforts should be made to ensure proper quality assurance methods are put in place when using clinically validated AI tools, including ways to measure algorithm performance, such as accuracy, drift, generalizability, and bias.

What is required for clinical validation prior to using AI for diagnostic purposes?

Guidelines are available on how best to validate WSI and quantitative image analysis for clinical use [89,90]. Unfortunately, the same cannot be currently said for AI algorithm validation, with definite gaps present in the overall clinical validation process for even some FDA-regulated clinical algorithms [91]. While no official regulatory checklists are currently available for AI clinical validation, there are general guidelines one can follow until these are formally developed [20]. Table 8.1 provides general proposed criteria for AI algorithm validation that pathology laboratories can use when beginning their validation process.

What is the ideal workflow when implementing AI in clinical practice?

AI tools can be embedded at multiple points within AP laboratory workflows, with the intended use of the AI algorithm dictating when it should be run (Fig. 8.1A). For example, if the intended use of the AI algorithm is to triage surgical pathology cases (e.g., AI-detected cancer cases receive higher priority) or assist with rare-event screening (e.g., microorganism identification), it would be embedded within the

Table 8.1 Example of an AI validation study parameters for clinical practice.

Phase	Calibration	Analytical validation	Clinical validation
Rationale	Check that AI works as specified	Check that AI consistently performs as expected	Confirm that AI works safely with routine workflow
Number WSIs	300	2000	60–120

Note that one should review any clinical AI algorithm to be deployed with the site CLIA laboratory director prior to its formal use in patient care.
WSIs = whole slide images.

146 **CHAPTER 8** AI challenges in AP

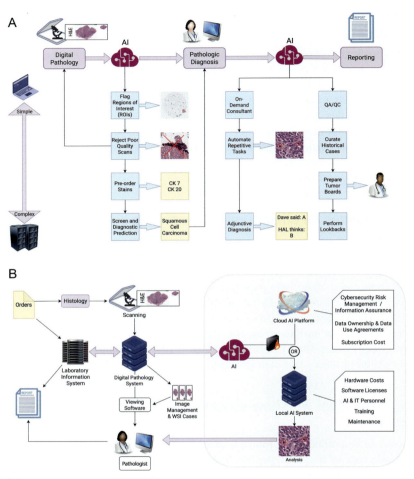

FIG. 8.1

(A) Embedding AI into AP clinical workflows. The top horizontal axis represents AI algorithm deployment opportunities before, during, or after the pathologist diagnostic interpretation process. These algorithms can be automatically executed in the background or manually triggered by the pathologist when needed. The left vertical axis represents how AI algorithms differ in relative complexity, ranging from relatively simple image analysis functions to complex rendering of adjunctive diagnoses as an "independent" observer. CK7 = cytokeratin 7, CK20 = cytokeratin 20, AI = artificial intelligence. (B) Proposed integration of AI into the AP laboratory IT architecture. In this schema, the Digital Pathology System is connected to a separate server running AI analysis pipelines, which can either be local/on-premises or integrated remotely as a cloud-based solution. WSI data is submitted to this server, running the algorithms and returning AI results/predictions to the digital pathology system and the pathologist. AI algorithm findings are interpreted by the pathologist and subsequently incorporated into the pathology report. WSI = whole slide imaging, AI = artificial intelligence, IT = information technology.

preanalytic workflows and run prior to pathologists receiving cases for review. AI tools can also be triggered to perform specific tasks during the analytic/interpretation process (e.g., counting mitotic figures for tumor grading), either by the pathologist manually or as part of an automated workflow. Finally, AI software can also be programmed to run in the background, continually performing quality assurance checks (e.g., histology quality control) or other important tasks (e.g., cytologic-histologic correlation). Creating a successful AI workflow to meet one's needs will ultimately depend on seamless interoperability between AI, DP, LIS, and other health information systems (e.g., electronic health record [EHR], enterprise picture archiving and communications system [PACS], etc.) (Fig. 8.1B).

What model do pathology laboratories use to pay AI vendors?

Unfortunately, given few pathology laboratories use AI for routine practice, little data exists regarding a "best" payment model for AI services. Current prevailing options include subscription, akin to cloud software as a service (SAAS), or pay-per-click (PPC) models. As with any formal contract, once an AI platform or system is chosen, the terms and payment conditions should be reviewed by both the supply chain/purchasing team and appropriate legal counsel.

What is the business use case for deploying AI?

There are multiple business use cases one can use to justify the deployment of AI in anatomic pathology. In general, three of the most common use cases are: (1) improved diagnostic accuracy and standardization of pathology reports; (2) improved pathologist efficiency and work satisfaction by automating nonvalue-added tasks; and (3) automated AP laboratory operations resulting in fewer manual handoffs, fewer FTEs needed for the same volume of work, and faster case turnaround times. Unfortunately, it is currently unknown whether improvements in efficiency and accuracy for pathologists will generate sufficient return on investment to drive the widespread adoption of AI tools. Ultimately, more real-world data is needed defining the added value AI tools bring to pathology practices before this question can be fully answered.

Should residents or fellows be allowed to use AI, or is this "cheating?"

Trainee de-skilling, where excessive reliance on AI assistance may provide less opportunities for trainees to develop skilled abilities to discern different types of histopathological lesions, is certainly a potential future concern with AI use. However, when used as an adjunctive tool (e.g., as an extra set of eyes) and not as a replacement for pathologists, AI has the ability to increase the efficiency and accuracy of clinical diagnoses. Further, AI has the potential to enrich pathology training, allowing residents and fellows to receive immediate feedback on their diagnoses and providing

CHAPTER 8 AI challenges in AP

centralized access to educational resources for additional study [92]. Therefore pathologists, starting early in residency, should be trained on the proper use, limitations, and pitfalls of AI tools similar to how they learn to wisely use special and immunohistochemistry stains, fluorescent in situ hybridization (FISH), molecular assays, and other ancillary studies. Finally, as digital pathology systems evolve, one can easily imagine a future for pathology trainees where AI-assistive tools can be customized to trigger only in specific instances, such as after a trainee has worked up a case and entered a provisional diagnosis.

Regulatory challenges

AI-based algorithms can easily give the illusion of being much more capable than what they really are, with most of these tools perceived as "black boxes" due to human difficulty in comprehending exactly how predictions are made. Even with methods such as gradient saliency maps [93] and filter visualization techniques that may help pathologists understand AI-based predictions, humans mostly cannot fully understand how millions of parameters contribute to a decision, leading to potential biases, misuse, and misdiagnoses. Therefore, a regulatory pathway for AI is required to promote safe, proper, and effective use of these algorithms while concurrently taking into account their intricacies and limitations. The issue of how to properly regulate AI will become increasingly important as more algorithms are put into clinical practice. Regulatory approval should aim to not only mitigate potential harm but also strike a proper balance between defining risk and benefits, developing effective validation standards, and promoting innovation [94]. The current landscape for AI device regulation is evolving, with regulatory bodies such as the FDA, the European Union (EU) CE, and the Centers for Medicare and Medicaid Services (CMS) not fully prepared for the influx of AI into medicine. Consequently, devices enabled with ML capabilities are being regulated under prior and arguably outdated standards for tests and medical devices [95].

FDA

Getting FDA device market approval, granting, or clearance is a long, expensive, and stringent process. The difficulty of submissions depends on the type of application the device falls into: premarket approval, De Novo granted, or 501(k) clearance. While the FDA has already approved several AI algorithms for medical use [1,15,25], it does recognize current regulations were not designed for machine learning and adaptive technologies, most of which would end up necessitating frequent, if not continuous, premarket review for algorithm modifications under current guidelines [96,97]. With this in mind, the FDA is currently drafting a set of new regulations to better address the safety and efficacy of AI-enabled devices in the market, in addition to creating a Software as a Medical Device (SaMD) Action Plan [98].

While there were discussions in Congress regarding how best to regulate laboratory-developed tests (LDTs), with the proposed Verifying Accurate Leading-Edge IVCT Development Act of 2021, or VALID Act, receiving the most attention, ultimately these efforts failed to pass [99,100]. In response to the lack of congressional action, the FDA instead issued a final rule on April 29, 2024 to amend it regulations and phase out its general enforcement discretion approach for LDTs, meaning LDTs will now be considered devices under the FDA [101]. The impact of this rule on AI model regulation within clinical laboratories is still not entirely fully understood, however, at a minimum, it means that an AI model included as a necessary component of an LDT, would be regulated under the LDT final rule guidelines. In some ways, the final rule simplifies how groups should approach AI model regulation since, if the model is not a fundamental component of an LDT, it should then be evaluated per current SaMD recommendations. Overall, the authors recommend consulting with organizational compliance and regulatory groups to understand how best to approach regulatory validation of an AI model for clinical use.

European Union Conformité Européenne

With a focus on health and safety, AI-capable devices can acquire EU CE marking for medical devices through the Medical Devices Regulation (EU) 2017/745 (MDR) and the In Vitro Diagnostic Medical Devices Regulation (EU) 2017/746 (IVDR) [102]. If obtained, CE mark approval means that an AI-based system can be legally marketed within the EU and European Free Trade Association (EFTA) member states. Both MDR (effective May 26, 2021) and IVDR (effective May 26, 2022) require scientific validity, analytical performance, and clinical performance to be demonstrated prior to approval [23]. Factsheets have been made available, based on one's activity area (e.g., medical device and in vitro device manufacturers, healthcare professionals, and institutions), to help groups navigate these regulations [103]. In order to address the growing impact of AI in Europe, the European Commission recently published a white paper on AI including broad proposals concerning AI regulation [104].

CMS/CLIA

In the United States, CMS regulates laboratory testing through the Clinical Laboratory Improvement Amendments of 1988 (CLIA) [105]. CLIA stipulates all tests performed on human-body-derived materials must undergo proper validation by the performing laboratory prior to their introduction for clinical use, regardless of the FDA status of the test [106]. Currently, there are no specific CLIA regulations for validating AI algorithms, nor does CMS/CLIA have a published opinion on whether AI algorithms fall under the category of LDTs. For AI algorithm validation, most pathologists would naturally look to typical laboratory validation methodologies, as defined by CLIA, CMS, and its deemed agencies (e.g., the College of American Pathologists). However, this is not the case for most (nonpathologist) clinicians and researchers since, from the authors' experiences, most nonlaboratorians are not familiar with CLIA and thus would not seek laboratory regulations for guidance.

In April 2022, the Clinical Laboratory Improvement Advisory Committee (CLIAC) CLIA Regulations Assessment Workgroup began initial discussions on how to incorporate the expanding use of laboratory data derived from human materials, including redefining the term "materials" to include image, genetic and protein, −omics, and other derivative data [107,108]. This is an interesting approach that warrants much future discussion and expert opinions, given that nonlaboratory use of this data to create novel interpretations affecting patient care (i.e., new patient "results") would require many new areas to fall under CLIA purview. Given the ambiguity that exists within the regulatory space not only now but also over the next 5–10 years, we recommend always consulting current FDA and CLIA regulations to better understand how AI algorithms should be validated.

Conclusion

AI holds tremendous potential in improving the practice of pathology by streamlining workflows (e.g., weeding out negative cases), reducing errors (e.g., screening for hard-to-find lesions, resolving atypical diagnoses), improving reproducibility (e.g., standardizing diagnostic decision-making), and conveying predictions not possible with a microscope alone. However, AI use for clinical work ideally should be affordable, practical, interoperable, explainable, generalizable, manageable, and reimbursable. Additionally, the domain expertise of pathologists is crucial for optimal expert system design and development [109]. With current limited clinical experience and uncertainty on how to embed AI technology into clinical practice, care must be taken when deploying AI to avoid unwanted consequences. Finally, only with pathologist buy-in, standardized recommendations and guidelines, a clearly defined regulatory framework, and the seamless interoperability of AI tools with current health information systems will AI be adopted and used safely in anatomic pathology laboratories.

Funding source

There was no external funding provided for this manuscript.

Disclosures/conflicts of interest

Jerome Y. Cheng has no disclosures.
Jacob T. Abel has no disclosures.
Ulysses G.J. Balis is on the advisory board of Inspirata.
Liron Pantanowitz is on the medical advisory board for Ibex and provides consultative services for Hamamatsu.
David S. McClintock is on the scientific advisory board for Epredia.
All authors attest that the disclosures listed are not directly related and have no conflicts of interest regarding this manuscript.

References

[1] Topol EJ. High-performance medicine: the convergence of human and artificial intelligence. Nat Med 2019;25:44–56. https://doi.org/10.1038/s41591-018-0300-7.

[2] Ahuja AS. The impact of artificial intelligence in medicine on the future role of the physician. PeerJ 2019;7, e7702. https://doi.org/10.7717/peerj.7702.

[3] Walsh F. AI "outperforms" doctors diagnosing breast cancer. BBC News; 2020.

[4] Leibowitz D. 'Sherlock Holmes' AI diagnoses disease better than your doctor. Medium: Study Finds; 2020. https://towardsdatascience.com/ai-diagnoses-disease-better-than-your-doctor-study-finds-a5cc0ffbf32. accessed May 2, 2022.

[5] Longoni C, Morewedge CK. AI can outperform doctors. So why Don't patients trust it? Harv Bus Rev 2019.

[6] Oh S, Kim JH, Choi S-W, Lee HJ, Hong J, Kwon SH. Physician confidence in artificial intelligence: an online Mobile Survey. J Med Internet Res 2019;21, e12422. https://doi.org/10.2196/12422.

[7] Voter AF, Meram E, Garrett JW, Yu J-PJ. Diagnostic accuracy and failure mode analysis of a deep learning algorithm for the detection of intracranial hemorrhage. J Am Coll Radiol 2021;18:1143–52. https://doi.org/10.1016/j.jacr.2021.03.005.

[8] He J, Baxter SL, Xu J, Xu J, Zhou X, Zhang K. The practical implementation of artificial intelligence technologies in medicine. Nat Med 2019;25:30–6. https://doi.org/10.1038/s41591-018-0307-0.

[9] Yousif M, McClintock DS, Yao K. Artificial intelligence is the key driver for digital pathology adoption. Clin Lab Int 2021;Dec2020/Jan2021:8–11.

[10] Yao K, Singh A, Sridhar K, Blau JL, Ohgami RS. Artificial intelligence in pathology: a simple and practical guide. Adv Anat Pathol 2020;27:385–93. https://doi.org/10.1097/PAP.0000000000000277.

[11] Pesapane F, Tantrige P, Patella F, Biondetti P, Nicosia L, Ianniello A, et al. Myths and facts about artificial intelligence: why machine- and deep-learning will not replace interventional radiologists. Med Oncol Northwood Lond Engl 2020;37:40. https://doi.org/10.1007/s12032-020-01368-8.

[12] Kiani A, Uyumazturk B, Rajpurkar P, Wang A, Gao R, Jones E, et al. Impact of a deep learning assistant on the histopathologic classification of liver cancer. NPJ Digit Med 2020;3:23. https://doi.org/10.1038/s41746-020-0232-8.

[13] Niazi MKK, Parwani AV. Gurcan MN: digital pathology and artificial intelligence. Lancet Oncol 2019;20 253-261.

[14] Sarwar S, Dent A, Faust K, Richer M, Djuric U, Van Ommeren R, et al. Physician perspectives on integration of artificial intelligence into diagnostic pathology. Npj Digit Med 2019;2:28. https://doi.org/10.1038/s41746-019-0106-0.

[15] Muehlematter UJ, Daniore P, Vokinger KN. Approval of artificial intelligence and machine learning-based medical devices in the USA and Europe (2015–20): a comparative analysis. Lancet Digit Health 2021;3:e195–203. https://doi.org/10.1016/S2589-7500(20)30292-2.

[16] Lee W-S, Ahn SM, Chung J-W, Kim KO, Kwon KA, Kim Y, et al. Assessing concordance with Watson for oncology, a cognitive computing decision support system for Colon Cancer treatment in Korea. JCO Clin Cancer Inform 2018;1–8. https://doi.org/10.1200/CCI.17.00109.

[17] Beede E, Baylor E, Hersch F, Iurchenko A, Wilcox L, Ruamviboonsuk P, et al. A human-centered evaluation of a deep learning system deployed in clinics for the detection of

152 CHAPTER 8 AI challenges in AP

diabetic retinopathy. In: Proceedings of the 2020 CHI Conference on Human Factors in Computing Systems. New York, NY, USA: Association for Computing Machinery; 2020. p. 1–12.

[18] LeCun Y, Bengio Y, Hinton G. Deep learning. Nature 2015;521:436–44. https://doi.org/10.1038/nature14539.

[19] Fu W, Menzies T. Easy over hard: A case study on deep learning. In: Proc. 2017 11th Jt. Meet. Found. Softw. Eng. - ESECFSE 2017. Paderborn, Germany: ACM Press; 2017. p. 49–60. https://doi.org/10.1145/3106237.3106256.

[20] Tan B. How to validate AI algorithms in anatomic pathology. Coll Am Pathol 2019. https://www.cap.org/member-resources/clinical-informatics-resources/how-to-validate-ai-algorithms-in-anatomic-pathology. [Accessed 1 May 2022].

[21] Center for devices and radiological. In: Good machine learning practice for medical device development: guiding principles. FDA; 2021.

[22] Harrington SG, Johnson MK. The FDA and artificial intelligence in radiology: defining new boundaries. J Am Coll Radiol JACR 2019;16:743–4. https://doi.org/10.1016/j.jacr.2018.09.057.

[23] García-Rojo M, De Mena D, Muriel-Cueto P, Atienza-Cuevas L, Domínguez-Gómez M, Bueno G. New European Union regulations related to whole slide image scanners and image analysis software. J Pathol Inform 2019;10:2. https://doi.org/10.4103/jpi.jpi_33_18.

[24] Lew M, Wilbur DC, Pantanowitz L. Computational cytology: lessons learned from pap test computer-assisted screening. Acta Cytol 2021;65:286–300. https://doi.org/10.1159/000508629.

[25] Food and Drug Administration. FDA authorizes software that can help identify prostate cancer. FDA; 2021. https://www.fda.gov/news-events/press-announcements/fda-authorizes-software-can-help-identify-prostate-cancer. [accessed May 3, 2022].

[26] Center for Devices and Radiological Health. Breakthrough Devices Program FDA; 2022.

[27] Business Wire. FDA Grants Breakthrough Designation to Paige, 2019. AI https://www.businesswire.com/news/home/20190307005205/en/FDA-Grants-Breakthrough-Designation-Paige.AI. [Accessed 18 August 2020].

[28] PR Newswire. Ibex granted FDA breakthrough device designation n.d. https://www.prnewswire.com/il/news-releases/ibex-granted-fda-breakthrough-device-designation-301308843.html (accessed May 3, 2022).

[29] Mindpeak's breast cancer cell detection software receives CE-IVD mark. Med Device Netw 2021. https://www.medicaldevice-network.com/news/mindpeak-breast-cancer-software/. [accessed May 3, 2022].

[30] PR Newswire. Lunit AI solution for PD-L1 expression analysis receives CE mark n.d. https://www.prnewswire.co.uk/news-releases/lunit-ai-solution-for-pd-l1-expression-analysis-receives-ce-mark-835925368.html (accessed May 3, 2022).

[31] Business Wire. Paige achieves CE Marks for breast Cancer detection and prostate Cancer grading and quantification AI-based digital diagnostics., 2020, https://www.businesswire.com/news/home/20201208005343/en/Paige-Achieves-CE-Marks-for-Breast-Cancer-Detection-and-Prostate-Cancer-Grading-and-Quantification-AI-based-Digital-Diagnostics. [Accessed 3 May 2022].

[32] PR Newswire. Ibex obtains CE mark for AI-powered breast Cancer detection n.d. https://www.prnewswire.com/il/news-releases/ibex-obtains-ce-mark-for-ai-powered-breast-cancer-detection-845153050.html (accessed May 3, 2022).

References   **153**

[33] NS Medical Devices. Roche launches digital pathology algorithm for non-small cell lung cancer., 2020, https://www.nsmedicaldevices.com/news/roche-pathology-algorithm-nsclc/. accessed May 3, 2022.

[34] NS Medical Devices. Aiforia gets CE-IVD marking for new AI model for breast cancer diagnostics. NS Med Devices 2022. https://www.nsmedicaldevices.com/news/aiforia-ai-model-breast-cancer-diagnostics/. [accessed May 3, 2022].

[35] Saha M, Chakraborty C, Racoceanu D. Efficient deep learning model for mitosis detection using breast histopathology images. Comput Med Imaging Graph 2018;64:29–40. https://doi.org/10.1016/j.compmedimag.2017.12.001.

[36] Yousif M, Huang Y, Sciallis A, Kleer CG, Pang J, Smola B, et al. Quantitative image analysis as an adjunct to manual scoring of ER, PgR, and HER2 in invasive breast carcinoma. Am J Clin Pathol 2021;aqab206. https://doi.org/10.1093/ajcp/aqab206.

[37] Yousif M, van Diest PJ, Laurinavicius A, Rimm D, van der Laak J, Madabhushi A, et al. Artificial intelligence applied to breast pathology. Virchows Arch 2021. https://doi.org/10.1007/s00428-021-03213-3.

[38] Niazi MKK, Senaras C, Pennell M, Arole V, Tozbikian G, Gurcan MN. Relationship between the Ki67 index and its area based approximation in breast cancer. BMC Cancer 2018;18:867. https://doi.org/10.1186/s12885-018-4735-5.

[39] Janowczyk A, Zuo R, Gilmore H, Feldman M, Madabhushi A. HistoQC: an open-source quality control tool for digital pathology slides. JCO Clin Cancer Inform 2019;3:1–7. https://doi.org/10.1200/CCI.18.00157.

[40] Kim D, Pantanowitz L, Schüffler P, Yarlagadda DK, Ardon O, Reuter V, et al. (Re) defining the high-power field for digital pathology. J Pathol Inform 2020;11:33. https://doi.org/10.4103/jpi.jpi_48_20.

[41] Cree IA, Tan PH, Travis WD, Wesseling P, Yagi Y, White VA, et al. Counting mitoses: SI(ze) matters! Mod Pathol 2021;34:1651–7. https://doi.org/10.1038/s41379-021-00825-7.

[42] Campanella G, Hanna MG, Geneslaw L, Miraflor A, Silva VWK, Busam KJ, et al. Clinical-grade computational pathology using weakly supervised deep learning on whole slide images. Nat Med 2019;25:1301–9. https://doi.org/10.1038/s41591-019-0508-1.

[43] Bera K, Schalper KA, Rimm DL, Velcheti V, Madabhushi A. Artificial intelligence in digital pathology—new tools for diagnosis and precision oncology. Nat Rev Clin Oncol 2019;16:703–15. https://doi.org/10.1038/s41571-019-0252-y.

[44] Jones AD, Graff JP, Darrow M, Borowsky A, Olson KA, Gandour-Edwards R, et al. Impact of pre-analytical variables on deep learning accuracy in histopathology. Histopathology 2019;75:39–53.

[45] Wu M, Wang S, Pan S, Terentis AC, Strasswimmer J, Zhu X. Deep learning data augmentation for Raman spectroscopy cancer tissue classification. Sci Rep 2021;11:23842. https://doi.org/10.1038/s41598-021-02687-0.

[46] Hipp JD, Sica J, McKenna B, Monaco J, Madabhushi A, Cheng J, et al. The need for the pathology community to sponsor a whole slide imaging repository with technical guidance from the pathology informatics community. J Pathol Inform 2011;2. https://doi.org/10.4103/2153-3539.83191.

[47] Tomczak K, Czerwińska P, Wiznerowicz M. The cancer genome atlas (TCGA): an immeasurable source of knowledge. Contemp Oncol Poznan Pol 2015;19 68-77.

[48] Hartman D, JeroenAWMVD L, Gurcan M, Pantanowitz L. Value of public challenges for the development of pathology deep learning algorithms. J Pathol Inf 2020;11.

[49] Business Wire. Pramana launches digital pathology as a service platform, offering comprehensive and scalable glass slide digitization for health systems., 2022, https://www.businesswire.com/news/home/20220316005227/en/Pramana-Launches-Digital-Pathology-as-a-Service-Platform-Offering-Comprehensive-and-Scalable-Glass-Slide-Digitization-for-Health-Systems. [Accessed 3 May 2022].

[50] Cleveland Clinic. PathAI and Cleveland clinic announce collaboration to build digital pathology infrastructure and evolve use of AI-powered pathology algorithms in research and clinical care. Clevel Clin Newsroom 2022. https://newsroom.clevelandclinic.org/2022/03/10/pathai-and-cleveland-clinic-announce-collaboration-to-build-digital-pathology-infrastructure-and-evolve-use-of-ai-powered-pathology-algorithms-in-research-and-clinical-care/. [accessed May 3, 2022].

[51] Roth H, Chitale P, Roopa M. Fast AI assisted annotation and transfer learning powered by the Clara train SDK. Nvidia Dev Blog 2019. https://developer.nvidia.com/blog/annotation-transfer-learning-clara-train/. [accessed July 4, 2020].

[52] Yan J, Chen H, Li X, Yao J. Deep contrastive learning based tissue clustering for annotation-free histopathology image analysis. Comput Med Imaging Graph 2022;97, 102053. https://doi.org/10.1016/j.compmedimag.2022.102053.

[53] Hughes AJ, Mornin JD, Biswas SK, Beck LE, Bauer DP, Raj A, et al. Quanti.Us: a tool for rapid, flexible, crowd-based annotation of images. Nat Methods 2018;15:587–90. https://doi.org/10.1038/s41592-018-0069-0.

[54] Deng J, Dong W, Socher R, Li L-J, Li K, Fei-Fei L. ImageNet: A large-scale hierarchical image database. In: 2009 IEEE Conf. Comput. Vis. Pattern Recognit. Miami, FL: IEEE; 2009. p. 248–55. https://doi.org/10.1109/CVPR.2009.5206848.

[55] Visiopharm. Visiopharm Research Solutions., 2022, https://visiopharm.com/research-pathology-image-analysis-software/. accessed May 3, 2022.

[56] Indica Labs. HALO. Indica Labs; 2022. https://indicalab.com/halo/. [Accessed 3 May 2022].

[57] Supervisely. Supervisely Image Labeling Toolbox. Supervisely n.d. https://supervise.ly/labeling-toolbox/images (accessed May 3, 2022).

[58] Bankhead P, Loughrey MB, Fernández JA, Dombrowski Y, McArt DG, Dunne PD, et al. QuPath: open source software for digital pathology image analysis. Sci Rep 2017;7:16878. https://doi.org/10.1038/s41598-017-17204-5.

[59] Witowski J, Choi J, Jeon S, Kim D, Chung J, Conklin J, et al. MarkIt: a collaborative artificial intelligence annotation platform leveraging Blockchain for medical imaging research. Blockchain Healthc Today 2021. https://doi.org/10.30953/bhty.v4.176.

[60] Lucas A. Introducing Piximi annotator, an easily accessible, user-friendly annotator tool. Broad Inst Imaging Platf Carpent-Singh Lab Cimini Lab Blog 2021. https://carpenter-singh-lab.broadinstitute.org/blog/introducing-piximi-annotator. [Accessed 3 May 2022].

[61] Eklund M, Kartasalo K, Olsson H, Ström P. The importance of study design in the application of artificial intelligence methods in medicine. NPJ Digit Med 2019;2.

[62] Rashidi HH, Tran NK, Betts EV, Howell LP, Green R. Artificial intelligence and machine learning in pathology: the present landscape of supervised methods. Acad Pathol 2019;6, 2374289519873088. https://doi.org/10.1177/2374289519873088.

[63] Morid MA, Borjali A, Del Fiol G. A scoping review of transfer learning research on medical image analysis using ImageNet. Comput Biol Med 2021;128, 104115. https://doi.org/10.1016/j.compbiomed.2020.104115.

[64] Komura D, Ishikawa S. Machine learning approaches for pathologic diagnosis. Virchows Arch Int J Pathol 2019;475:131–8. https://doi.org/10.1007/s00428-019-02594-w.

References 155

[65] Rajamohan S. Are GPUs really expensive? Databricks: Benchmarking GPUs for Inference on the Databricks Clusters; 2021. https://databricks.com/blog/2021/12/15/are-gpus-really-expensive-benchmarking-gpus-for-inference-on-the-databricks-clusters.html. accessed May 3, 2022.

[66] Butterworth W. Why are graphics cards so expensive? Cold Wire., 2022, https://www.thecoldwire.com/why-are-graphics-cards-so-expensive/. [Accessed 3 May 2022].

[67] Schmidt CMD. Anderson breaks with IBM Watson, raising questions about artificial intelligence in oncology. JNCI. J Natl Cancer Inst 2017;109:djx113. https://doi.org/10.1093/jnci/djx113.

[68] Visiopharm. Oncotopix discovery—AI deep learning for pathology tissue image analysis., 2022, https://visiopharm.com/oncotopix-discovery/. accessed May 3, 2022.

[69] DeePathology. DeePathology™ STUDIO 2022. https://deepathology.ai/our-solutions/deepathology-studio/ (accessed May 3, 2022).

[70] Aiforia. Aiforia Create., 2022, https://www.aiforia.com/aiforia-create. accessed May 3, 2022.

[71] Tosun AB, Pullara F, Becich MJ, Taylor DL, Fine JL, Chennubhotla SC. Explainable AI (xAI) for anatomic pathology. Adv Anat Pathol 2020;27:241–50. https://doi.org/10.1097/PAP.0000000000000264.

[72] Volynskaya Z, Evans AJ, Asa SL. Clinical applications of whole-slide imaging in anatomic pathology. Adv Anat Pathol 2017;24:215–21. https://doi.org/10.1097/PAP.0000000000000153.

[73] Zarella MD, Bowman D, Aeffner F, Farahani N, Xthona A, Absar SF, et al. A practical guide to whole slide imaging: a White paper from the digital pathology association. Arch Pathol Lab Med 2018;143:222–34. https://doi.org/10.5858/arpa.2018-0343-RA.

[74] Hanna MG, Parwani A, Sirintrapun SJ. Whole slide imaging: technology and applications. Adv Anat Pathol 2020;27:251–9. https://doi.org/10.1097/PAP.0000000000000273.

[75] Chen P-HC, Gadepalli K, MacDonald R, Liu Y, Kadowaki S, Nagpal K, et al. An augmented reality microscope with real-time artificial intelligence integration for cancer diagnosis. Nat Med 2019;25:1453–7. https://doi.org/10.1038/s41591-019-0539-7.

[76] McClintock DS, Abel JT, Cornish TC. Whole slide imaging hardware, software, and infrastructure. In: Parwani AV, editor. Whole slide imaging Curr. Appl. Cham: Springer International Publishing; 2022. p. 23–56. https://doi.org/10.1007/978-3-030-83332-9_2. Future Dir.

[77] Retamero JA, Aneiros-Fernandez J, del Moral RG. Complete digital pathology for routine histopathology diagnosis in a multicenter hospital network. Arch Pathol Lab Med 2020;144:221–8. https://doi.org/10.5858/arpa.2018-0541-OA.

[78] Stathonikos N, Nguyen TQ, Spoto CP, Verdaasdonk MAM, Van DPJ. Being fully digital: perspective of a Dutch academic pathology laboratory. Histopathology 2019;75:621–35. https://doi.org/10.1111/his.13953.

[79] Evans AJ, Salama ME, Henricks WH, Pantanowitz L. Implementation of whole slide imaging for clinical purposes: issues to consider from the perspective of early adopters. Arch Pathol Lab Med 2017;141:944–59. https://doi.org/10.5858/arpa.2016-0074-OA.

[80] Ajami S, Bagheri-Tadi T. Barriers for adopting electronic health records (EHRs) by physicians. Acta Inform Medica 2013;21:129–34. https://doi.org/10.5455/aim.2013.21.129-134.

[81] Orly Ardon PDMBA, MD VER, MD MH, BFA LC. Digital pathology operations at an NYC tertiary Cancer center during the first 4 months of COVID-19 pandemic response. Acad Pathol 2021;8. https://doi.org/10.1177/23742895211010276.

[82] Lujan GM, Savage J, Shanaah A, Yearsley M, Thomas D, Allenby P, et al. Digital pathology initiatives and experience of a large academic institution during the coronavirus disease 2019 (COVID-19) pandemic. Arch Pathol Lab Med 2021. https://doi.org/10.5858/arpa.2020-0715-SA.

[83] Langmead B, Nellore A. Cloud computing for genomic data analysis and collaboration. Nat Rev Genet 2018;19:208–19. https://doi.org/10.1038/nrg.2017.113.

[84] Office for Civil Rights. Guidance on HIPAA & cloud computing, 2016. HHSGov https://www.hhs.gov/hipaa/for-professionals/special-topics/health-information-tech nology/cloud-computing/index.html. accessed May 3, 2022.

[85] Anastasijevic D. Mayo Clinic selects Google as strategic partner for health care innovation, cloud computing. Mayo Clin News Netw 2019. https://newsnetwork. mayoclinic.org/discussion/mayo-clinic-selects-google-as-strategic-partner-for-health-care-innovation-cloud-computing/. [accessed May 3, 2022].

[86] Landi H. UC Davis health taps AWS to accelerate digital health innovation, with an eye toward health equity. Fierce Healthc 2021. https://www.fiercehealthcare.com/tech/uc-davis-health-taps-aws-to-accelerate-digital-health-innovation-eye-toward-health-equity. [accessed May 3, 2022].

[87] Miliard M. HCA enters new partnership with Google cloud. Healthc IT News 2021. https://www.healthcareitnews.com/news/hca-enters-new-partnership-google-cloud. [accessed May 3, 2022].

[88] Matthews D. Supercharge your data wrangling with a graphics card. Nature 2018;562:151–2. https://doi.org/10.1038/d41586-018-06870-8.

[89] Pantanowitz L, Sinard JH, Henricks WH, Fatheree LA, Carter AB, Contis L, et al. Validating whole slide imaging for diagnostic purposes in pathology: guideline from the College of American Pathologists Pathology and Laboratory Quality Center. Arch Pathol Lab Med 2013;137:1710–22. https://doi.org/10.5858/arpa.2013-0093-CP.

[90] Evans AJ, Brown RW, Bui MM, Chlipala EA, Lacchetti C, Milner DA, et al. Validating whole slide imaging Systems for Diagnostic Purposes in pathology. Arch Pathol Lab Med 2022;146:440–50. https://doi.org/10.5858/arpa.2020-0723-CP.

[91] Ebrahimian S, Kalra MK, Agarwal S, Bizzo BC, Elkholy M, Wald C, et al. FDA-regulated AI algorithms: trends, strengths, and gaps of validation studies. Acad Radiol 2022;29:559–66. https://doi.org/10.1016/j.acra.2021.09.002.

[92] Wells A, Patel S, Lee JB, Motaparthi K. Artificial intelligence in dermatopathology: diagnosis, education, and research. J Cutan Pathol 2021;48:1061–8. https://doi.org/10.1111/cup.13954.

[93] Pasa F, Golkov V, Pfeiffer F, Cremers D, Pfeiffer D. Efficient deep network architectures for fast chest X-ray tuberculosis screening and visualization. Sci Rep 2019;9.

[94] Allen TC. Regulating artificial intelligence for a successful pathology future. Arch Pathol Lab Med 2019;143:1175–9. https://doi.org/10.5858/arpa.2019-0229-ED.

[95] Minssen T, Gerke S, Aboy M, Price N, Cohen G. Regulatory responses to medical machine learning. J Law Biosci 2020;7:lsaa002. https://doi.org/10.1093/jlb/lsaa002.

[96] Food and Drug Administration. Artificial intelligence and machine learning in software as a medical device. FDA 2021. https://www.fda.gov/medical-devices/software-medi cal-device-samd/artificial-intelligence-and-machine-learning-software-medical-device. [accessed May 3, 2022].

[97] Food and Drug Administration. Artificial intelligence and machine learning (AI/ML)-enabled medical devices. FDA 2021. https://www.fda.gov/medical-devices/software-

medical-device-samd/artificial-intelligence-and-machine-learning-aiml-enabled-medi cal-devices. [accessed May 3, 2022].

[98] Food and Drug Administration. Artificial intelligence/machine learning (ai/ml)-based: software as a medical device (SaMD) action plan., 2021, https://www.fda.gov/media/ 145022/download. [accessed May 3, 2022].

[99] Bassett SE, Gaba MM. FDA Finalizes Rule Regulating Laboratory Developed Tests. Nat Law Rev; 2024. https://natlawreview.com/article/fda-finalizes-rule-regulating-lab oratory-developed-tests/. [Accessed 13 May 2024].

[100] Parkins K, Is FDA. LDT surveillance set to improve as VALID act heads to resolution? Med device Netw., 2022, https://www.medicaldevice-network.com/analysis/fda-ldt-surveillance-set-to-improve-as-valid-act-heads-to-resolution/. [Accessed 3 May 2022].

[101] Food and Drug Administration. Medical Devices; Laboratory Developed Tests. Federal Register 2024. https://www.federalregister.gov/documents/2024/05/06/2024-08935/ medical-devices-laboratory-developed-tests/. [Accessed 13 May 2024].

[102] European Commission. Medical devices—new regulations—overview., 2022, https:// ec.europa.eu/health/medical-devices-new-regulations/overview_en. [Accessed 3 May 2022].

[103] European Commission. Medical devices—new regulations—publications., 2022, https://ec.europa.eu/health/medical-devices-new-regulations/publications_en. [Accessed 3 May 2022].

[104] European Commission. White paper on artificial intelligence, a European approach to excellence and trust; 2020.

[105] Centers for Disease Control and Prevention. Clinical Laboratory Improvement Amendments (CLIA)., 2022, https://www.cdc.gov/clia/index.html. accessed May 3, 2022.

[106] Definitions. Code of Federal Regulations, title 42, chapter IV, subchapter G, subpart a, section 493.2. Natl Arch 2022. https://www.ecfr.gov/cgi-bin/text-idx?node=pt42.5. 493&rgn=div5#se42.5.493_12. [accessed May 3, 2022].

[107] Centers for Disease Control and Prevention. clinical laboratory improvement advisory committee (CLIAC)—LIAC meeting., 2022, https://www.cdc.gov/cliac/upcoming-meeting.html. accessed May 2, 2022.

[108] Stang H, CDC/DDPHSS/CSELS/DLS. Clinical Laboratory Improvement Advisory Committee (CLIAC). CLIA Regulations Assessment Workgroup; 2022. April 1. Meeting Summary Report 2022.

[109] Huss R, Coupland SE. Software-assisted decision support in digital histopathology. J Pathol 2020;250:685–92. https://doi.org/10.1002/path.5388.

CHAPTER

Ethics of AI in pathology: Current paradigms and emerging issues

9

Chhavi Chauhan[a,*] and Rama R. Gullapalli[b,c,*]

[a]*Director of Scientific Outreach, American Society of Investigative Pathology, Rockville, MD, United States,* [b]*Department of Pathology, University of New Mexico, Albuquerque, NM, United States,* [c]*Department of Chemical and Biological Engineering, University of New Mexico, Albuquerque, NM, United States*

Introduction

Healthcare is inherently data-centric, encompassing various data-generating subdomains such as insurance, pharmacy, administration, healthcare institutions, and different specialties of actual patient clinical practice [1]. Vast amounts of information are generated at each level of healthcare with the potential to provide unique insights into how medicine is practiced at scale [2]. Artificial intelligence (AI)-enabled clinical workflows have tremendously improved our ability to collect healthcare data [3]. However, large-scale data analytics across healthcare subdomains mentioned above are lagging [4,5]. Computational algorithms based on principles of machine learning (ML) and natural language processing (NLP) are expected to automate big data analytics, identify patterns to improve our understanding of healthcare processes, and improve efficiencies of healthcare delivery [1,2,6].

Although data-generating sources within healthcare are vast. AI researchers have been mostly focused on the data generated in the context of routine clinical work. Clinicians generate vast amounts of unstructured data (e.g., clinical notes during patient encounters) [1]. However, especially in the developed nations, clinicians rely heavily on radiology and pathology to guide the diagnosis, prognosis, therapeutics, and management of each patient [7–10]. Radiologists are adept in the use of technology (e.g., computerized tomography (CT) and magnetic resonance imaging (MRI)), and they use it as key drivers of the practice of radiology [8–11]. Technological innovation is also playing an increasingly dominant role in the practice of pathology. Both radiology and pathology are image-intensive specialties making extensive

[*]Contributed equally.

Artificial Intelligence in Pathology. https://doi.org/10.1016/B978-0-323-95359-7.00009-1
Copyright © 2025 Elsevier Inc. All rights are reserved, including those for text and data mining, AI training, and similar technologies.

160 **CHAPTER 9** AI ethics in pathology

use of image data for patient care via specialist-generated interpretations. Though radiology is farther along the path of digitization and management of medical images, pathology is increasingly moving along the same path [8,10–13]. The heavy reliance on images and digitization makes these two healthcare specialties most attractive to AI researchers to test emerging ideas of imaging AI research. Imaging AI algorithms have seen the greatest amount of research and advances over the past decade. The confluence of abundant imaging data, ever-increasing cheap and powerful computational capacity, and advancing algorithmic AI research make radiology and pathology prime targets for disruptive innovation of healthcare AI applications over the next decade [7,9,14,15].

The second arm of the practice of pathology (in addition to image-intensive anatomical pathology) is the area of clinical laboratory medicine. Automation in clinical laboratory medicine has been well underway for many decades resulting in vastly improved efficiency in patient test result delivery. In emerging fields such as precision medicine, there is a lot of interest in the use of genomic and other forms of "omic" data for both diagnosis and prognosis using information at a molecular level [16]. The new frontier of "omic" technologies is a true "big-data" specialty with vast amounts of "omic" patient data generated in each encounter [17,18]. The field of bioinformatics focuses on algorithmic computational methods to manage and interpret such "omic" data in various clinical settings. AI researchers are highly interested in using AI-based methods to understand "omic" data in the context of patient healthcare [16,19]. Perhaps the ultimate challenge in the use of AI-enabled healthcare is to synthesize both imaging and genomic data from a patient to provide novel insights into clinical outcomes and management [17,18,20]. Many such efforts are currently underway.

Although the potential of AI-based algorithms to effectively manage and interpret big data in healthcare is considerable, there are significant downsides to using such a powerful technology without the necessary ethical and moral safeguards [5,21–24]. There is increasing unease with the unrestricted use of AI in healthcare especially with regards to ethical issues such as patient privacy, exacerbation of race and gender inequities, and patient safety outcomes. The broader field of AI ethics is focused on the use of AI technologies to ensure development in an ethically and morally appropriate manner to benefit society at large. Issues surrounding AI technology misuse have common themes across specialties and reflect more specific specialty-centric concerns. Thus, the development of AI ethical guidelines *requires* the participation of domain experts (e.g., practicing pathologists) to develop specialty-centric guidelines for the ethical use of AI technologies. Key participants in enabling ethical AI technologies include AI researchers, pathologists, clinicians, institutional administration, professional societies, and finally, the patients themselves. The AI ethics paradigm and the participants involved in this interactive process of development are illustrated in Fig. 9.1. Also, some of the key definitions associated with topics of AI and ethics discussed in this paper are listed in Table 9.1 to better inform the reader.

Introduction **161**

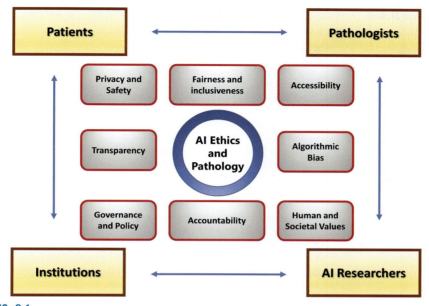

FIG. 9.1
Key issues and participants involving the study of AI ethics in pathology. The various topics shown at the center of this diagram are key emerging topics of significance to understand AI ethics in detail. These topics are common across various fields of study and also unique within the context of each specialty. Interactions between the various participants (patients, AI researchers, pathologists, and institutions) is key to developing a comprehensive framework of understanding AI ethics in pathology.

Table 9.1 A brief description of the key terms used in this review explaining artificial intelligence (AI), ethics, and pathology.

Terminology	Explanation
Artificial intelligence	Intelligence emulated or simulated by the use of technological means. Computational machinery is used to achieve intellectual autonomy and independence of thought similar to that seen in humans
Algorithm	A set of step-wise commands in order to accomplish a specific task/goal/objective. In AI, algorithms are the programming code that enable the functionality of an objective task and are key to emulating intelligence in an artificial manner
Bias	Discrimination in favor or against a set of outcomes in a particular setting. In AI ethics, this often refers to the ability of the AI algorithm to discriminate against individuals, groups, or populations based on the design of the original algorithm

Continued

162 CHAPTER 9 AI ethics in pathology

Table 9.1 A brief description of the key terms used in this review explaining artificial intelligence (AI), ethics, and pathology—*Cont'd*

Terminology	Explanation
Big data	"Big data" refers to data produced by an automated and repetitive technological process. Big data may be quantified in terms of the abundance of the data size generated. In pathology, some examples of big data are (1) a digital pathology whole slide image repository, or (2) databases containing complete blood counts across a population and time
Data privacy	The moral, legal, and ethical expectations to maintain confidentiality of data collected from either individuals or nonindividual resources. In pathology, institutions responsible for the collection of patient laboratory data are tasked with the responsibility to ensure data privacy at an individual and population level
Data-shifts	A concept referring to the change in the data distribution between training and real-world datasets in AI algorithm development
Digital pathology	An emerging paradigm of pathology focused on digitization of traditional glass-based slides read by pathologists. Digitized slide data can be stored, viewed, and shared in real-time leading to enhanced efficiency of the sign-out process
Ethics	A branch of philosophy studying the concepts of right and wrong human behavior in a systematized manner
Machine learning	Computational algorithms that are capable of automated learning processes through iterative feedback of data without (or with minimal) human intervention
Underspecification	Failure to specify adequate details in the context of a training set of an AI algorithm

While writing this manuscript, we identified an article with a similar thematic focus examining the role of AI ethics in pathology and laboratory medicine [25]. The article by Jackson et al. discusses AI ethics from a traditional bioethicist's perspective, relating to the core principles of bioethics as laid down in the Belmont report [25] (also see https://www.hhs.gov/ohrp/regulations-and-policy/belmont-report/read-the-belmont-report/index.html for a full PDF version of the Belmont report; last accessed—08/01/24). In contrast, this chapter mainly focuses on AI ethics from the perspective of ongoing developments in the area of AI research and algorithmic decision-making (ADM) and imagining its potential impact on the future practice of pathology. Thus, we feel these articles are complementary. This chapter informs the practicing pathologist about the topics of current concern in AI ethics within the broader field of AI research and ADM and educates them how these AI ethics issues are likely to impact the practice of pathology in the future. Hopefully, this piece encourages the interested pathologists to get involved in this emerging area to help guide the future development of AI algorithms in the practice of pathology.

In this chapter, we first discuss the issues of AI ethics from the perspective of an AI researcher interested in AI-enabled pathology. We then address issues of AI ethics and research pertinent to a practicing pathologist including risks of AI to the practice of pathology. We then discuss AI ethics issues relevant at a professional society or institutional level to guide the safe development and deployment of AI in pathology. Finally, we discuss some of the latest developments in this field in the past few years.

Ethical AI study designs in pathology

The current section reviews issues in AI pathology research studies based on principles of ethical AI design. There are a multitude of imaging-based pathology AI studies underway, and it is worthwhile to remember the key ethical issues involved must be reviewed *prior* to initiating such studies.

Inclusive AI design and bias

Pathology is highly data-centric, making use of both clinical and phenotypic (histomorphological) data elements to enable the traditional practice of pathology. Yet, there is an increasing appreciation for the need to place a classical pathology expert diagnosis in a broader context by integrating additional patient data elements as part of routine diagnostic workflow (e.g., molecular biomarker information at an individual and population level) [26–28]. There is also a need to include additional data elements such as lifestyle and socioeconomic data to improve research categories such as cohort description, methods applied, and patient outcomes within the domains of pathology and precision medicine. Such complex, cross-domain data handling and integration is perfectly suited for the use of AI algorithms. A key strength of AI and ADM is their ability to enable integrative cross-domain analyses of diverse research datasets, which are perhaps not amenable easily to the human mind. However, one must be careful while conducting such studies as the complexity of these cross-domain datasets may lead to introduction of bias which can manifest in two major ways [29,30].

These include biases inherent to the AI algorithm itself and biases arising within the datasets used for the purpose of training the AI algorithm. The former does not usually have ethical implications, rather, it is necessary to understand the inner workings of an AI algorithm appropriately. For example, the popular *k-means* clustering algorithm works best when the data form clusters that are roughly spherical and similar in size; however, sometimes they may not [31]. This is not an ethical dilemma; it is simply a function of how the algorithm works. Similarly, DL networks typically have tunable variables (known as hyperparameters) integral to the AI algorithm that must be assigned on an ad hoc basis by human beings as a part of the AI research study. These are conscious choices on the part of the AI researcher. But in some cases, ingrained, and sometimes unconscious biases may creep in with the potential to result in cascading downstream effects unforeseen by the AI researchers.

164 CHAPTER 9 AI ethics in pathology

From an AI ethics point of view, the key critical issues relating to algorithmic bias are mainly in the context of datasets used for a research study [29,30,32]. Both sample choice and valuation play a role in this regard. For example, if a dataset has category imbalance (i.e., a study that is composed mainly of adult white males due to factors such as sample availability and socioeconomic factors of healthcare access), then the results of the AI algorithm trained on such data may not be accurate when implemented on the population as a whole. Results from such a homogenous research study might inadvertently disadvantage a minority subpopulation. A second problem is related to underspecification (Table 9.1), which describes a phenomenon in which an AI-training dataset is not provided with all of the necessary parameters. For example, if the genetics of a population were a key factor in categorization of histology images, not including those details would lead to an incompletely trained AI model. Underspecification can thus lead to faulty correlations in predicting clinical outcomes. Another example is that when in trying to weigh the importance of various factors in determining the extent of disease, one might consider that a subpopulation that spends less on medical care might be a healthier one. However, it is just as likely, if not more so, that the subpopulation spending less on medical care is from a lower socioeconomic group, and just simply cannot afford costly care. They are not healthier; they simply do not have access to appropriate care. Issues such as these mandate a deep understanding on the part of the AI researcher seeking to train DL algorithms to improve the practice of pathology. It also befits pathologists to be aware of such issues that may result in a skewed interpretation while utilizing an AI algorithm in pathology practices.

In 2016, Arkansas approved the use of an algorithm-based program designed by InterRAI, a nonprofit coalition of health researchers from around the world, to determine the care hours needed by patients with limited mobility (source: https://www.theverge.com/2018/3/21/17144260/healthcare-medicaid-algorithm-arkansas-cerebral-palsy; last accessed—08/02/24). However, the AI algorithm was limited in its utility as it assigned variable scores for people with similar disabilities and made several erroneous decisions in calculating care hours needed, resulting in life-altering outcomes for hundreds of patients. There was no explanation offered to the patients as the standards for use were neither clearly defined nor disclosed to all stakeholders to identify and rectify these errors in a timely fashion. AI researchers must consider such high-level factors related to inclusivity and algorithmic bias while designing AI studies using pathology-based data.

Algorithmic bias, and statistical bias in general, is an ill-understood topic. It is perhaps unrealistic to expect a busy practicing pathologist to be well-versed in the various nuances of algorithmic and/or statistical bias. The solution to the issue of algorithmic bias may thus fall primarily on the shoulders of regulatory agencies (e.g., the US Food and Drug Administration (FDA)) now identifies Software as a Medical Device (SaMD) (https://www.fda.gov/medical-devices/digital-health-center-excellence/software-medical-device-samd; last accessed—08/01/24) to approve sustainable AI-enabled workflows. However, pathologists must realize the potential of bias to manifest itself over long-term use of unregulated AI-enabled pathology workflows. AI researchers and vendors must partner with pathologists to obtain practical perspectives to identify and remedy potential sources of long-term bias in AI pathology algorithms. There are multiple complementary efforts underway in professional pathology organizational committees (e.g., Digital Pathology Association (DPA) and College of American Pathology (CAP)) tasked with understanding the use of AI within the practice of pathology.

Race in ethical AI design

Another key variable to consider while designing AI workflows in pathology is the impact of race. AI model performance deterioration may occur due to issues related to data shifts, faulty correlations, and underspecification (Table 9.1) that limit the eventual utility of the AI algorithms in pathology classification studies. These limitations become even more striking in healthcare data incorporating complexities such as underlying physiological effects and genetic factors of disease predisposition. These complexities are further compounded when these data are analyzed without due consideration of race and ethnicity.

Though historically race has been considered to be a social construct without any biological basis, and evidence suggests that it is tied to genetics [33]. Work presented by AI researchers like Joy Boulamwini and Dr. Timnit Gebru have made it evident that the AI systems and the defined parameters should be tested intersectionally with race, to determine their efficacy and broad utility beyond the use in training datasets alone [34]. During the global COVID-19 pandemic, the Centers for Disease Control and Prevention (CDC) had placed emphasis on health equity considerations in racial and ethnic minorities (APM Research Lab, St. Paul, Minnesota). The COVID Tracking Project confirms that COVID-19 has affected Black, Indigenous, Hispanic, and other minorities at higher rates (https://covidtracking.com/race; last accessed—08/01/24). These recent findings make it imperative to include race and ethnicity datapoints in datasets used to train AI algorithms in the healthcare sector for broad applicability of any clinically facing AI model.

Optum, a subsidiary of insurance giant UnitedHealth Group, designed an application to identify high-risk patients with untreated chronic diseases. The automated categorization algorithm however, was noted to discriminate against Black patients based on the cost of an individual patient's past treatments [35]. AI-based methods have been extensively deployed for cancer staging, especially in breast cancer. The FDA has granted clearance to these applications without a need to publicly disclose how extensively their tools have been tested on people of color. Thus AI-based applications have the potential to exacerbate disparities in clinical outcomes in breast cancer, a disease which is 46% more likely to be fatal for Black women [36]. This example provides an illustration of the potential of the cascading damaging effects a poorly researched and designed AI algorithm can have with real-world consequences on minority communities in particular.

Stakeholder concerns: Consent and awareness

In the absence of any defined guardrails and recommendations for the use of AI-based algorithms in clinical practice, issues related to patient consent and awareness have come to the forefront. Although, historically, there have been rigorous standards in place monitored by institutional review boards (IRBs) to protect patients with respect to data privacy, there appears to be a stunning lack of regulation and even basic guidelines on how to train an AI algorithm on available pathology datasets. For example, if a researcher trains an AI model on annotated data from a pathology slide archive and launches an independent AI company based on this model, do they own rights to all of the profits generated from it? How should the pathologists who worked on generating the annotation for each patient based on their expertise, be compensated? This is an issue that has not been dealt with yet. Ethically, one would imagine that at least a part of the profit should be shared toward the continued maintenance of the pathology archive and/or support the department whose pathologists performed initial annotation to generate a viable AI-based product of the company.

There are other pressing concerns that need a thorough discussion in the context of stakeholder concerns. For example, are patients made aware that an AI-based platform may have affected a pathologist's decision? Should the patients be informed about the use of AI-aided decision-making? Are the patients offered a choice to approve or reject the regimen based on their informed consent for the use of an AI-based model? Are the pathologists familiar with, and do they completely understand the parameters and limitations of an AI-based algorithm? Are AI-based algorithms audited for bias, fairness, transparency, ethics, and risk mitigation, and if so, how often? Who are the governing bodies overseeing these audits?

In developing and underdeveloped countries, physicians often form informal consultation groups over various social media platforms and apps that often have loose privacy settings. These platforms and apps may allow data sharing to variable degrees within the platform/app as well as with third-party vendors. Such data-sharing practices put patient data safety and privacy at a high risk. There is often no regulation regarding how patient data are shared within an AI-enabled app that is often used without patient consent. How does one address concerns of de-identification of patient data and patient data privacy concerns in such cases? AI researchers must consider such questions from pathologist and patient perspectives while designing research studies before AI-enabled algorithms can become mainstream within the discipline of pathology.

Risks of AI in pathology and to pathologists—Real or imagined?

Pathologists in current clinical practice are anxious to understand the scope and impact of AI algorithms. Questions such as "Will AI replace pathologists?" and "How impactful is AI in enabling patient diagnoses?" are commonplace and reviewed often (with justified concern) by pathologists [37,38]. In this section, we review some of these issues from a practicing pathologist's perspective, focusing on AI ethics and the important role pathologists have to play in the future of AI algorithm development.

An intensely studied aspect of AI ethics is risk assessment and evaluation of the dangers associated with implementation of AI. The popular press and social media are key drivers in fueling the imagination of the public about the future of AI, often in apocalyptic terms. Emerging AI-based technologies such as driverless cars, automated facial recognition, and AI-based deep fakes are indeed a cause for concern on the economic and moral outlook of our society as a whole [39]. Yet, the reality is that AI is here to stay in one form or another with all of the attendant risks associated with it. A key requirement of the latest iteration of AI, deep learning (DL), is the need for vast amounts of training data for eventual implementation. By training on large amounts of raw data, the algorithmic performance of DL workflows is much better compared to that achieved by previous ADM approaches. In contrast to previous ML algorithms, DL algorithms are able to work on both structured (e.g., laboratory data) and unstructured (e.g., pathologist reports) data to create ADM models capable of attaining high levels of prediction accuracy.

A commonly held notion is that laboratory-generated data influences 70% of all clinical decisions [40]. Although the accuracy of the aforementioned claim is contested, it is undeniable that laboratory-generated data (both clinical and anatomical) makes up a significant portion of the *quantitative* data associated with a patient's electronic health record (EHR) [40]. The ready availability of a significant amount of structured and unstructured clinical data in pathology archives and databases is highly attractive to AI researchers (and potential unscrupulous actors) to leverage laboratory-generated data for purposes of benefit (and harm) constituting a potential risk to patient safety. Pathologists, as custodians of laboratory data, will be at the center of heated debates on issues of patient data ownership for purposes of AI research in the future. The profession of pathology must be ready for this battle.

Assessing risk outcomes associated with AI research implementation is a key concern of AI ethicists. In fact, there are specific institutes dedicated to enabling such risk assessments and are populated by subject matter experts from different fields (e.g., Machine Intelligence Research Institute—https://intelligence.org; last accessed—08/01/24). As AI research into pathology accelerates, there is a significant opportunity for practicing pathologists to contribute toward such risk assessments to understand the broader implications of AI algorithms within pathology and patient management. Some of these risks (with a specific focus on pathology) are discussed here. We focus on two key issues related to AI risks in pathology—(i) risk potential of AI to the pathologist workforce and (ii) risk impact of AI on the practice of pathology itself.

Underestimating the risks of AI to pathology

Research into medical applications of AI is proceeding rapidly. Radiology is perhaps at the forefront of many of these AI imaging initiatives. However, other case studies of AI healthcare applications, such as AI-enabled natural language processing (NLP) in clinical electronic health record (EHR) text evaluation and AI-based analysis of whole slide images (WSI) in digital pathology, are also moving forward [1,41–43]. Economic incentives of low-cost healthcare also drive commercial AI research efforts. Workforce automation is a key economic driver for many industry-driven AI initiatives to reduce the overall costs.

Traditionally, pathologists have relied heavily on experience in rendering diagnoses. To the extent pathologists are unwilling to move on from simply rendering a diagnosis based on visual impression, the risk of AI over the next decade to such practitioners is nonnegligible. AI techniques such as DL are beginning to outperform humans in certain image-based tasks, particularly those that involve a quantitation component. Pathologists must thus expand the scope of their practices and be more integrated within the overall clinical care of the patient. This includes adopting accessory techniques such as molecular, clinical, and epidemiological data to provide comprehensive diagnoses for each patient. Future pathology practitioners must develop skills to synthesize information from multiple sources to provide *integrated* prognostic and even established high-level therapeutic NCCN guidelines within their reports. Humans have the unique ability to synthesize information across different domains of knowledge with relative ease. In contrast, cross-domain integration is an ability that AI algorithms lack at the current time. And it is this shortcoming of AI that human pathologists must capitalize on to improve the scope of pathology practice and stay uniquely relevant to the practice of medicine.

At the pathology practice level, it is also inevitable that the analytical instruments used by pathologists and laboratories will adopt increasing degrees of AI-enabled automation. Some examples of such AI-enabled automation would include providing first-pass oversight of the diagnostic algorithmic outputs, AI-based quality assurance of laboratory data, and automated assessments in high-volume clinical laboratory tests. In the majority of these instances, such developments would be driven primarily by the companies responsible for the instrument development. This may limit the impact an end-user pathologist or a laboratory has directly on the AI-enabled instrumentation development process. However, as an end-user customer, ideally, a pathologist could significantly influence the adoption of such AI-enabled laboratory instrumentation into clinical practice by adopting due diligence for AI technologies and risk assessments based on principles of ethical AI. Thus, an awareness of issues surrounding algorithmic bias and ethical AI should be kept at the forefront while evaluating instruments and technologies in pathology practices. Active development of the guidelines of ethical AI and norms of assessment by professional organizations (e.g., College of American Pathologists) would raise awareness among pathologists that could help mitigate the risks of adopting AI-enabled technologies into clinical practice in the future. Pathologists and administrators must be aware of the subtle, unanticipated risks posed by AI algorithms in issues related to patient data privacy and the potential of AI-enabled technologies to (unconsciously) deviate from established pathology practice guidelines. The need for a cautious eye on quality and process control by pathologists is unlikely to be automated away anytime soon.

Overestimating the risks of AI to pathology

AI has enabled some truly impressive advances in the automation of narrowly defined tasks, particularly through the use of DL-based approaches. However, it would be a stretch to say that AI is "human-like" in its capabilities at the current time. In fact, many experts are also of the opinion (perhaps pejoratively) that DL is no more than an extremely efficient statistical means to fit data. True human-like AI (also known as strong AI/AI-hard/AI-complete), capable of emulating human-like awareness and decision-making capabilities, is unlikely to become a reality for decades to come [44,45]. Within the context of pathology, AI algorithm capable of replacing the skills of a highly trained human pathologist is highly unlikely to come to pass anytime soon. Current DL techniques perform exceedingly well in addressing narrowly defined and well-formed questions in pathology with strict boundaries of performance [44,45]. Pathologists must be wary of the "hype" and "oversell" commonly associated with AI research studies, while assessing the claims made by AI researchers. The hype associated with AI has been a well-known issue since the 1960s. AI research has passed through multiple boom and bust cycles when the ambitious goals of AI researchers failed to pan out [44,45]. While it feels as if the recent advances in computer hardware, networking, and data storage capacity may have allowed the field to turn a corner, deep learning techniques may still prove to be part of one such hype cycle playing itself out now. While AI technologies are undoubtedly improving in each successive cycle of development, there is still a very high bar that AI-driven technologies need to clear before they are ready for widespread application in patient healthcare.

It is instructive to review the study by Frey and Osborne [46] which assessed the impact of AI technology on workforce displacement in over 700 professions in the United States. Physicians working in healthcare were the 15th least likely profession to be impacted by AI automation with an assigned low probability score of 0.0042. This finding is not unexpected as the nature of work performed by physicians is complex, interactive, cross-domain, and multifaceted. Pathologists must thus actively seek skills and expertise to enable cross-domain relevancy in medicine. An oft-repeated quote regarding this issue is worth remembering—"AI will not replace pathologists. However, pathologists who know nothing about AI will be replaced in future."

The laboratory workflow processes will likely see increased automation with incorporation of AI algorithms at various steps; however, the pace of change will be at best incremental. Instrument vendors in healthcare are by nature cautious. Regulatory FDA oversight will also ensure that evolution will be gradual. Labor-intensive and uncompensated steps in laboratory workflow are the areas where AI-enabled automation will be implemented first to improve overall process efficiencies. Equally, one may predict newer job opportunities within AI-enabled pathology to develop, deploy, and maintain these automated AI workflows in the future. However, the specifics of such development are not yet visible.

In summary, the profession of pathology and laboratory medicine must be well informed of the potential risks associated with AI. The only thing predictable about the AI-enabled future of pathology automation is that it will be unpredictable. By adopting a proactive stance toward these technological developments within the AI space, pathologists can be at the forefront of mitigating the risks posed by AI while benefiting from the potential advantages it could confer. Awareness of issues of AI ethics will ensure the balanced and informed viewpoint of pathologists is incorporated into the development of AI-enabled technologies in the field of pathology.

Institutional frameworks to enable ethical AI in pathology

Spending on healthcare worldwide was estimated at approximately $9 trillion in 2014 and is projected to grow to approximately $24 trillion by 2040 [47]. The United States alone spent approximately $3.8 trillion, accounting for nearly 17.7% of its national gross domestic product (GDP). The United States also had the highest healthcare expenditures per person ($\sim$$11,582) in the entire world which is nearly double compared to the second most expensive country for healthcare (see links at https://www.ama-assn.org/about/research/trends-health-care-spending; last accessed—08/01/24). Naturally, there is an increasing push for automation to reduce healthcare costs. The economic incentives of lowering such costs are directly aligned with the potential of AI to help with this process.

Pathologists and lab medicine can expect to be squarely in the middle of the upcoming scramble to mine patient healthcare data at scale to enable AI workflows in healthcare. Although the ultimate value of such approaches remains to be determined, it will not impede the desire to acquire patient data for research and commercial purposes. The quantitative (clinical pathology) and (semi) structured (anatomical pathology) data formats are highly attractive to an AI and ML researchers to assess the efficacy of AI algorithms in healthcare. An important emerging question pertains to the question of ownership of patient data. Ultimately, who "owns" patient data?—Is it the patients themselves? Is it the institution where the data is held? Do pathologists who generate and curate the extensive data residing in the institutional databases have any intellectual property rights over the potential payoffs of collaborative AI algorithm development? Is it even ethical to consider patient data as something that is "ownable" with initiatives to mine it using AI?

Conundrums such as these are likely to continue to confront pathology practice over the next decade. Professional organizations such as College of American Pathologists (CAP) and American Society of Clinical Pathology (ASCP) have a key role to play in guiding the ethical development of AI in a manner that is appropriate to the practice of pathology. In the next subsections, we describe a basic framework of three key principles governing AI ethics within the specialty of pathology.

Transparency

A key element driving AI and ADM research ethics since the early 1960s is the idea of radical transparency [48,49]. Algorithmic transparency may be defined as the idea of open availability of all information pertaining to the working of an AI algorithm. AI algorithmic transparency enables the interactive and relational assessment (moral and ethical) of the way an AI algorithm functions and eventually impacts the end-user. Transparency, more generally, is defined as the robust and open availability of all information content related to the function and interactions within a system, including both human and nonhuman information content [49,50]. In healthcare-associated AI research, it is important to implement the idea of radical transparency due to the significant impact of ADM on a patient's healthcare [48,51]. In pathology, an active area of research interest is the use of DL algorithms as agents to aid the anatomical pathologist's workflow. Questions such as "Can DL AI algorithms identify a disease process as efficiently as a human pathologist?" underpins much of the AI research efforts into pathology. Ironically, though, much of the inner workings of DL algorithms remain unknown [39]. It may come as a surprise to many pathologists that the manner in which AI-DL neural networks work is still a mystery [39]. When training a DL algorithm on a set of images (anatomical pathology or otherwise) or any other data, the precise imaging data features that a DL algorithm uses to identify and discriminate between different target categories are obscure. Thus, in essence, DL algorithms are a black box in their capabilities at the current time. In light of this, one must ponder when, if ever, we might be confident enough in trusting an ADM process with life-and-death decisions that are commonplace in the healthcare domain? Pathologists make such critical calls on a daily basis. Will we ever trust any such call made by an AI algorithm? This is an issue that remains to be answered. What is indisputable, though, is that AI and ADM algorithms demonstrate an ever-increasing capability to identify and predict patterns with accuracy that is near human (or even better) across a variety of fields [2]. It may well turn out that the performance characteristics of DL algorithms are so well-established in the future that we may implicitly trust them without a detailed understanding of their inner workings. However, a transparent development process is the key to achieving general acceptance of the AI technologies in the pathology workflow.

Beyond the use of AI and ADM for mere "diagnostic" purposes, there are multiple pathology domains where issues of AI transparency are likely to play an important role in the decade ahead. AI transparency will be critical while seeking to implement algorithmically enabled clinical workflow in a health institution. Although one may rely on the FDA to provide overall regulatory supervision, it is institutions and pathologists who will eventually be responsible and liable for real-world outcomes. Thus, it is incumbent on the industry to provide transparency of the ADM process during the development and implementation phases of an AI algorithm. Due to the potential for AI to constantly "learn" as a part of the workflow, developers, and vendors must make two-way communication of the AI performance a routine part of their normal implementation protocols with end-users. As the AI and

ADM protocols are upgraded iteratively, radical transparency must also be maintained to protect patient health. Another scenario where one can envision the need for transparency is in the data used to test and train as a part of the AI development process. AI trained on a local set of data may not necessarily translate at a global implementation level. Transparency of such information will inform the pathologist of potential variability in the performance of the ADM and AI model in the local context and allow them to adjust accordingly. Equally, pathologists must be trained to recognize and deal with such issues. Over the next decade, as our understanding of the use and benefits of healthcare AI expands, our notions and expectations of AI and ADM transparency will also evolve in tandem.

Accountability

A natural corollary of transparency in AI is accountability. Accountability pertains to both human and nonhuman factors and their interactions alike. For additional details on this topic, refer to the article by Kroll [52]. Current AI research initiatives in pathology focus on the eventual use case of AI algorithms as an "assistant" in the normal pathology workflow. This may be attributed to the complexity of a normal pathology workflow process, and it reflects the reluctance of AI researchers and industry to accept full "accountability" in the final decision-making process. Accountability is a shared transactional concept involving multiple entities such as the AI algorithm, the human responsible for the healthcare decision process (e.g., clinician and/or pathologists), and finally, the institution implementing the AI-enabled workflow process. The eventual goal of a shared accountability process is to assign appropriate "answerability" as a part of the normal healthcare delivery. Accountability may be either desired or undesired in the form of a "reward" for the beneficial outcome or "blame" for a nonbeneficial outcome, respectively. However, the key is that there needs to be a formal accountability process in place when considering the implementation of AI-enabled workflows, including pathology. This includes formal documentation of the accountability hierarchy at an institution, as well as detailed oversight specifying who is responsible for what and what outcomes are anticipated due to the implementation and use of algorithmic healthcare AI [49,51,53,54]. Additionally, periodic review and updating of the institutional AI accountability protocols are mandatory to reflect the current state of knowledge of the healthcare AI processes, which itself is constantly evolving. The "black box" nature of the neural networks was alluded to in the previous section. In healthcare, where patient safety is paramount, such a black box scenario of an AI algorithm mandates the need for clearly defined hierarchies of accountability to ensure safe patient outcomes over the long term. In seeking to establish the use of AI technologies within the field of pathology, one must establish the different tiers of accountability associated with the use of AI algorithms within an institution. This would include the AI algorithm vendors, the human pathologists using AI technologies, and the institutions adopting AI themselves. Both physicians and institutions must be accountable for the use of AI in an ethical manner and share an equitable burden for the successful long-term use of AI/ADM in patient healthcare [49,51,53,54].

Governance

The third leg of the proposed framework for AI ethics in pathology relates to governance. The potential impact of AI and ADM tools creates enormous financial incentives for research and commercial interests in healthcare. AI has the potential to change the practice of healthcare delivery in the next decade, but it can also pose temptations for unethical entities to take advantage of shortcuts. Establishing and enforcing rules underpinning the governance of AI and policy will be critical to the moral and ethical implementation of AI and ADM in healthcare. Rules guiding the governance will thus need to be implemented at multiple scales—national, professional, and institutional. At a national level, the European Union (EU) has been proactive in instituting rules governing data ownership and privacy, in marked contrast to the practice in the United States where regulation is highly lax. The rules implemented by the EU in May 2018 are part of a framework known as the General Data Protection Regulation (GDPR; see https://gdpr.eu/; last accessed—06/24/2021). The GDPR framework is directed toward countering the overwhelming power of the large tech giants which are at the leading edge of AI and ADM research. Similar policies are likely to be adopted globally over the next decade to ensure data privacy rights while harnessing the benefits of AI in a fair and egalitarian manner. More pertinent to the current chapter is the role of professional pathology societies in guiding the development and implementation of AI and ADM.

As domain experts, AI researchers, engineers, and scientists have been at the forefront of assessing the ethical aspect of AI and ADM and the wider impact of the technology. A majority of AI researchers and engineers are professionally affiliated with organizations such as the Institute for Electrical and Electronics Engineers (IEEE). IEEE has created a Global Initiative on Ethics of Autonomous and Intelligent Systems under the title Ethically Aligned Design (see https://ethicsinaction.ieee.org/ and https://standards.ieee.org/content/dam/ieee-standards/standards/web/documents/other/ead1e.pdf; last accessed—08/01/24). The science and ethics of AI are frequently represented at cutting-edge technical engineering conferences such as the Association for Advancement of AI (AAAI) and the ACM Conference on Fairness, Accountability, and Transparency (ACM FAccT). However, there is increasing realization that individuals who are involved in the development of AI technology (i.e., engineers and AI researchers) also cannot simultaneously be the arbiters of the ethics of these technologies and that it will be necessary to incorporate the voices of the end-users of these technologies into the development process. For example, physicians may be well positioned to understand the long-term impact of AI and ADM usage on patient outcomes within an AI-enabled clinical decision support (CDS) system [29,42]. Similarly, a radiologist may be better positioned to understand the weaknesses of an AI algorithm while incorporating clinical information into an AI-enhanced radiological technique [14]. Due to the interdisciplinary and transformative nature of AI and ADM technology, non-AI technical expert voices (physicians, lawyers, and lawmakers) are required to participate in building a framework of AI ethics to provide domain-specific expertise [5,55].

Our colleagues in radiology at the American College of Radiology (ACR) have been at the forefront of launching AI-centric initiatives in the practice of radiology. ACR AI-LAB is an ACR initiative to educate radiologists in the use and implementation of AI tools [14]. This course aims to empower radiologists with the basic knowledge to participate directly in the creation, validation, and the use of healthcare AI. The College of American Pathologists (CAP) also has similar early-phase initiatives within pathology AI through the creation of various AI and technology committees. The mandate of the CAP AI committee is to create a broad pathology-centric AI strategy. Additionally, the committee aims to provide subject-matter expertise to CAP councils and enable the creation of AI laboratory standards in pathology. We propose that AI ethics *must* be a core component of the mandate of this committee moving forward. As more of AI and ADM technologies are incorporated into pathology workflows, pathologists must be at the table to guide the development of these technologies in a manner that aligns with the core principles of the profession of pathology. Finally, institutions employing pathologists (i.e., universities and clinical practices) must also be actively engaged in helping to build a framework of AI ethics meaningful in a local context. This would involve initiatives such as the implementation of professional norms and of creating procedural guidelines for the adoption of ethical AI workflows within an institution. Effectiveness of ethical AI initiatives depends on the awareness of its importance, periodic review and oversight, and reinforcement of the saliency of this issue to employees (clinical and nonclinical healthcare workers) in general.

In summary, we have discussed three foundational frameworks to enable and guide ethical AI in specialty of pathology—Transparency, Accountability, and Governance. These three core principles represent a starting point for adoption of AI- and ADM-based initiatives in pathology at an institutional level. As the science of AI evolves, pathologists must review and adopt additional measures to enable the ethical AI usage in pathology practice in a manner that benefits the patients. Also, as end-users of powerful data-centric technologies of AI and ADM and as custodians of structured patient data repositories within the field of medicine, pathologists are in a strong position to drive the adoption of ethical healthcare AI in the future.

Recent developments in the use of AI in pathology

The public release of OpenAI's ChatGPT in November 2023 has paved the way for the widespread use of large language models (LLMs) in all domains, and pathology is no exception. The major concerns with these LLMs have been the unreliable source of data used for training, the underlying biases in the training dataset, and the lack of accountability as many data sources were used without proper consent or attribution leading to a trail of copyright lawsuits (https://www.bakerlaw.com/services/artificial-intelligence-ai/case-tracker-artificial-intelligence-copyrights-and-class-actions/; last accessed on 08/01/24). As discussed previously and given the focus of pathology on the ethical principles of transparency, accountability, and governance, several academic and healthcare institutions have now resorted to building their own LLMs on real-world patient data-enriched datasets from their own healthcare systems. Though cost and often time intensive, these responsible efforts will prove to be more valuable and will remain sustainable over time. Also, given that the training datasets may be enriched in real-world data, the tools may be able to draw more reliable trends and make highly relevant predictions for the demographic populations they may serve.

The release of the European Union's AI Act (EU AI Act) has now provided a much-needed ethical framework as well as some metrics to ensure that all AI tools and systems adhere to ethical principles so that they can be deployed responsibly (https://artificialintelligenceact.eu/; last accessed on 08/01/24). The EU AIA came into effect today (08/01/24) and it would be interesting to see its impact on all domains, especially digital pathology.

As of May 13, 2024, the FDA has approved six AI/ML-enabled medical devices in the field of pathology (https://www.fda.gov/medical-devices/software-medical-device-samd/artificial-intelligence-and-machine-learning-aiml-enabled-medical-devices; last accessed on 08/01/24). These include Paige Prostate (Paige.AI), Tissue of Origin Test Kit-FFPE (Cancer Genetics, Inc. and Pathwork Diagnostics, Inc.), Pathwork Tissue Of Origin Test Kit-Ffpe (Pathwork Diagnostics), Pathwork Tissue of Origin Test (Pathwork Diagnostics Inc.), and PAPNET Testing System (PAPNET Testing System). Several associations and societies are tirelessly working to institute standards and better define processes to generate a blueprint to make FDA approval for AI/ML-enabled medical devices in the field of pathology easier.

The fact that we have not heard of any massive, unprecedented outcomes of using AI/ML in the pathology domain is a testament to the responsible approach the community has been taking by learning from the lessons from other AI tools that mis performed before in the healthcare domain to ensure that these new tools adhere to strict standards of ethics and responsible AI.

Conclusions

AI research and the ethical implications of AI have become areas of great interest across different scientific fields, including healthcare. In this chapter, we have highlighted some of the ethical issues that a researcher needs to consider while conducting AI research in pathology. These include factors such as race, gender, and ethnicity which play a key role in pathology AI research designs and outcomes. Multiple examples are described demonstrating that neglecting these factors leads to a downstream exacerbation of existing inequities in healthcare delivery. This is particularly important in pathology which serves as the "big data" repository of quantitative (and imaging) data measures of patient progress in the practice of medicine and has a central role in personalized patient care. We reviewed potential "risk" scenarios associated with the use of AI in the field of pathology. Even though AI pathology is in its infancy, it behooves the profession to be aware of the potential risks posed by AI workflows to the practice of pathology. By improving awareness of such risks, pathologists can help guide the careful development of these technologies to benefit patients while minimizing potential downsides. Finally, we discuss three key foundational principles of AI ethics for professional organizations to adapt to enable the development of AI within the field of pathology—Transparency, Accountability, and Governance. This framework merely represents a starting point for the development of ethical AI in pathology at an institutional and organizational level. As the impact of AI-enabled workflows in pathology continues over the next decade, more elements need to be added to the pathology AI ethics framework in accordance with the specific needs of the field of pathology. Pathologists have a critical role in enabling AI-based workflows in the laboratory and must have a seat at the table to guide the development and implementation of ethical AI and ADM within the practice of pathology. AI- and ADM-based workflows may create incredibly powerful new approaches to the practice of medicine. Pathologists must leverage this once-in-a-generation opportunity to be key drivers of this emerging paradigm shift within the practice of medicine.

Acknowledgment

We thank Dr. Stanley Cohen for critically reviewing select sections of this chapter.

Disclosures

None declared.

Funding

R.R.G. is supported by the Centers of Biomedical Research Excellence (CoBRE) grant number 1P20GM130422 from the National Institute of General Medical Sciences (NIGMS).

References

[1] Rajkomar A, Oren E, Chen K, Dai AM, Hajaj N, Hardt M, Liu PJ, Liu X, Marcus J, Sun M, Sundberg P, Yee H, Zhang K, Zhang Y, Flores G, Duggan GE, Irvine J, Le Q, Litsch K, Mossin A, Tansuwan J, Wang D, Wexler J, Wilson J, Ludwig D, Volchenboum SL, Chou K, Pearson M, Madabushi S, Shah NH, Butte AJ, Howell MD, Cui C, Corrado GS, Dean J. Scalable and accurate deep learning with electronic health records. NPJ Digit Med 2018;1:18.
[2] Rajkomar A, Dean J, Kohane I. Machine learning in medicine. N Engl J Med 2019;380:1347–58.
[3] Beam AL, Kohane IS. Big data and machine learning in health care. JAMA 2018;319:1317–8.

References 177

[4] Bukowski M, Farkas R, Beyan O, Moll L, Hahn H, Kiessling F, Schmitz-Rode T. Implementation of eHealth and AI integrated diagnostics with multidisciplinary digitized data: are we ready from an international perspective? Eur Radiol 2020;30:5510–24.

[5] Vayena E, Salathe M, Madoff LC, Brownstein JS. Ethical challenges of big data in public health. PLoS Comput Biol 2015;11:e1003904.

[6] Cohen IG, Amarasingham R, Shah A, Xie B, Lo B. The legal and ethical concerns that arise from using complex predictive analytics in health care. Health Aff (Millwood) 2014;33:1139–47.

[7] Madabhushi A, Lee G. Image analysis and machine learning in digital pathology: challenges and opportunities. Med Image Anal 2016;33:170–5.

[8] Niazi MKK, Parwani AV, Gurcan MN. Digital pathology and artificial intelligence. Lancet Oncol 2019;20:e253–61.

[9] Shen D, Wu G, Suk HI. Deep learning in medical image analysis. Annu Rev Biomed Eng 2017;19:221–48.

[10] Stenzinger A, Alber M, Allgauer M, Jurmeister P, Bockmayr M, Budczies J, et al. Artificial intelligence and pathology: from principles to practice and future applications in histomorphology and molecular profiling. Semin Cancer Biol 2022;84:129–43.

[11] Miotto R, Wang F, Wang S, Jiang X, Dudley JT. Deep learning for healthcare: review, opportunities and challenges. Brief Bioinform 2018;19:1236–46.

[12] Abels E, Pantanowitz L, Aeffner F, Zarella MD, van der Laak J, Bui MM, Vemuri VN, Parwani AV, Gibbs J, Agosto-Arroyo E, Beck AH, Kozlowski C. Computational pathology definitions, best practices, and recommendations for regulatory guidance: a white paper from the digital pathology association. J Pathol 2019;249:286–94.

[13] Komura D, Ishikawa S. Machine learning methods for histopathological image analysis. Comput Struct Biotechnol J 2018;16:34–42.

[14] Langlotz CP, Allen B, Erickson BJ, Kalpathy-Cramer J, Bigelow K, Cook TS, Flanders AE, Lungren MP, Mendelson DS, Rudie JD, Wang G, Kandarpa K. A roadmap for foundational research on artificial intelligence in medical imaging: from the 2018 NIH/RSNA/ACR/the academy workshop. Radiology 2019;291:781–91.

[15] Louis DN, Feldman M, Carter AB, Dighe AS, Pfeifer JD, Bry L, Almeida JS, Saltz J, Braun J, Tomaszewski JE, Gilbertson JR, Sinard JH, Gerber GK, Galli SJ, Golden JA, Becich MJ. Computational pathology: a path ahead. Arch Pathol Lab Med 2016;140:41–50.

[16] Min S, Lee B, Yoon S. Deep learning in bioinformatics. Brief Bioinform 2017;18:851–69.

[17] Diao JA, Wang JK, Chui WF, Mountain V, Gullapally SC, Srinivasan R, Mitchell RN, Glass B, Hoffman S, Rao SK, Maheshwari C, Lahiri A, Prakash A, McLoughlin R, Kerner JK, Resnick MB, Montalto MC, Khosla A, Wapinski IN, Beck AH, Elliott HL, Taylor-Weiner A. Human-interpretable image features derived from densely mapped cancer pathology slides predict diverse molecular phenotypes. Nat Commun 2021;12:1613.

[18] Djuric U, Zadeh G, Aldape K, Diamandis P. Precision histology: how deep learning is poised to revitalize histomorphology for personalized cancer care. NPJ Precis Oncol 2017;1:22.

[19] Castaneda C, Nalley K, Mannion C, Bhattacharyya P, Blake P, Pecora A, Goy A, Suh KS. Clinical decision support systems for improving diagnostic accuracy and achieving precision medicine. J Clin Bioinforma 2015;5:4.

178 CHAPTER 9 AI ethics in pathology

[20] Bhargava R, Madabhushi A. Emerging themes in image informatics and molecular analysis for digital pathology. Annu Rev Biomed Eng 2016;18:387–412.

[21] Ananny M. Toward an ethics of algorithms. Sci Technol Hum Values 2015;41:93–117.

[22] Boddington P, O'Sullivan B, Wooldridge M. Towards a code of ethics for artificial intelligence. Cham, Switzerland: Springer; 2017.

[23] Brey PAE. Anticipatory ethics for emerging technologies. NanoEthics 2012;6:1–13.

[24] Ienca M, Ferretti A, Hurst S, Puhan M, Lovis C, Vayena E. Considerations for ethics review of big data health research: a scoping review. PLoS One 2018;13: e0204937.

[25] Jackson BR, Ye Y, Crawford JM, Becich MJ, Roy S, Botkin JR, de Baca ME, Pantanowitz L. The ethics of artificial intelligence in pathology and laboratory medicine: principles and practice. Acad Pathol 2021;8. 2374289521990784.

[26] Fu Y, Jung AW, Torne RV, Gonzalez S, Vöhringer H, Shmatko A, Yates LR, Jimenez-Linan M, Moore L, Gerstung M. Pan-cancer computational histopathology reveals mutations, tumor composition and prognosis. Nat Can 2020;1:800–10.

[27] Kather JN, Heij LR, Grabsch HI, Loeffler C, Echle A, Muti HS, Krause J, Niehues JM, Sommer KAJ, Bankhead P, Kooreman LFS, Schulte JJ, Cipriani NA, Buelow RD, Boor P, Ortiz-Bruchle NN, Hanby AM, Speirs V, Kochanny S, Patnaik A, Srisuwananukorn A, Brenner H, Hoffmeister M, van den Brandt PA, Jager D, Trautwein C, Pearson AT, Luedde T. Pan-cancer image-based detection of clinically actionable genetic alterations. Nat Can 2020;1:789–99.

[28] Moore DA, Young CA, Morris HT, Oien KA, Lee JL, Jones JL, Salto-Tellez M. Time for change: a new training programme for morpho-molecular pathologists? J Clin Pathol 2018;71:285–90.

[29] Gianfrancesco MA, Tamang S, Yazdany J, Schmajuk G. Potential biases in machine learning algorithms using electronic health record data. JAMA Intern Med 2018;178:1544–7.

[30] Parikh RB, Teeple S, Navathe AS. Addressing bias in artificial intelligence in health care. JAMA 2019;322(24):2377–8. https://doi.org/10.1001/jama.2019.18058.

[31] Cohen S. Artificial intelligence and deep learning in pathology. Amsterdam: Elsevier; 2021 [chapter 5].

[32] Rajkomar A, Hardt M, Howell MD, Corrado G, Chin MH. Ensuring fairness in machine learning to advance health equity. Ann Intern Med 2018;169:866–72.

[33] Gannon M. Race is a social construct, scientists argue. Scientific American; 2016.

[34] Buolamwini J, Gebru T. Gender shades: intersectional accuracy disparities in commercial gender classification. In: Sorelle AF, Christo W, editors. Proceedings of machine learning research: PMLR. Proceedings of the 1st conference on fairness, accountability and transparency; 2018. p. 77–91.

[35] Obermeyer Z, Powers B, Vogeli C, Mullainathan S. Dissecting racial bias in an algorithm used to manage the health of populations. Science 2019;366:447–53.

[36] Pacilè S, Lopez J, Chone P, Bertinotti T, Grouin JM, Fillard P. Improving breast cancer detection accuracy of mammography with the concurrent use of an artificial intelligence tool. Radiol Artif Intell 2020;2, e190208.

[37] Allen TC. Regulating artificial intelligence for a successful pathology future. Arch Pathol Lab Med 2019;143:1175–9.

[38] Jackups Jr R. Deep learning makes its way to the clinical laboratory. Clin Chem 2017;63:1790–1.

[39] O'Neil C. Weapons of math destruction: how big data increases inequality and threatens democracy. Crown Publishing Group; 2016.

[40] Hallworth MJ. The '70% claim': what is the evidence base? Ann Clin Biochem 2011;48:487–8.

[41] Hoyt RE, Snider D, Thompson C, Mantravadi S. IBM Watson analytics: automating visualization, descriptive, and predictive statistics. JMIR Public Health Surveill 2016;2:e157.

[42] Miotto R, Li L, Kidd BA, Dudley JT. Deep patient: an unsupervised representation to predict the future of patients from the electronic health records. Sci Rep 2016;6:26094.

[43] Zarella MD, Bowman D, Aeffner F, Farahani N, Xthona A, Absar SF, Parwani A, Bui M, Hartman DJ. A practical guide to whole slide imaging: a white paper from the digital pathology association. Arch Pathol Lab Med 2019;143:222–34.

[44] Larson EJ. The myth of artificial intelligence : why computers can't think the way we do. Cambridge, Massachusetts: The Belknap Press of Harvard University Press; 2021.

[45] Mitchell M. Artificial intelligence : a guide for thinking humans. 1st ed. New York: Farrar, Straus and Giroux; 2019.

[46] Frey CB, Osborne MA. The future of employment: how susceptible are jobs to computerisation? Technol Forecast Soc Chang 2017;114:254–80.

[47] Dieleman JL, Campbell M, Chapin A, Eldrenkamp E, Fan VY, Haakenstad A, Kates J, Li Z, Matyasz T, Micah A, Reynolds A, Sadat N, Schneider MT, Sorensen R, Abbas KM, Abera SF, Kiadaliri AA, Ahmed MB, Alam K, Alizadeh-Navaei R, Alkerwi AA, Amini E, Ammar W, CAT A, Atey TM, Avila-Burgos L, Awasthi A, Barac A, Berheto TM, Beyene AS, Beyene TJ, Birungi C, Bizuayehu HM, NJK B, Cahuana-Hurtado L, Castro RE, Catalia-Lopez F, Dalal K, Dandona L, Dandona R, Dharmaratne SD, Dubey M, Faro A, Feigl AB, Fischer F, JRA F, Foigt N, Giref AZ, Gupta R, Hamidi S, Harb HL, Hay SI, Hendrie D, Horino M, Jürisson M, Jakovljevic MB, Javanbakht M, John D, Jonas JB, Karimi SM, Khang Y-H, Khubchandani J, Kim YJ, Kinge JM, Krohn KJ, Kumar GA, Leung R, HMA ER, MMA ER, Majeed A, Malekzadeh R, Malta DC, Meretoja A, Miller TR, Mirrakhimov EM, Mohammed S, Molla G, Nangia V, Olgiati S, Owolabi MO, Patel T, AJP C, Pereira DM, Perelman J, Polinder S, Rafay A, Rahimi-Movaghar V, Rai RK, Ram U, Ranabhat CL, Roba HS, Savic M, Sepanlou SG, BJT A, Tesema AG, Thomson AJ, Tobe-Gai R, Topor-Madry R, Undurraga EA, Vargas V, Vasankari T, Violante FS, Wijeratne T, Xu G, Yonemoto N, Younis MZ, Yu C, Zaidi Z, El Sayed Zaki M, CJL M. Future and potential spending on health 2015–40: development assistance for health, and government, prepaid private, and out-of-pocket health spending in 184 countries. Lancet 2017;389:2005–30.

[48] Botkin JR. Transparency and choice in learning healthcare systems. Learn Health Syst 2018;2:e10049.

[49] Diakopoulos N. Accountability in algorithmic decision making. Commun ACM 2016;59:56–62.

[50] Diakopoulos N. Transparency. In: Oxford handbook of ethics and AI, vol. 1; 2020. p. 197–213.

[51] Vayena E, Blasimme A, Cohen IG. Machine learning in medicine: addressing ethical challenges. PLoS Med 2018;15:e1002689.

[52] Kroll JA. Accountability in computer systems. In: The Oxford handbook of ethics of AI, vol. 1. Oxford Academic; 2020. p. 180–96. https://doi.org/10.1093/oxfordhb/9780190067397.013.10.

[53] McCradden MD, Joshi S, Anderson JA, Mazwi M, Goldenberg A, Zlotnik Shaul R. Patient safety and quality improvement: ethical principles for a regulatory approach to bias in healthcare machine learning. J Am Med Inform Assoc 2020;27:2024–7.

[54] Wieringa M. What to account for when accounting for algorithms. In: Proceedings of the 2020 conference on fairness, accountability, and transparency; 2020. p. 1–18.

[55] Vayena E, Tasioulas J. Adapting standards: ethical oversight of participant-led health research. PLoS Med 2013;10:e1001402.

Applications

PART II

CHAPTER

Applications of artificial intelligence for image enhancement in pathology

10

Tanishq Abraham[a], Austin Todd[b], Daniel A. Orringer[c], and Richard Levenson[d]

[a]*Department of Biomedical Engineering, University of California, Davis, CA, United States,* [b]*UT Health, San Antonio, TX, United States,* [c]*Associate Professor, Department of Surgery, NYU Langone Health, New York City, NY, United States,* [d]*Professor and Vice Chair, Pathology and Laboratory Medicine, UC Davis Health, Sacramento, CA, United States*

Introduction

This chapter will describe and discuss opportunity for artificial intelligence (AI) to contribute to image quality improvements and to simplify and accelerate imaging methods. There are two major goals for adding AI to imaging tasks. The first is direct image enhancement and conversion to different imaging modes (e.g., autofluorescence to simulated H&E) for direct visual "consumption." The second is the eventual deployment of AI-based classification and quantitative approaches. We will concentrate on the former and give one example of the latter, which demonstrates the real-world application of AI in intraoperative surgical guidance in brain tumor surgery. This approach, which has already entered the clinic, accomplishes not only improved patient care via accelerated, high-quality tumor-site imaging but also can reintegrate pathology into surgical decision-making and reinforce collaborative relationships. This will be explored in detail in the latter section of this chapter.

The basic underpinning of all the approaches to be discussed below is that there is information content present in images that is hard to discern with direct (human) visual observation, but that can be elicited using AI tools that have been trained to associate image type 1 with image type 2. This requires some degree of image content stereotyping, in that the relationships have to be established within a context of relatively limited spatial, textural, and biological variability. While transfer learning from "irrelevant" sources, such as images of cats and dogs, can prime a learning network to be sensitive to image primitives such as edges and textures, it will not do a good job predicting image content for histology specimens without specific and focused exemplars. To take this argument to the limit, the more random noise or extreme variation is present, the poorer AI predictive or conversion tools will perform. It is impossible to take an image consisting of pure noise, downsize it, and then predict its true high-resolution content via AI. Fortunately, histology exhibits

183

Artificial Intelligence in Pathology. https://doi.org/10.1016/B978-0-323-95359-7.00010-8
Copyright © 2025 Elsevier Inc. All rights are reserved, including those for text and data mining, AI training, and similar technologies.

a limited repertoire—only so many colors are anticipated, nuclei have a reasonable degree of self-similarity, the background is often mostly white or mostly black, and so on.

Various image conversion tasks have been identified as worth pursuing in histology, and these are described below.

Common machine learning tasks

In histopathology, there are several tasks that can be assisted through the application of deep learning technologies. Most of these problems fall into four classes: classification, segmentation, image translation, and image style transfer. Here, we will briefly define these tasks and explain how problems in histopathology can be formulated as common machine learning tasks. In all cases, we focus on image tasks, as this is most relevant to histopathology.

Classification

Image classification refers to the task of assigning a label to an image. Much of the field of histopathology is comprised of various classification tasks. This is because histopathology is mainly focused on assigning a diagnosis based on the review of slide-based microscopy. Automatic classification of tissue structures and subtypes can also be extremely useful to augment and improve the histopathology workflow.

As a central problem in computer vision and machine learning, the methodologies for solving classification problems have been broadly explored in both academic and well-funded commercial enterprises, and considerable progress has been made. The leading algorithms for image classification are convolutional neural networks (CNNs), which have demonstrated better-than-human performance on various benchmark datasets [1–6]. However, their real-world performance across novel institutions and differently curated collections remains to be determined. We will briefly cover an example application of deep learning–based classification in histology used for intraoperative surgical guidance in Deep learning for computationally aided diagnosis in histopathology section.

Segmentation

Image segmentation refers to the task of assigning labels to specific regions of an image. It can also be seen as a pixel-level classification task. Segmentation of subcellular structures, such as nuclei and cytoplasm and membrane compartments, can be useful for automating common tasks such as cell enumeration (via nuclei counting), determination of intracellular locations of molecular markers, and is important for the analysis of subcellular morphological features such as nuclear size, eccentricity, and chromatin texture [7]. Automatic segmentation can also be used to help pathologists recognize tissue components by delineating different tissue types.

Commonly used deep learning methodologies **185**

While segmentation in pathology is not a focus of this chapter, some image microscopy problems can be reformulated as segmentation problems. More importantly, the most common deep learning architecture for segmentation, U-nets, is often utilized for image enhancement tasks, as we will see later.

Image translation and style transfer

Image translation refers to the conversion of one image representation to another image representation. The goal is to learn the transformation between the input and output images. This can be applied to a wide variety of applications, such as converting between night and day images, winter and summer images, or more useful tasks like converting satellite images to maps. In computational pathology, image-to-image translation has been explored for stain color normalization (Section "Stain color normalization") and conversion between different image modalities (Section "Mode switching"). The most common models used for image translation tasks are generative adversarial networks or their variants (Section "Generative adversarial networks and their variants"). Most model architectures for image translation tasks require paired-image datasets. However, in some cases, obtaining paired datasets may be difficult or even impossible. For example, when converting between image modalities, it may be impossible to obtain pixel-scale paired images using different image modalities. Therefore, models like CycleGANs (Section "Generative adversarial networks and their variants") have been developed to solve the problem of unpaired image-to-image translation.

Another task related to that is style transfer [8,9]. Style transfer focuses on transferring the style of a single reference image to the content of the input image in order to generate an output image. Style transfer can be thought of as a single-sample version of image-to-image translation. Style transfer also has potential applications in computational pathology, such as for stain-color normalization, the task being to convert an input image to have the color gamut and structural color assignments that mimic a target, reference image. This is demonstrated in Ref. [10].

Commonly used deep learning methodologies
Convolutional neural networks

A CNN belongs to a class of deep learning algorithms most commonly used for image tasks such as image classification (Fig. 10.1A). The building blocks of a CNN are learned convolutional kernels [11,12]. These convolutional kernels are applied to the image and their subsequent representations in the neural network in order to extract learned features. The convolutional kernels are developed during the training process, and updated through backpropagation. A key advantage of convolutional layers is their shift invariance [11]. As the convolutional kernel is applied over the entire image, shifts in the input shift the output in an equivalent manner.

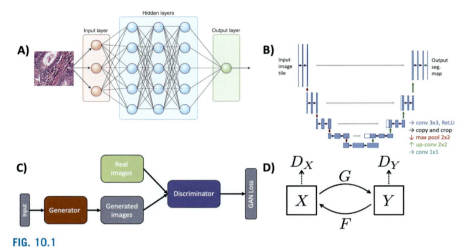

FIG. 10.1

(A) Convolutional neural networks take in images with an input layer, extract features with convolutional blocks in the hidden layers, and often output a label probability for classification tasks. (B) U-nets comprise a contracting feature-extraction path and an expanding up-sampling path, finally outputting a segmentation map. (C) Generative adversarial networks utilize both a generator and discriminator network. The generator network takes in an input and generates an image. The discriminator tries to classify between real and generated images. The discriminator tries to maximize correct classification in the form of a GAN loss, while generated images try to minimize it. (D) CycleGANs contain two sets of generators and discriminators to convert between two domains in both directions.

(A) Adapted with permission from Topol EJ. High-performance medicine: the convergence of human and artificial intelligence. Nat Med 2019;25(1):44–56. https://doi.org/10.1038/s41591-018-0300-7; Shaban MT, Baur C, Navab N, Albarqouni S. Staingan: stain style transfer for digital histological images. In: 2019 IEEE 16th international symposium on biomedical imaging (ISBI 2019); 2019. pp. 953–956. https://doi.org/10.1109/ISBI.2019.8759152; (B) Adapted with permission from Ronneberger O, Fischer P, Brox T. U-net: convolutional networks for biomedical image segmentation. In: Medical image computing and computer-assisted intervention–MICCAI 2015, Cham; 2015. pp. 234–241. https://doi.org/10.1007/978-3-319-24574-4_28; (D) Adapted with permission from Zhu JY, Park T, Isola P, Efros AA. Unpaired image-to-image translation using cycle-consistent adversarial networks. In: 2017 IEEE international conference on computer vision (ICCV), Venice; 2017. pp. 2242–2251. https://doi.org/10.1109/ICCV.2017.244.

This is an especially useful property for microscopy. The stacking of convolutional layers and other layers, such as fully connected layers and pooling layers, are used to build CNNs.

A variety of CNN architectures have been developed, but some network architectures are more commonly used due to their generalizability and effectiveness over a wide variety of tasks. Some of the most common architectures include VGG nets [4], ResNets [5], and Inception nets [6].

For many computer vision tasks, there is little data available for successful training of CNNs from scratch. However, the knowledge of models trained on larger datasets can be transferred for application to learning cycles applied to a smaller, focused dataset [13]. This takes advantage of the fact that many of the convolutional layers derived from large-dataset-derived models serve to extract general, low-level features. Therefore, only some of the layers of models that were pretrained on larger datasets are subsequently retrained on the smaller dataset. This process is known as transfer learning.

U-nets

A wide variety of algorithms and deep learning architectures [14,15] are used for image segmentation, but undoubtedly the most common neural network architecture for image segmentation is the U-net [16] (Fig. 10.1B). Ronneberger et al. originally introduced U-nets specifically for biomedical segmentation, but this approach has now been used across a variety of segmentation problems. This work builds upon existing fully convolutional architectures [14] by adding up sampling operations with many feature channels to the contracting subnetworks. As a result, there is an expansive path that is symmetric to the contracting path, leading to a U-shaped architecture. This model was tested successfully across various cell and more complex tissue segmentation tasks [16].

While U-nets were originally developed for segmentation, they have been used for other paired image tasks such as achieving computational super-resolution [17], and are commonly employed as components in some generative adversarial networks (GANs) (described below). With some 10,000 citations already reported in Google Scholar, the U-net architecture is clearly one of the most influential deep learning architectures developed to date.

Generative adversarial networks and their variants

A GAN is a deep learning method that was pioneered by Goodfellow et al. [18] and leverages multiple neural networks working "against" each other to transform images of different classes (Fig. 10.1C). The basic premise is that one neural network (the "generator") is trained to take images of one class and transform them into a second class, for example, taking zebras and turning them into horses. The second neural network (the "discriminator") is trained to discriminate real images of horses from fake images generated by the first network.

This is done cyclically until the images generated cannot be classified accurately as generated or real by the discriminator. The potential for this method has been continually explored and expanded since its original description in 2014; it has found many applications, such as in super-resolution [19] and artistic endeavors [20]. One of the early limitations of this general methodology was that the data had to be pixel-registered to facilitate training. In other words, the starting and target mode images needed to contain exactly the same scene. This is tractable in situations where a slide is imaged under one condition, and then reimaged perhaps after additional staining steps. In this case, the scenes being viewed intrinsically have pixel-registered content. This requirement makes the creation of large datasets for complex applications costly and time-consuming, if it is possible at all. CycleGAN [21] is a variation of the traditional method that aimed to correct this problem by adding cyclic consistency between the conversions, allowing the training to be completely unsupervised with spatially unpaired (but contextually paired) images (Fig. 10.1D). This is accomplished by doubling the number of neural networks and training the network to not only convert from, for example, zebras to horses but also from zebras to horses and then back to zebras. By forcing the reconstruction of the original image from the converted image, the training is almost pseudo-supervised, permitting unpaired image training. Applications of this class of deep learning tools (mostly) in histopathology are described in more detail below.

Common training and testing practices
Dataset preparation and preprocessing

Often, when applying deep learning to microscopy, ground truth and input images are collected separately. For example, images of the same sample may be taken on separate microscopes with different imaging modalities. In this case, the images need to be aligned to obtain a paired-image dataset. Many image coregistration approaches have been developed for accurate image coregistration [22]. Often, a global registration algorithm is used for rough coregistration of the image, while a subsequent local coregistration algorithm is used to account for local deformations. Often, accurate coregistration is essential for the performance of an algorithm on a paired dataset. Note that coregistration is not required for unpaired image translation tasks, however.

Not infrequently, datasets are quite small and only represent a small sample of the kinds of images to which a trained model could be applied. A neural network model generalizes well across a diverse set of samples if there is a large and diverse training set. If the dataset is small, the model might perform well on the training set but may not perform well on unseen data. This phenomenon is known as overfitting. In order to overcome this challenge, two methods are commonly used. Transfer learning, as discussed in the Convolutional neural networks section, can be used to decrease the amount of data required for training [13]. Additionally, data augmentation is a common approach for improving the utility of small datasets. This involves the creation of synthetic but representative examples for the dataset. More specifically, transformations like rotation, flipping, distorting, color variation, etc., are used to create new images and augment the original image dataset [23]. Data augmentation is therefore a useful technique for limiting overfitting.

Loss functions

In order to train a neural network, the weights (i.e., the trainable parameters of the network) of the neural network are adjusted to minimize a user-defined loss function that best represents the task that the neural network is required to solve. For classification tasks, the most commonly used loss function is a cross-entropy loss. When comparing predicted and ground-truth images in image-mapping tasks, the cost function is instead usually based on per-pixel differences, such as mean absolute error (L1 loss) or mean squared error (L2 loss). Other networks can also have specialized loss functions for training, such as GANs (adversarial loss) and CycleGANs (cyclic loss). Loss functions can also be augmented with regularization terms (e.g., weight decay, identity loss) in order to prevent overfitting. In many cases, careful selection of the loss function is critical in optimizing the performance of the neural network.

Metrics

Quantitative metrics are required in order to reliably evaluate the performance of a model. For classification tasks, the most common metrics includes accuracy and area under the curve of the receiver operating characteristic (AUC-ROC). For segmentation tasks, the most common metrics are intersection over union and Dice scores. However, for image translation and style transfer tasks, the most appropriate metric is often dependent on the application. For applications in microscopy, metrics such as structural similarity index (SSIM), peak signal-to-noise ratio (PSNR), and Pearson correlation coefficient are often used to compare the structural content between the predicted and ground truth images.

Deep learning for microscopy enhancement in histopathology
Stain color normalization

A major problem in the field of computational pathology and computer-aided diagnosis (CAD) is variability between individual stained whole-slide (WSI) microscopy images. Different scanners, staining procedures, and institutional environments can lead to wide variance in staining appearance. Image analysis methods find it difficult to cope with these variations in image appearance. In fact, a major contributor to CAD algorithm performance is proper deployment of a standardization algorithm [24]. Consequently, a significant body of work has been dedicated to methodologies intended to reduce variability between and within microscopy image datasets. Collectively, these algorithms are designated as stain color normalization methods (Fig. 10.2A).

One of the most popular algorithms is color deconvolution [25,26]. This algorithm attempts to determine a stain matrix, which describes how much each stain contributes to the final pixel color. However, these stain matrices do not generalize well to images from other labs and scanning platforms. Other common approaches use clustering-based algorithms. However, these require that clusters to be proportionately represented in the images. Both of these approaches are color-based and do not incorporate spatial information. However, recent algorithms, especially those that employ deep learning, have begun to utilize contextual information to improve normalization.

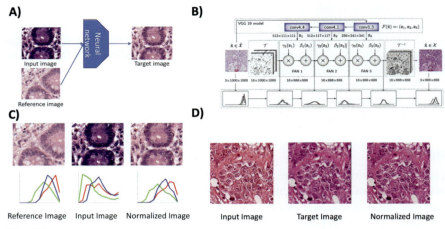

FIG. 10.2

(A) Stain color normalization is often formulated as trying to change the color of the input image to be similar to the reference image of desired H&E color. (B) An end-to-end neural network was developed with a two-path architecture. A feature extractor path utilizes a VGG19, while a transformer network utilizes feature-aware normalization units to transform the input image into a normalized image. (C) The StaNoSA algorithm successfully adjusted the color distribution of the input image to match the reference image, with accurate coloring of subcellular structures. (D) A GAN-based approach was used to successfully normalize H&E images and reduce interscanner variability. Additionally, using GAN-based normalization as preprocessing also improved classification results on a tumor dataset.

(A) Adapted with permission from Janowczyk A, Basavanhally A. Madabhushi A, Stain normalization using sparse AutoEncoders (StaNoSA): application to digital pathology. Comput Med Imag Graph 2017;57:50–61. https://doi.org/10.1016/j.compmedimag.2016.05.003; (B) Adapted from permission from Janowczyk A, Basavanhally A, Madabhushi A. Stain normalization using sparse AutoEncoders (StaNoSA): application to digital pathology. Comput Med Imag Graph 2017;57:50–61. https://doi.org/10.1016/j.compmedimag.2016.05.003; (C) Adapted with permission from Bug D, et al. Context-based normalization of histological stains using deep convolutional features. In: Deep learning in medical image analysis and multimodal learning for clinical decision support, Cham; 2017. pp. 135–142. https://doi.org/10.1007/978-3-319-67558-9_16; (D) Adapted with permission from Shaban MT, Baur C, Navab N, Albarqouni S. Staingan: stain style transfer for digital histological images. In: 2019 IEEE 16th international symposium on biomedical imaging (ISBI 2019); 2019, pp. 953–956. https://doi.org/10.1109/ISBI.2019.8759152.

Here, we will highlight a few studies that incorporate deep learning into new stain normalization algorithms. For a more thorough review on both traditional and deep learning–based stain color normalization algorithms, please refer to Ref. [27, 28].

In Janowczyk et al. [25], an unsupervised deep learning approach termed Stain Normalization using Sparse AutoEncoders (StaNoSA) was developed to overcome some of the challenges with traditional stain color normalization methods (Fig. 10.2B). A sparse autoencoder is trained on the dataset, and the embeddings of the template were clustered with K-means clustering and the input image embeddings were assigned to the K clusters. Then, a traditional histogram matching algorithm was applied per cluster. Three datasets were used to evaluate the StaNoSA algorithm. A breast biopsy slide dataset was used to study how StaNoSA could decrease intrascanner and interscanner variation. A gastrointestinal biopsy slide dataset was used to study how StaNoSA could decrease variability between sites, which was simulated with different staining protocols. The third dataset is a subset of the second with nuclei annotations to test how StaNoSA could improve object detection in pathology. StaNoSA was also compared with four state-of-the-art techniques. StaNoSA was demonstrated to bring the interscanner error to a similar range of the intrascanner error. It also decreased variability between staining protocols with better performance than other tested techniques. Finally, StaNoSA had the best Dice score for nuclei segmentation using a simple thresholding method.

While the previous study utilized a neural network to improve normalization, a neural network architecture can also be used for end-to-end normalization processes, as done in Bug et al. [29]. Their network utilized two parallel network paths: a main transformation path and a feature extractor path (Fig. 10.2C). The feature extractor path is based on a pretrained VGG-19 architecture. The transformation path is composed of novel network blocks termed Feature Aware Normalization (FAN) units, which are inspired by batch normalization and long short-term memory units (LSTMs). The features from each block in the feature extractor network were passed into the FAN unit to gradually rescale and shift the image (similar to batchnorm) through addition and multiplication gates (similar to LSTM). As there are no paired data from H&E images, noise is deliberately added to the reference image. The noise model is constructed using normal distribution data with variances defined based on principal component analysis. To train and test the model, the staining protocol was varied by either increasing or decreasing hematoxylin concentration, eosin concentration, and slice thickness. The changes in color deviation were measured with the sum of squared differences between the color histogram (SSDH). Influence on texture was measured with the structural dissimilarity index (SDSIM). The magnitude of perceivable colors (color volume) was measured by the product of the standard deviations of the LAB color-space channels. In terms of SSDH and color volume, the proposed method significantly outperformed previous approaches. When assessed using alternative metrics, other methods performed equally well. However, these other methods required similar tissue distributions in the input and reference images, and therefore also required careful selection of a reference image. The proposed method overcomes these challenges, while providing an end-to-end algorithm to perform stain normalization, with the caveat that the introduction of contextual information in the conversion model did introduce some undesirable textural changes.

Generative models have also been applied for histopathological stain color normalization. Zanjani et al. [30] introduce fully unsupervised generative models for stain-color normalization. The GAN model includes a generator network, a discriminatory network, and an auxiliary network. The images are converted to CIELAB color space, and the lightness channel is passed into the generator, and only the chromatic channels are generated by the GAN. The auxiliary network is based on the InfoGAN architecture [31] in order to maximize the information present in the generated image.

The variational autoencoder (VAE) model [32] is inputted with the H&E image in the HSD color space, with the encoder being trained to segregate the input image into tissue channels. The parameters of the color distribution are computed per tissue class. The decoder network reconstructs the chromatic channels of the input image. The deep convolutional Gaussian mixture model (DCGMM) used by these authors was a fully convolutional network jointly optimized with the parameters of a Gaussian mixture model (GMM). The DCGMM fits the GMM to the pixel-color distribution while being conditioned on the tissue classes. The detection of the tissue classes is done by exploiting CNN embeddings in a similar fashion to the output of the encoder of a VAE. The algorithms were evaluated with a normalized median intensity (NMI) measure to assess color constancy of the normalized images. Out of the three models, and compared with previous state-of-the-art results, the DCGMM had highest performance, with the lowest standard deviation and coefficient of variation. While these models do not make assumptions about H&E image content, they do rely on other assumptions such as various hard-coded thresholds and nuclei shape fitting that may not be applicable to other tissue types.

The stain color normalization problem can also be reformulated as an image translation task. For example, in Shaban et al. [33], a CycleGAN was trained to reduce interscanner variability (Fig. 10.2D). The stain normalization task was reformulated as image translation between images from two different scanners. The CycleGAN, termed StainGAN in this study, outperformed other methods in terms of SSIM, FSSIM, Pearson correlation coefficient, and PSNR. StainGAN was also applied as a preprocessing step for cancer classification on the Camelyon16 dataset. The StainGAN improved the AUC-ROC of classification when compared with no normalization and alternative normalization preprocessing steps. The use of a GAN also eliminates human error that can arise when an expert is called upon to manually select a reference slide for normalization. While this application was tasked to decrease interscanner variability, similar models could also be used for interlaboratory variability.

Tellez et al. [34] also developed a network-based stain normalization algorithm by formulating it as an image translation problem. The authors trained a U-net architecture to reconstruct the original images from inputted augmented images. The model was trained on a multiorgan, multisite dataset. They also performed a systematic multicenter evaluation of several other stain color augmentation and normalization algorithms. A total of four classification tasks spanning different centers were used to determine the effect of data augmentation and stain normalization on the performance of these tasks. These tasks are mitotic figure detection, tumor metastasis detection, prostate epithelium detection, and colorectal cancer tissue type detection. Basic, morphology-based, brightness and contrast, HSV-based, and H&E-to-DAB (HED)-based augmentation methods were tested. In terms of stain normalization, starting points of no transformation, grayscale transformation, and a chromatic distribution–based technique were tested. The augmentation and stain color normalization methods were used as preprocessing for a simple patch CNN classifier. A CNN classifier was trained for each combination of organ, color normalization, and data augmentation method studied. The performance of the classifier was evaluated with the AUC-ROC metric. A global ranking of the methods based on their AUC-ROC score for each task and organ was established. Based on these rankings, it was determined that stain color augmentation was crucial for the performance of the CNN classifier. The novel deep learning–based stain color normalization also improved the results. Importantly and somewhat surprisingly, using the deep learning normalization method did not alleviate the need for color augmentation. Utilizing stain color normalization alone is insufficient to achieve high classification accuracy.

Mode switching

Traditionally, microscopy with H&E (or related) staining or fluorescent labeling requires time-consuming and potentially variable labeling procedures. In addition, staining or labeling can typically only be used for ex vivo applications. Another major disadvantage is the damage to the sample that staining and/or labeling can induce. Additionally, the act of imaging itself can degrade the specimen via forms of photodamage.

Many novel microscopy techniques have been developed in order to overcome these challenges. However, the traditional H&E-stained histopathology images are most accessible to histopathologists. Additionally, the images obtained with these new techniques may be harder to interpret and susceptible to a variety of artifacts, making these techniques more difficult for histopathologists to adopt. Therefore, converting novel modalities to resemble H&E-stained slides (or vice versa) may allow for a wider adoption of these modalities. For example, Fig. 10.3A demonstrates how quantitative phase imaging is an imaging modality that is hard for the uninitiated to interpret, but still can be introduced in a histopathologist's workflow by utilizing neural networks to convert these images to H&E-like results. We will describe several reports on how deep learning techniques are used for such mode of conversion. Most of these reports utilize GAN architectures and their variants, discussed in Section "Image translation and style transfer".

Quantitative phase imaging (QPI) is a relatively recent label-free imaging technique with lower phototoxicity that can be deployed on various platforms (including portable platforms). In order to bridge the gap between QPI and standard histopathology, Rivenson et al. [35] used a GAN to convert phase-based microscopy images to histological images (Fig. 10.3A–B). Three models were trained for H&E for skin tissue, Jones' stain for kidney tissue, and Masson's trichrome for liver tissue. The models were then applied to similar images that were not present in the training set. They utilize a custom global coregistration method and elastic image registration for local feature registration. A U-net architecture is used for the generator, with H&E images converted to YCbCr color space passed as the input. With fast inference time, and by removing the need for staining, this technology has the potential to reduce labor and time costs. While lens-free holographic microscopy was used, this algorithmic approach could be applied to other QPI techniques.

In Ref. [36], the authors sought to convert confocal microscopy images to virtually stained H&E images for easier interpretation by pathologists and surgeons.

194 CHAPTER 10 Applications of artificial intelligence

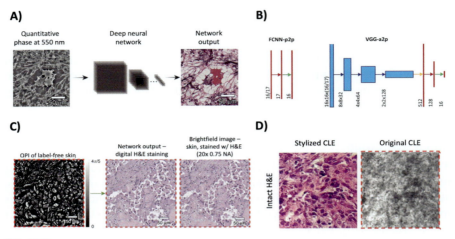

FIG. 10.3

(A) Often, novel image modalities can be difficult to interpret. However, deep neural networks can be used to convert these images to H&E-like images for better interpretation. (B) A fully convolutional neural network and pixel-based VGG classification were tested for converting multiphoton microscopy images to H&E-stained images. (C) Quantitative phase imaging is an example of a difficult-to-interpret imaging modality that has other benefits over H&E images, such as lower phototoxicity and portability. A GAN was trained to convert the QPI images to digital H&E stained images. (D) Confocal laser endomicroscopy allows for in vivo real-time analysis of tissue but is difficult to interpret. Instead of using traditional image translation algorithms like GANs, neural style transfer was used to digitally stain CLE images. GAN, generative adversarial network; QPI, quantitative phase imaging; CLE, confocal laser endomicroscopy.

(A) Adapted with permission from Rivenson Y, Liu T., Wei Z., Zhang Y., de Haan K, Ozcan A. PhaseStain: the digital staining of label-free quantitative phase microscopy images using deep learning. Light Sci Appl 2019; 8(1):1–11. https://doi.org/10.1038/s41377-019-0129-y (B) Adapted with permission from Rivenson Y, Liu T, Wei Z, Zhang Y, de Haan K, Ozcan A. PhaseStain: the digital staining of label-free quantitative phase microscopy images using deep learning. Light Sci Appl 2019;8(1):1–11. https://doi.org/10.1038/s41377-019-0129-y (C) Adapted with permission from Borhani N, Bower AJ, Boppart SA, Psaltis D. Digital staining through the application of deep neural networks to multi-modal multi-photon microscopy. Biomed Opt Express 2019;10 (3):1339–1350.https://doi.org/10.1364/BOE.10.001339; (D) Adapted with permission from Izadyyazdanabadi M, et al., Fluorescence image histology pattern transformation using image style transfer. Front Oncol 2019;9. https://doi.org/10.3389/fonc.2019.00519

Confocal microscopy can be operated in both reflectance (RCM) and fluorescence (FCM) modes. Both types of confocal microscopy can be used to digitally generate images resembling H&E. First, a fully convolutional network is used to despeckle the RCM images. This network is trained with artificially contaminated single-channel H&E images, as appropriate despeckled RCM images are not available. The RCM and FCM images are linearly combined to form an RGB image, by assigning RCM images an eosin-like color, and FCM images a hematoxylin-like color. This naively converted image is then passed to the CycleGAN to improve the overall mode conversion. Importantly, it was demonstrated that despeckling is essential for functional mode switching. Without it, the CycleGAN generated unwanted artifacts.

Another attractive microscopy modality with high potential applicability is multiphoton microscopy (MPM). MPM uses multiple near-infrared photons that arrive nearly simultaneously at excitable molecules to induce fluorescence, with a potential for less photodamage and deeper tissue observation than is provided by both one-photon excitation fluorescence, and even tissue staining. In addition, since there are many endogenous fluorophores, it also allows for in vivo tissue imaging, which is not typically possible with tissue staining. However, the high-dimensional data produced by MPM is hard for histopathologists to interpret.

Therefore, Borhani et al. [37] demonstrate techniques for deploying deep learning for mapping label-free MPM to H&E-stained brightfield images. First, a modal coregistration method applies an affine transformation to the H&E image to align the MPM and H&E images. This is done at a global and local level. Color reduction to 16 colors by k-means clustering was performed to decrease the complexity of the required modal transfer DNN architectures. Two models were developed: a pixel-to-pixel classification model and an area-to-pixel classification model (Fig. 10.3C). The pixel-to-pixel model is a two-layer, fully connected classifier, which maps one pixel from MPM to its respective H&E pixel. The area-to-pixel classification model uses a VGG network to map an image tile from MPM to the respective central pixel for H&E. These models performed well when trained and applied to rat liver samples. The area-to-pixel models performed better than pixel-to-pixel models as they also consider spatial and structural information. Coregistration was also demonstrated to be important for accurate results. When the liver-trained model was applied to mouse ovary and rat mammary tumor tissues, the model performed poorly. This indicates that this model can only be applied to datasets similar to the source dataset and must be retrained for other tissue types. Nevertheless, this novel approach that does not require GAN architectures could increase accessibility for MPM microscopy.

In certain settings, staining procedures can result in irreversible modifications or damage; this is usually regarded as disadvantageous as it can preclude further analysis of the tissue. However, this is particularly true in the special case of sperm cell selection for in vitro fertilization (IVF) due to the potential cytotoxicity of staining. Interferometric phase microscopy (IPM) is currently used for sperm evaluation, as it is compatible with available medical microscopes which are modified with an additional wavefront sensor to allow for holographic microscopy. This combination provides quantitative topological maps of the cells without the need for stains. However, morphological evaluation of cells is most effectively accomplished using chemical staining of cell organelles. A method, "HoloStain," was therefore developed to convert quantitative phase images and phase gradient images, which are extracted from the digital holograms, to virtually stain brightfield images using DCGANs [38]. For each cell, an ensemble of four images was used to train the model: a stain-free quantitative phase image, two stain-free phase gradient images, and the target-stained brightfield image. An advantage of HoloStain is that cells that were out of focus could be virtually stained and refocused. This is possible since digital reconstruction of the holograms can be done so that unfocused cells can be brought into focus. In order to demonstrate the usefulness of the algorithm, sperm cells were classified by an embryologist using data from brightfield images (no staining), stain-free quantitative phase images, one of the stain-free gradient images, and HoloStain images. The HoloStain method had the highest F1 score compared with the alternative three methods.

While most research studies employ GANs, it is also possible to use image-style transfer algorithms as an alternative approach, as demonstrated by Izadyyazdanabadi et al. [10]. In particular, the authors sought to convert confocal laser endomicroscopy (CLE) images to H&E. CLE is a real-time, in vivo interoperative imaging technique with high potential to assist neurosurgeons during surgery. The authors developed a transformation tool that can remove common CLE artifacts, add histological patterns, and avoid removing critical features or adding unreal features to images (Fig. 10.3D). Unfortunately, due to problems in exact colocalization and intrinsic tissue movements, it is difficult to obtain spatially registered H&E-CLE image pairs. Instead, the authors used an image-style transfer algorithm using a pretrained VGG19 architecture and a content and style loss. One advantage of style transfer is that it only requires a single image of the desired style (H&E). Five neurosurgeons independently judged the stylized CLE images as having a significantly higher diagnostic quality than the original CLE images.

The potential for these methods is quite high, as well as being quite close in terms of implementation, with multiple groups working on clinical trials. The quality of images acquired from these methods greatly exceeds the current gold standard for real-time pathology, frozen sections, which introduce artifacts that can render evaluation challenging. By converting images from fluorescence color scheme to one immediately familiar to the pathologist, the odds of adoption increase dramatically. These imaging modalities have the potential to become a new gold standard of real-time pathology in the near future; hence the widespread interest in accurate color mapping.

In silico labeling

Deep learning has supported the development of methods that can extract information from other image modalities in order to reconstruct brightfield stained images. However, can molecularly stained images, highlighting fluorescently labeled cellular structures be reconstructed from stain-free brightfield images (Fig. 10.4A)? Recent research has suggested these kinds of transformations are indeed possible. We refer to such work as in-silico labeling, as inspired by Ref. [39].

Rana et al. [40] demonstrated how it is possible to use cGANs to computationally stain images of unstained paraffin-embedded WSIs. Both staining and destaining models were trained using traditional GAN and L1 loss functions as well as utilizing an additional regularization term based on Pearson's correlation coefficient in order to prevent tiling noise. The models could be used for further downstream tasks such as tumor diagnosis.

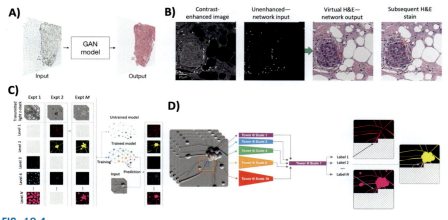

FIG. 10.4

(A) New research has demonstrated that it is possible to convert unstained images to digitally stained images using deep learning. (B) A GAN-based model was successfully applied to convert unstained autofluorescence images to H&E virtually stained images. (C) A multitask model was trained to convert unlabeled transmitted-light images to fluorescently labeled images. GAN, generative adversarial network.

(A) Adapted with permission from Rana A, Yauney G, Lowe A, Shah P. Computational histological staining anddestaining of prostate core biopsy RGB images with generative adversarial neural networks, In: 2018 17th IEEE international conference on machine learning and applications (ICMLA), Orlando, FL; 2018. pp. 828–834. https://doi.org/10.1109/ICMLA.2018.00133 (B) Adapted with permission from Rivenson Y, et al., Virtual histological staining of unlabelled tissue-autofluorescence images via deep learning. Nat Biomed Eng 2019;3(6):466–477. https://doi.org/10.1038/s41551-019-0362-y; (C) Adapted with permission from Christiansen EM, et al., In silico labeling: predicting fluorescent labels in unlabeled images. Cell 2018;173 (3):792–803. https://doi.org/10.1016/j.cell.2018.03.040.

Rivenson et al. [41] instead utilized autofluorescence generated by exciting unstained histology sections using a DAPI excitation/emission filter set to reconstruct stained images using GANs. They trained on salivary gland tissue stained with H&E, kidney tissue stained with Jones stain, and liver and lung stained with Masson's trichrome. The GAN model performed very well converting the images, with considerable detail generated and accurate histopathological features captured in the virtual images (Fig. 10.4B). The virtually stained images were also evaluated by pathologists, and no significant difference between the virtually stained and regular brightfield images was observed. This indicates that the virtually stained images can provide similar diagnostic utility. The variation in the colorization of the virtually stained images is lower than that of the histologically stained images. Therefore, intrinsic staining standardization is an interesting positive by-product of this approach. The authors also demonstrated transfer learning by using it to rapidly train the model for application on thyroid tissue sections. This technology has the potential to be applied to non–slide-based methods, which would allow for impressive label-free imaging without the need to retrain pathologists. The added ability to generate

198 CHAPTER 10 Applications of artificial intelligence

special stains increases the applicability, as many of these stains require hours of skilled technician time and expensive reagents.

While histochemically stained images can give useful information about the cellular and extracellular components, fluorescent labels can give even more detailed information about desired subcellular structures. Hence, there is tremendous interest in obtaining fluorescently labeled images in cell biology and pathology. Ounkomol et al. [42] developed a U-net architecture to convert 3D transmitting light and 2D electron microscopy (EM) to 3D fluorescence microscopy and 2D immunofluorescence (IF) microscopy. Importantly, the predictions for the different structure models can be combined into a 3D multichannel fluorescent image prediction from a single 3D TL image. Another application of image modality conversion is to facilitate automatic registration between images. EM and IF images of ultrathin brain sections are commonly collected as part of an array tomography protocol. However, as they are collected from two different microscopes, they are not inherently aligned, and manual registration is tedious and time-consuming. The authors trained a model on manually registered EM and immunofluorescence images labeled for myelin basic protein (MBP-IF). The MBP-IF prediction, and therefore the original EM image, was registered with the target MBP-IF image using traditional intensity-matching techniques. While EM is not routinely used for clinical diagnosis outside of renal biopsies, converting images to immunofluorescence stains has great utility as these stains are often expensive, cumbersome, and take hours to complete. While in its early stages with further validation and interest, this technology could be very useful for pathologists.

In Christiansen et al. [39], a more universal and flexible architecture was developed to address a similar problem. A multitask, multiscale deep neural network was trained on paired sets of fluorescently labeled 2D images and transmitted light z-stack images (Fig. 10.4C). These z-stack images include images from various focal planes, with the hypothesis that even out-of-focus planes contain information useful for accurately predicting the fluorescent labels. The images were obtained from various labs, samples, and even imaging modalities. This allows for the training of a generalizable network through multitask learning. When tasks are similar in multitask learning, common features can be learned to improve the performance of the model. The training set included human motor neurons derived from induced pluripotent stem cells, primary murine cortical cultures, and breast cancer cell lines. The fluorescent labels used were Hoechst and DAPI for labeling nuclei, propidium iodide (PI) for assessing cell viability, and CellMask for labeling plasma membrane. Some cells were also immunolabeled with antibodies against TuJ1 (labeling for neurons), the Islet1 protein (labeling for motor neurons), and for MAP2 (labeling dendrites). The deep neural network had multiscale input with five paths for each of these scales (Fig. 10.4D). The scale with the finest detail stays at the same scale, while the other scales have contracting and upsampling convolution operations, inspired by a U-net architecture. The final network outputs a discrete probability distribution over 256 intensity values (all possible 8-bit pixel values) for all the pixels in the image. It reads z-stacks of either brightfield, phase-contrast, or differential interference contrast images.

The network was tested for various tasks. It was first used to predict Hoechst or DAPI fluorescence in order to label cell nuclei. Qualitatively, the in silico labels and the true fluorescent labels looked identical. The Pearson correlation coefficient between the true and predicted pixel intensities was 0.87, indicating the model accurately predicted the nuclear regions. Similar experiments were done to predict propidium iodide (PI) fluorescence, which is used to detect dead cells (since live cells can exclude the dye). The network also performed well-predicting immunolabels for TuJ1 protein, which labels neurons. The neural network was also able to successfully predict independent axon and dendrite labels, which was unexpected as it was unclear whether the transmitted-light images contained enough information to be able to distinguish whether a neurite extending from a cell was an axon or dendrite. In order to demonstrate the flexibility of the model, transfer learning was used to learn how to predict a different label from a different cell type from data from a single well. Specifically, transfer learning was used to predict CellMask (cell membrane label) fluorescence from differential interference contrast images from cancer cells. The model was trained from data from a single training well but was regularized by simultaneously training on previous conditions. The network had a Pearson correlation score of 0.95. This suggests this model could continue to learn and be adapted for other tasks, when provided with more training examples.

While fluorescent microscopy is not commonly used in routine clinical pathology applications, deep learning techniques used for fluorescence microscopy could also be applied in histopathology. In particular, Christiansen et al. [39] suggested a way to solve a key problem in deep learning for mode conversion. Most mode conversion methods are limited by constraints imposed by differences in magnification, staining, modality, tissue types, and other factors. However, Christansen et al. used a multitask and multiscale network that was generalizable to various tissue types, magnification scales, and even various modalities. Using transfer learning, these universal models could also be further trained for more specific tasks with very additional little data. The development of similar models for conversion to H&E images could greatly facilitate the adoption of novel imaging modalities for use in histopathology.

Super-resolution, extended depth-of-field, and denoising

In order to generate high-quality WSIs, it has been necessary to use high-numerical-aperture lenses that can capture high-resolution images. The downside is that these lenses have restricted fields of view, which greatly increases the scanning time and shallow depths of field (Fig. 10.5A). While used increasingly for improving the resolution of standard (macro) photography images [19], the application of deep learning methods to increase the resolution of microscope images is quite recent. Wang et al. [43] used a training set consisting of pixel-matched images taken with a $10\times/0.4$ NA lens paired with images taken with a $20\times/0.75$ NA lens. The use of paired data forces the GAN to concentrate on improving the resolution of the image and confers high-quality transformations with little artifacts or errors (Fig. 10.5B). The benefit is a field of view that is four times larger, while allowing for a similar

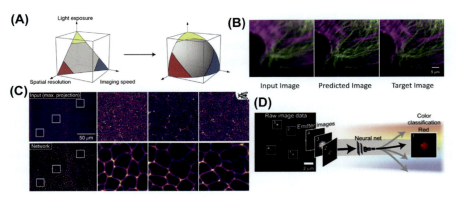

FIG. 10.5

(A) In traditional microscopy, there are key trade-offs between exposure, resolution, and speed. However, application of deep learning can augment this design space and diminish these trade-offs. (B) A GAN-based approach was successfully applied to convert diffraction-limited fluorescence images to super-resolved images. (C) A U-net architecture was successfully utilized for image denoising of fluorescent microscopy images. (D) Even with grayscale cameras, deep learning can be used to determine the color of emitters, as the point spread function is determined by the wavelength. GAN, generative adversarial network.

(A) Adapted with permission from Weigert M, et al. Content-aware image restoration: pushing the limits of fluorescence microscopy. Nat Methods 2018;15(12):1090–1097. https://doi.org/10.1038/s41592-018-0216-7; (B) Adapted with permission from Wang H, et al. Deep learning enables cross-modality super-resolution in fluorescence microscopy. Nat Methods 2019;16(1):103–110. https://doi.org/10.1038/s41592-018-0239-0; (C) Adapted with permission from Weigert M, et al. Content-aware image restoration: pushing the limits of fluorescence microscopy. Nat Methods 2018;15(12):1090–1097. https://doi.org/10.1038/s41592-018-0216-7; (D) Adapted with permission from Hershko E, Weiss LE, Michaeli T, Shechtman Y. Multicolor localization microscopy and point-spread-function engineering by deep learning. Opt Express 2019;27(5):6158–6183. https://doi.org/10.1364/OE.27.006158.

reduction in scanning time. This has major implications not only for speeding up scanning but also for allowing larger fields of view in time-lapse studies, allowing for more data to be collected from a single experiment. Another benefit is that the depth of field of the 10× lens is greater, so the focus and sharpness of the generated images can actually be better than the ground truth images themselves, meaning that they are not only equivalent but superior in terms of quality. This is not limited to brightfield as the group has also published similar deep learning algorithms to convert brightfield images from a 40× lens to those better than a 100× oil immersion lens even though the latter was used as the target for the transformation [44]. All the same benefits apply from the previous fluorescent paper with the added benefit that a vast majority of current clinical pathology is done with brightfield imaging. Taking this a step further the group showed it was possible to achieve image quality similar to that of a 20× microscope lens with a smartphone lens, with the added benefit of increased depth of field achieved by the lower NA lens. The algorithm here was not only improving the spatial resolution, but was able to correct the color distortions

Deep learning for microscopy enhancement in histopathology **201**

from the low-quality lens/camera to that of similar quality from a benchtop microscope [45]. This has even further reaching potential as incredibly low-cost hardware could generate much higher quality diagnostic images, allowing for better care in resource-poor nations.

An example of the use of deep learning to increase depth of field is an implementation of a CNN to allow for rapid automated diagnosis of blood films for malaria [46]. Blood films are thick samples, and the interpretation of scanned images requires the collection of multiple focal planes merged in a way that can be interpreted by either a human pathologist or an artificial neural network (ANN). Manescu et al. [46] trained a network with a three-plane input that generated an image close to the quality obtained with fourteen different focus plane images. It convincingly outperformed traditional extended depth-of-field algorithms, which could greatly increase the speed at which automated scanning and diagnosis of malaria could be made.

Increasing depth of field, however, is not always beneficial; there are many methods that allow for the reduction of depth of field to allow for axially confined resolution. Such techniques include structured-illumination microscopy (SIM), confocal microscopy, and computational deconvolution. These methods allow for the acquisition of high-resolution images from thick samples without physically sectioning tissue. Zhang et al. [47] implemented a CNN to reduce the optical depth of widefield fluorescence images without having to employ the optical methods just listed. The largest advantage is the reduced complexity of the microscope and the reduction of artifacts. Stripe artifacts that can be seen in structured illumination images are common and can distract from the content of the image; deep learning can help generate similarly optically thinned images while retaining comparable quality. Application of a traditional deconvolution method also eliminated stripe artifacts but the resulting image appeared is significantly thicker and lacked the sharpness of the AI-generated counterpart. While not currently in clinical use, light sheet microscopy for pathology has been explored in Ref. [48] and proved especially useful in the evaluation of prostate biopsies. The ability to capture 3D datasets allowed a more accurate assessment of Gleason scoring (a prognostic factor for prostate tumors), which could in turn lead to a better quality of care for patients.

Beyond increasing resolution or affecting depth of field, another contribution of AI that can markedly improve image quality is the reduction of noise. Images with low signals can be difficult to interpret; this can make in vivo imaging particularly challenging since the data typically must be acquired quickly and with low (safe) light levels. Paired data (low and high noise versions) for developing noise reduction models are relatively easy to acquire or synthesize, meaning that large and representative datasets can be generated quickly. For example, Weigert et al. [49] successfully demonstrated the use of a convolutional neural net trained on multiphoton images using very low laser power and low exposure times as input with the target output being the high-quality images taken with normal exposure times and laser powers (Fig. 10.5C). While not immediately relevant to pathology as it is practiced today, this concept could be applied to all of the previous fluorescent methods to help speed them up. If any of these techniques are to replace the frozen section, they must either offer better-quality histology or be faster. In addition, these techniques could be used to improve label-free imaging methods, possibly allowing for in vivo microscopic inspection of tumor margins allowing for even better resection margins than obtained via frozen section, as there is no tissue lost due to sectioning or embedding.

This general approach can also be applied to standard super-resolution techniques, such as photoactivated localization microscopy (PALM) or stochastic optical reconstruction microscopy (STORM). Both methods rely on the acquisition of thousands of images to reconstruct structure below the diffraction limit of light microscopy. By training an ANN, Ouyang et al. [50] decreased the number of images required for a PALM reconstruction from 60,000 to a mere 500, representing a difference in acquisition time from 10 min to 9 s. While not currently applicable to pathology, there are some applications of electron microscopy that may be able to be replaced by techniques such as STORM/PALM; and as the field evolves and our understanding of disease processes increases, there may be a place for these techniques in clinical pathology.

Increased acquisition time requirements arise from a desire to maximize signal-to-noise ratios; many fluorescence image sessions require serial imaging with different sets of filters in order to separate multiple labels. The result is a large increase in imaging times and light exposure to the sample, with attendant potential for phototoxicity or photobleaching. By training a neural net on filter-acquired images, Hershko et al. [51] have shown that it is possible to use the chromatic aberration differences related to the use of different wavelengths of light to separate multiple labels from a single image (Fig. 10.5D). The implication of this method is much the same as described previously, namely, the potential for more rapid collection of data, and a reduction in imaging time.

Deep learning for computationally aided diagnosis in histopathology

In the section above, we described the possibility of using AI tools for improving the quality and information salience of images via AI-assisted tools. However, such tools are still regarded with some skepticism on the regulatory side, as they are seen as premature for deployment in high-impact clinical settings. The number of approved AI tools—with or without AI-enhanced images—for pathology is still very small, and for intraoperative guidance, even smaller. The notion is that for direct-to-AI-interpretation, the need for an intermediate step of AI enhancement may be diminished, since extraction of information and reduction of the impact of noise and distortion is implicitly performed by the classifier itself. An example from the neuropathology domain is given below, emphasizing as well the opportunity to combine AI tools with novel rapid histology technology.

A rationale for AI-assisted imaging and interpretation

Serving the US population that grew 8% from 2007 to 2017, the pathology workforce contracted 17% during the same time period [52]. Within certain subspecialties, workforce shortages are even more severe. Approximately 800 neuropathologists cover the 1400 centers performing brain tumor surgery [53]. Worsening deficits in neuropathology are expected with a vacancy rate of approximately 42% in neuropathology fellowships [54]. In neuropathology, the workforce shortage is felt in a number of ways. Although academic centers tend to be well staffed with board-certified neuropathologists, community hospitals rarely have one on staff. Consequently, nonexpert pathologists are compelled to provide services that could better be provided by a more specialized practitioner. Although hard to quantify, lack of neuropathology expertise probably leads to delays in intraoperative diagnosis, unnecessary surgical intervention (where a lymphoma or germinoma is misdiagnosed as a tumor where surgery might be required), and a limitation on how pathology can be used to formulate surgical strategy.

Histologic data have been recognized as central to decision-making in brain tumor surgery since at least 1930 [55]. There are numerous decisions that must be made by surgeons at the time of an operation on a brain lesion. First and foremost, surgeons rely on pathologists to ensure the tissue they have biopsied is indeed lesional and can be used to establish a diagnosis. Beyond that, differentiating neoplastic from nonneoplastic processes in the brain is essential, as is differentiating surgical and nonsurgical tumors. With respect to the surgical candidates, differentiating between primary and metastatic tumors, low- and high-grade gliomas, circumscribed and diffuse tumors all have specific implications for decision-making.

While the importance of histologic data in brain tumor surgery is now well established, in modern surgical practice, a chasm remains between the operating room and neuropathology. In most centers, the frozen section lab is off-site or a substantial distance from the operating table, creating a logistic barrier for surgeons interested in viewing slides with the pathologists. Moreover, the process for creating high-quality pathology slides is time- and labor-intensive. Consequently, pathologists will call into an OR with a diagnosis 30–45 min after a specimen is collected. The inefficiency of this process isolates surgeons from histologic data and limits the way it can be used to inform surgical decision-making.

Given the challenges in the modern neuropathology workflow, AI holds great promise in several regards: (1) verifying that lesional tissue has been collected, (2) assisting in tissue diagnosis, (3) helping to define surgical endpoints in tumors without gross boundaries, and (4) in the future, predicting actionable and diagnostic molecular markers. In fixed tissue, AI has been applied to the classification of epithelial vs stromal regions in breast and colon cancer [56], differentiating necrotic and nonnecrotic regions [57], and differentiating high- and low-grade gliomas [58]. That said, the application of AI for assisting in diagnosis only addresses the challenges associated with establishing a diagnosis in neuropathology. With conventional histologic workflows, many of the barriers related to tissue transport and processing

204 **CHAPTER 10** Applications of artificial intelligence

remain. To fully leverage the opportunity created by rapid advances in AI-assisted pathologic diagnosis, marriage with a technique for rapid histology is essential.

Approaches to rapid histology interpretations

To this end, several methods for slide-free histology have been proposed and validated to varying degrees. MUSE, stimulated Raman histology (SRH), and light-sheet microscopy are among the most promising tools for imaging unprocessed or minimally processed biological specimens (Fig. 10.6).

While numerous techniques for slide-free histology have been proposed, only SRH has been incorporated into an AI-based diagnostic pipeline validated for brain tumor diagnosis in a clinical setting. Hollon et al. adapted the Inception-V2 CNN to diagnose the 10 most common brain tumors using SRH images and deployed their software on a bedside SRH imager. In a head-to-head clinical trial of 278 patients, Hollon et al. demonstrated that an autonomous diagnostic pathway utilizing SRH and a CNN could predict diagnosis with performance that was noninferior to board-certified neuropathologists interpreting conventional histologic slides in the same patients [59]. The accuracy of both arms was approximately 94% and interestingly the errors in both arms were entirely nonoverlapping. The authors of this study therefore conclude that the SRH + CNN pathway can be used as a stand-alone method for brain tumor diagnosis when pathology resources are unavailable and as a method to improve diagnostic accuracy in neuropathology, even in well-staffed centers. Although the authors limited trial inclusion to common types of brain tumors, they also offered a means of differentiating common and unusual histologic classes in an effort to make their system universal to application in all central nervous system neoplasms. The algorithm was also used to develop a semantic segmentation tool capable of depicting the areas of a histologic image most likely to contain lesional tissue (Fig. 10.7). The authors also conclude that given the features that their CNN used to classify brain tumors, similar algorithms could be developed for tumors of other organs in the future.

It is interesting to speculate whether AI-based image enhancement prior to the classification learning step would have improved the performance of the latter.

Future prospects

The entire field of histopathology is experiencing rapid developments across all domains. Opportunities and challenges include:

(a) The wider deployment of conventional whole-slide scanning capabilities that can finally enable the entry of slide interpretation into the digital realm.
(b) The use of AI to enhance the performance of slide-scanning by interposing color normalization, denoising, sharpening, and depth-of-field enhancements with specialized AI tools prior to classification steps.

Future prospects **205**

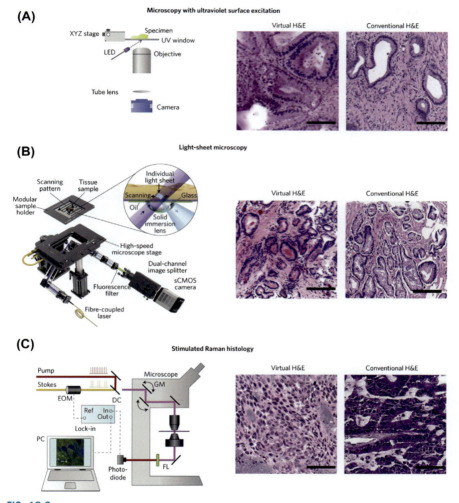

FIG. 10.6

Methods for slide-free histology including (A) microscopy with ultraviolet surface excitation, (B) light sheet microscopy, and (C) stimulated Raman histology (SRH), all of which can be performed on minimally processed or unprocessed tissue to generate diagnostic histologic images without the need for the resources of a conventional pathology lab. Each technique relies on optical, rather than physical sectioning of imaged tissue, simplifying the workflow required to generate a diagnostic histologic image. SRH has unique value as a modality for bedside histology as no staining, clearing, or handling of the tissue is required to generate a histological image. In SRH, fresh specimens can be placed into an imager at the bedside, which rapidly creates a virtual H&E microscopic image through relatively simple spectral band combination and recoloring. To this end, academic–industrial partnership has resulted in the first FDA-registered imager, now employed in several operating rooms across the United States.

Adapted with permission from Orringer DA, Camelo-Piragua S. Fast and slide-free imaging. Nat Biomed Eng, 2017;1(12):926–928. https://doi.org/10.1038/s41551-017-0172-z.

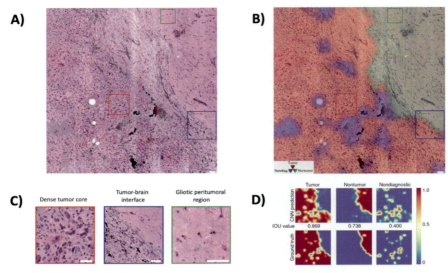

FIG. 10.7

Semantic segmentation. (A) An SRH mosaic image of a specimen from a glioblastoma patient.(B) Tiles from the mosaic are fed into a semantic segmentation algorithm and classified by a convolutional neural network (CNN) which then assigns a numerical value, depicted by color (*red, green*, or *blue*) corresponding to the likelihood that a given region contains tumor, nondiagnostic tissue, or nonlesional tissue, as illustrated in (C). The entire mosaic can be predicted as a heatmap for each of the 10 categories, 3 of which are shown here (D). SRH, stimulated Raman histology; CNN, convolutional neural networks (CNN).

Adapted with permission from Hollon TC, et al. Near real-time intraoperative brain tumor diagnosis using stimulated Raman histology and deep neural networks. Nat Med 2020;26(1):52–58. https://doi.org/10.1038/s41591-019-0715-9.

(c) The use of AI to dramatically simplify optomechanical requirements for scanners by correcting for lower raw scan performance with computational tools, thus decreasing instrumentation costs, scan times, and inadequate scan performance issues.
(d) Combining novel slide-free imaging hardware and AI to maximize both information content and similarities with conventional H&E-stained slide appearance.
(e) Getting pathologists, surgeons, administrators, the FDA, and payors comfortable with the above.

These are exciting times, as pathology finally seems poised to take advantage of notable advances in microscopy and computation.

Acknowledgement

The authors would like to acknowledge the diligent and skilled assistance of Nathan Anderson in the preparation of the revised version of this chapter.

References

[1] Russakovsky O, et al. ImageNet large scale visual recognition challenge. Int J Comput Vision 2015;115(3):211–52. https://doi.org/10.1007/s11263-015-0816-y.

[2] Cireşan D, Meier U, Masci J, Schmidhuber J. Multi-column deep neural network for traffic sign classification. Neural Netw 2012;32:333–8. https://doi.org/10.1016/j.neunet.2012.02.023.

[3] Krizhevsky A, Sutskever I, Hinton GE. ImageNet classification with deep convolutional neural networks. In: Pereira F, Burges CJC, Bottou L, Weinberger KQ, editors. Advances in neural information processing systems 25. Curran Associates, Inc; 2012. p. 1097–105.

[4] Simonyan K, Zisserman A. Very deep convolutional networks for large-scale image recognition. ArXiv14091556 Cs; 2015.

[5] He K, Zhang X, Ren S, Sun J. Deep residual learning for image recognition. In: 2016 IEEE conference on computer vision and pattern recognition (CVPR); 2016. p. 770–8. https://doi.org/10.1109/CVPR.2016.90.

[6] Szegedy C, et al. Going deeper with convolutions. In: 2015 IEEE conference on computer vision and pattern recognition (CVPR); 2015. p. 1–9. https://doi.org/10.1109/CVPR.2015.7298594.

[7] Gurcan MN, Boucheron LE, Can A, Madabhushi A, Rajpoot NM, Yener B. Histopathological image analysis: a review. IEEE Rev Biomed Eng 2009;2:147–71. https://doi.org/10.1109/RBME.2009.2034865.

[8] Gatys LA, Ecker AS, Bethge M. A neural algorithm of artistic style. ArXiv150806576 Cs Q-Bio; 2015.

[9] Gatys LA, Ecker AS, Bethge M. Image style transfer using convolutional neural networks. In: 2016 IEEE conference on computer vision and pattern recognition (CVPR), Las Vegas, NV, USA; 2016. p. 2414–23. https://doi.org/10.1109/CVPR.2016.265.

[10] Izadyyazdanabadi M, et al. Fluorescence image histology pattern transformation using image style transfer. Front Oncol 2019;9. https://doi.org/10.3389/fonc.2019.00519.

[11] Goodfellow I, Bengio Y, Courville A. Deep learning. The MIT Press; 2016.

[12] LeCun Y, Bengio Y, Hinton G. Deep learning. Nature 2015;521(7553):436–44. https://doi.org/10.1038/nature14539.

[13] Yosinski J, Clune J, Bengio Y, Lipson H. How transferable are features in deep neural networks? In: Ghahramani Z, Welling M, Cortes C, Lawrence ND, Weinberger KQ, editors. Advances in neural information processing systems 27. Curran Associates, Inc; 2014. p. 3320–8.

[14] Long J, Shelhamer E, Darrell T. Fully convolutional networks for semantic segmentation. In: 2015 IEEE conference on computer vision and pattern recognition (CVPR); 2015. p. 3431–40. https://doi.org/10.1109/CVPR.2015.7298965.

[15] He K, Gkioxari G, Dollár P, Mask GR. R-CNN. In: 2017 IEEE international conference on computer vision (ICCV); 2017. p. 2980–8. https://doi.org/10.1109/ICCV.2017.322.

[16] Ronneberger O, Fischer P, Brox T. U-net: convolutional networks for biomedical image segmentation. In: Medical image computing and computer-assisted intervention – MICCAI 2015; 2015. p. 234–41. https://doi.org/10.1007/978-3-319-24574-4_28. cham.

[17] Fang L, et al. Deep learning-based point-scanning super-resolution imaging. BioRxiv 2019. preprint.

[18] Goodfellow I, et al. Generative adversarial nets. In: Ghahramani Z, Welling M, Cortes C, Lawrence ND, Weinberger KQ, editors. Advances in neural information processing systems 27. Curran Associates: Inc; 2014. p. 2672–80.

[19] Ledig C, et al. Photo-realistic single image super-resolution using a generative adversarial network. In: 2017 IEEE conference on computer vision and pattern recognition (CVPR); 2017. p. 105–14. https://doi.org/10.1109/CVPR.2017.19.

[20] Radford A, Metz L, Chintala S. Unsupervised representation learning with deep convolutional generative adversarial networks. ArXiv151106434 Cs; 2016.

[21] Zhu J-Y, Park T, Isola P, Efros AA. Unpaired image-to-image translation using cycle-consistent adversarial networks. In: 2017 IEEE international conference on computer vision (ICCV), Venice; 2017. p. 2242–51. https://doi.org/10.1109/ICCV.2017.244.

[22] de Haan K, Rivenson Y, Wu Y, Ozcan A. Deep-learning-based image reconstruction and enhancement in optical microscopy. Proc IEEE 2020;108(1):30–50. https://doi.org/10.1109/JPROC.2019.2949575.

[23] Perez L, Wang J. The effectiveness of data augmentation in image classification using deep learning. ArXiv171204621 Cs; 2017.

[24] Ciompi F, et al. The importance of stain normalization in colorectal tissue classification with convolutional networks. In: 2017 IEEE 14th international symposium on biomedical imaging (ISBI 2017); 2017. p. 160–3. https://doi.org/10.1109/ISBI.2017.7950492.

[25] Janowczyk A, Basavanhally A, Madabhushi A. Stain normalization using sparse Auto-Encoders (StaNoSA): application to digital pathology. Comput Med Imag Graph 2017;57:50–61. https://doi.org/10.1016/j.compmedimag.2016.05.003.

[26] Ruifrok AC, Johnston DA. Quantification of histochemical staining by color deconvolution. Anal Quant Cytol Histol 2001;23(4):291–9.

[27] Azevedo Tosta TA, de Faria PR, Neves LA, do Nascimento MZ. Computational normalization of H&E-stained histological images: progress, challenges and future potential. Artif Intell Med 2019;95:118–32. https://doi.org/10.1016/j.artmed.2018.10.004.

[28] Roy S, Kumar Jain A, Lal S, Kini J. A study about color normalization methods for histopathology images. Micron 2018;114:42–61. https://doi.org/10.1016/j.micron.2018.07.005.

[29] Bug D, et al. Context-based normalization of histological stains using deep convolutional features. In: Deep learning in medical image analysis and multimodal learning for clinical decision support; 2017. p. 135–42. Cham https://doi.org/10.1007/978-3-319-67558-9_16.

[30] Zanjani FG, Zinger S, Bejnordi BE. Histopathology stain-color normalization using generative neural networks. p. 16.

[31] Chen X, Duan Y, Houthooft R, Schulman J, Sutskever I, Abbeel P. InfoGAN: interpretable representation learning by information maximizing generative adversarial nets. In: Lee DD, Sugiyama M, Luxburg UV, Guyon I, Garnett R, editors. Advances in neural information processing systems 29. Curran Associates, Inc.; 2016. p. 2172–80.

[32] Kingma DP, Welling M. Auto-encoding variational bayes. ArXiv13126114 Cs Stat; 2014.

[33] Shaban MT, Baur C, Navab N, Albarqouni S. Staingan: stain style transfer for digital histological images. In: 2019 IEEE 16th international symposium on biomedical imaging (ISBI 2019); 2019. p. 953–6. https://doi.org/10.1109/ISBI.2019.8759152.

[34] Tellez D, et al. Quantifying the effects of data augmentation and stain color normalization in convolutional neural networks for computational pathology. Med Image Anal 2019;58, 101544. https://doi.org/10.1016/j.media.2019.101544.

[35] Rivenson Y, Liu T, Wei Z, Zhang Y, de Haan K, Ozcan A. PhaseStain: the digital staining of label-free quantitative phase microscopy images using deep learning. Light Sci Appl 2019;8(1):1–11. https://doi.org/10.1038/s41377-019-0129-y.

[36] Combalia M, Perez-Anker J, García-Herrera A. Digitally stained confocal microscopy through deep learning. In: Proceedings of the 2nd international conference on medical imaging with deep learning (PMLR); 2019. p. 121–9. 102.

[37] Borhani N, Bower AJ, Boppart SA, Psaltis D. Digital staining through the application of deep neural networks to multi-modal multi-photon microscopy. Biomed Opt Express 2019;10(3):1339–50. https://doi.org/10.1364/BOE.10.001339.

[38] Nygate YN, et al. HoloStain: holographic virtual staining of individual biological cells. p. 20.

[39] Christiansen EM, et al. In silico labeling: predicting fluorescent labels in unlabeled images. Cell 2018;173(3):792–803. https://doi.org/10.1016/j.cell.2018.03.040.e19.

[40] Rana A., Yauney G., Lowe A., Shah P.. Computational histological staining and destaining of prostate core biopsy RGB images with generative adversarial neural networks. In: 2018 17th IEEE international conference on machine learning and applications (ICMLA), Orlando, FL. 2018. p. 828–834 https://doi.org/10.1109/ICMLA.2018.00133.

[41] Rivenson Y, et al. Virtual histological staining of unlabelled tissue-autofluorescence images via deep learning. Nat Biomed Eng 2019;3(6):466–77. https://doi.org/10.1038/s41551-019-0362-y.

[42] Ounkomol C, Seshamani S, Maleckar MM, Collman F, Johnson GR. Label-free prediction of three-dimensional fluorescence images from transmitted-light microscopy. Nat Methods 2018;15(11):917–20. https://doi.org/10.1038/s41592-018-0111-2.

[43] Wang H, et al. Deep learning enables cross-modality super-resolution in fluorescence microscopy. Nat Methods 2019;16(1):103–10. https://doi.org/10.1038/s41592-018-0239-0.

[44] Rivenson Y, Göröcs Z, Günaydin H, Zhang Y, Wang H, Ozcan A. Deep learning microscopy. Optica 2017;4(11):1437–43. https://doi.org/10.1364/OPTICA.4.001437.

[45] Rivenson Y, et al. Deep learning enhanced mobile-phone microscopy. ACS Photonics 2018;5(6):2354–64. https://doi.org/10.1021/acsphotonics.8b00146.

[46] Manescu P, et al. Deep learning enhanced extended depth-of-field for thick blood-film malaria high-throughput microscopy. ArXiv190607496 Cs Eess; 2019.

[47] Zhang X, et al. Deep learning optical-sectioning method. Opt Express 2018;26 (23):30762–72. https://doi.org/10.1364/OE.26.030762.

[48] Reder NP, Glaser AK, McCarty EF, Chen Y, True LD, Liu JTC. Open-top light-sheet microscopy image atlas of prostate core needle biopsies. Arch Pathol Lab Med 2019;143(9):1069–75. https://doi.org/10.5858/arpa.2018-0466-OA.

[49] Weigert M, et al. Content-aware image restoration: pushing the limits of fluorescence microscopy. Nat Methods 2018;15(12):1090–7. https://doi.org/10.1038/s41592-018-0216-7.

[50] Ouyang W, Aristov A, Lelek M, Hao X, Zimmer C. Deep learning massively accelerates super-resolution localization microscopy. Nat Biotechnol 2018;36(5):460–8. https://doi.org/10.1038/nbt.4106.

[51] Hershko E, Weiss LE, Michaeli T, Shechtman Y. Multicolor localization microscopy and point-spread-function engineering by deep learning. Opt Express 2019;27(5): 6158–83. https://doi.org/10.1364/OE.27.006158.

[52] Metter DM, Colgan TJ, Leung ST, Timmons CF, Park JY. Trends in the US and Canadian pathologist workforces from 2007 to 2017. JAMA Netw Open 2019;2(5), e194337. https://doi.org/10.1001/jamanetworkopen.2019.4337.

[53] Orringer DA, et al. Rapid intraoperative histology of unprocessed surgical specimens via fibre-laser-based stimulated Raman scattering microscopy. Nat Biomed Eng 2017;1 (2):1–13. https://doi.org/10.1038/s41551-016-0027.

[54] Robboy SJ, et al. Pathologist workforce in the United States: I. Development of a predictive model to examine factors influencing supply. Arch Pathol Lab Med 2013;137 (12):1723–32. https://doi.org/10.5858/arpa.2013-0200-OA.

[55] Eisenhardt L, Cushing H. Diagnosis of intracranial tumors by supravital technique. Am J Pathol 1930;6(5):541–52. .7.

[56] Xu J, Luo X, Wang G, Gilmore H, Madabhushi A. A Deep convolutional neural network for segmenting and classifying epithelial and stromal regions in histopathological images. Neurocomputing 2016;191:214–23. https://doi.org/10.1016/j.neucom.2016.01.034.

[57] Sharma H, Zerbe N, Klempert I, Hellwich O, Hufnagl P. Deep convolutional neural networks for automatic classification of gastric carcinoma using whole slide images in digital histopathology. Comput Med Imaging Graph 2017;61:2–13. https://doi.org/10.1016/j.compmedimag.2017.06.001.

[58] Barker J, Hoogi A, Depeursinge A, Rubin DL. Automated classification of brain tumor type in whole-slide digital pathology images using local representative tiles. Med Image Anal 2016;30:60–71. https://doi.org/10.1016/j.media.2015.12.002.

[59] Hollon TC, et al. Near real-time intraoperative brain tumor diagnosis using stimulated Raman histology and deep neural networks. Nat Med 2020;26(1):52–8. https://doi.org/10.1038/s41591-019-0715-9.

CHAPTER 11

Foundation models and information retrieval in digital pathology

H.R. Tizhoosh

Kimia Lab, Mayo Clinic, Rochester, MN, United States

Introduction

The surge in the adoption of digital pathology has the potential to revolutionize medical diagnosis by allowing computerized analysis of tissue images [1–3]. Central to this technology is the digitization of formalin-fixed, paraffin-embedded (FFPE) tissue sections mounted on glass slides. This process converts physical tissue samples into high-resolution, gigapixel digital images called *whole slide images* (WSIs) [4,5]. These WSI files contain detailed patterns of tissue morphology, enabling the application of computer-vision algorithms in diagnostic pathology. Pathologists can now analyze tissue images seamlessly on computer screens at various magnifications [6]. This shift from light microscopes to digital displays allows for easier visual inspection of anatomic clues that may indicate specific diseases. Additionally, and in contrast to the physical sharing of glass slides, digital pathology facilitates fast and efficient sharing of tissue images for consultation and annotation, as well as quantitative analysis [7,8].

In recent years, the advent of deep learning and steady and rapid progress in artificial intelligence (AI) have been incessantly pushing forward the research in digital pathology. The availability of a myriad of AI models, among others, has enabled the application of many sophisticated operations in computational pathology, operations such as tissue segmentation, tumor detection, and other assistive actions for diagnosis [9–11]. In addition, digital pathology is offering new frameworks for consultations (i.e., securing second opinions) for complex and rare cases, and facilitating next-generation telemedicine [12,13]. Although in many cases, AI models are still unable to deliver acceptable diagnostic accuracy and efficiency for clinical utility, the perspectives of enhanced collaborations, efficient computer-aided analysis, and the development of a large set of quantitative methods are depicting a rather optimistic view of the future in histopathology [14]. However, as one of the major obstacles, the management of sheer volume and complexity of WSI data present a daunting challenge to the digital pathology community. Efficiently assembling

Artificial Intelligence in Pathology. https://doi.org/10.1016/B978-0-323-95359-7.00011-X
Copyright © 2025 Elsevier Inc. All rights reserved, including those for text and data mining, AI training, and similar technologies.

CHAPTER 11 Foundation models and information retrieval

well-organized (small or large) repositories of tissue images for information access is a critical task for the future of computational pathology [15–17]; information retrieval is experiencing a renaissance in light of all these developments.

Information retrieval

For decades, information retrieval (IR) has played a crucial role in structuring, identifying, and facilitating user-friendly access to information within complex clinical datasets [18,19]. IR systems are designed to efficiently match user queries (predominantly in the form of "text") with massive amounts of data to locate and retrieve the desired information [20,21]. The most fundamental and evident application of IR is searching in large archives of medical literature. Platforms like PubMed and Google Scholar are essential tools for researchers, clinicians, and even patients seeking information in research papers, clinical trials, and publications [22–24]. These platforms leverage IR technologies to identify relevant articles based on keywords, medical terminology, and other search criteria.

Perhaps the most common usage of IR systems involves searching Electronic Health Records (EHRs). IR functionalities allow us to efficiently search for and retrieve patient information from large hospital archives [25,26]. Timely and accurate diagnosis, effective and individualized treatment planning, and productive and timely research, all require access to patients' medical history and clinical data [27]. Search and retrieval, therefore, connect the more general category of decision support systems, a set of software tools to provide clinicians with relevant patient information at the point of care. By locating patient data in the context of relevant literature, identifying personalized treatment options, and predicting patient outcomes, IR can be a major contributor for paving the way for individualized medicine [21,28]. Finding and retrieving patients' unique characteristics in their medical history can then be used to customize the treatment plan. In context of research, drug discovery and development is another key application of IR systems. By searching vast and heterogeneous repositories of chemical compounds and biological information, search and retrieval can identify potential drug candidates and most promising development trajectories, accelerating the complex drug discovery process [29,30].

While the significant role of present and future IR systems in medical advancements is rather obvious, the implementation of sophisticated IR platforms has been experiencing formidable roadblocks. The rapid growth of multimodal patient data makes veracity, accuracy, and quality of big data a daunting challenge. Additionally, investigating user-friendly interfaces for complex IR queries, as an academically less alluring task, has not received enough attention. In the realm of digital pathology, with its emphasis on whole slide imaging, and with the existence of multiple H&E and IHC whole slide images for each patient, *image search* as a specific form of information retrieval has recently gained more traction among researchers.

Image search

Human vision is central to information gathering, processing, and analysis in most adults. This is especially true in medicine, where radiology and pathology images serve as vital diagnostic tools [31,32]. In pathology, the gold standard for diagnosing many diseases, particularly cancers (neoplasia), relies heavily on the visual inspection of tissue samples, whether digitally on a computer screen or analogously through the eyepiece of a light microscope [33,34]. Image search, as a form of visual information retrieval, offers a powerful new approach by bridging the gap between words, molecular data, and the visual patterns observed in tissue samples [16]. This technology, sometimes referred to as "reverse image search," allows pathologists to use IR systems and submit a *"tissue image as a query"* (and via typing words) and retrieve tissue samples with similar morphological features. This specific type of image search in an atlas (an indexed archive) is called Content-Based Image Retrieval (CBIR) (see Fig. 11.1), and it holds significant potential for various applications in histology, histopathology, and cytopathology [35–38].

Due to their gigapixel size, complex nature, and lack of labeled data, processing WSIs for search and retrieval purposes requires a *divide-and-conquer* approach (Fig. 11.2). This strategy allows for the construction of an "atlas" for a specific anatomical site (lung, breast, prostate, skin, etc.) and corresponding diagnoses (Fig. 11.3).

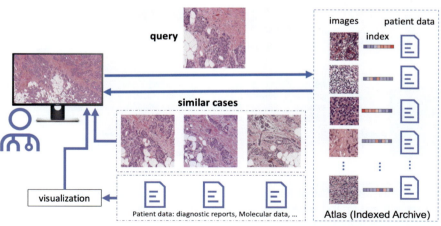

FIG. 11.1

CBIR in digital pathology. Pathologists can submit a query (a "patch" or a WSI) to an atlas (indexed archive of tissue images, patches, or WSIs). The atlas generally contains patient data or links to other archives to request metadata and multimodal patient data. Similar cases can be retrieved after search for query tissue and sent back to the pathologist's workstation for visualization.

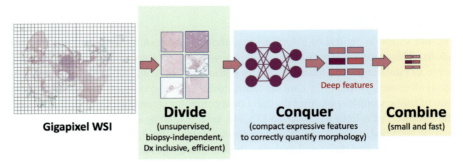

FIG. 11.2

Processing WSIs with their gigapixel dimensionality requires meticulous design of the "divide & conquer" concept. The divide, splitting the WSI into a small number of patches, has to be unsupervised, biopsy-independent, diagnostically inclusive, and efficient. The conquer is mostly performed by feature extraction from a properly designed deep network. A last stage of combine may use some notion of compression or encoding for fast processing and lean storage of the features [39].

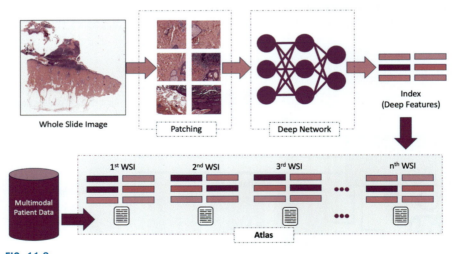

FIG. 11.3

An "atlas" for CBIR in histopathology can be constructed by indexing WSIs and storing indexes with links to other patient data. It is paramount that the index is compact and does not need excessive storage overhead.

The word "atlas" has been in use for centuries and generally means "*a collection of maps or charts, usually bound together… In addition to maps and charts, atlases often contain pictures, tabular data, facts about areas, and indexes of placenames keyed to coordinates of latitude and longitude or to a locational grid with numbers and letters along the sides of maps*" (Encyclopaedia

Britannica). In the context of information retrieval, the word atlas can be understood as follows:

> An **atlas** for a specific disease is a structured and indexed collection of patient data, well-curated to represent the spectrum of the disease diversity. To each patient entry in the atlas an outcome may be attached including but not limited to primary diagnosis, and successful treatment. The outcome is ideally free from variability. The indexing is a computerized process to create and use the atlas mainly consisting of patient data representations that can be easily and efficiently searched and matched to locate semantically, biologically, anatomically, clinically and genetically correct pattern similarities.

By facilitating the analysis and management of high-volume histological images through deep features for indexing (i.e., patient representation) and atlas creation, CBIR can significantly assist pathologists in their daily workflow. One application of CBIR in digital pathology involves quality assurance. For instance, CBIR can help identify potential discrepancies between a patient's tissue image and similar cases within the archive. More broadly, CBIR contributes to disease diagnosis by enabling the comparison of a query image with well-characterized cases in an atlas of WSIs with known diagnoses [17,40]. This retrieval of visually similar WSIs with confirmed diagnoses empowers pathologists with evidence-based information, allowing them to identify disease patterns with greater confidence [16]. The same atlas-building principles can be applied to develop evidence-based platforms for advanced triaging and treatment planning tools.

In recent years, several CBIR solutions have been proposed for histopathology. Among others, hashing-based image retrieval [41] (short HBIR), visual dictionary (or bag of visual words, BoVW, for instance [42]), SMILY [43], Yottixel [40], SISH [44,45], and RetCCL [46]. Recent examples also include HSDH (histopathology Siamese deep hashing) [47] and High-Order Correlation-Guided Self-Supervised Hashing-Encoding Retrieval (HSHR) [48].

Table 11.1 summarizes various CBIR methods for histopathology. While some, like BoVW, Yottixel, and SISH, function as complete search engines, others focus on specific aspects of information retrieval. For example, HBIR, SMILY, and HSHR primarily deal with image patches rather than entire WSIs. Additionally, HSDH and RetCCL are centered on specific deep learning models they propose. A critical challenge in histopathology search is dividing WSIs into manageable subimages (i.e., tiles or patches) for efficient and inclusive retrieval, independent of the original biopsy [39]. This key step (see Fig. 11.2) is often overlooked by CBIR solutions by prioritizing speed or accuracy over storage. SISH exemplifies this approach, boasting constant search time but demanding excessive storage requirements. In contrast, Yottixel's patching method, which creates a "mosaic" of image patches, has become a foundation for other CBIR approaches due to its effectiveness [40]. This method, combined with deep feature barcoding, enables both fast search and efficient storage (Fig. 11.4). We can generally postulate the following general rules:

- Image search will be very slow if no patch selection is applied.
- Image search will not be accurate if deep features emanate from a suboptimally trained model.

Table 11.1 Overview of CBIR solutions for histopathology (also see Refs. [39,49]).

	Patching	Features	Encoding	Searching	WSI matching	Speed	Memory need
HBIR	None	None	Hashing	Hashcode matching	No	Fast	Low
BoVW	Visual words	Any feature extraction method	Counting	Histogram matching	Yes	Fast	Very low
SMILY	None	Custom	None	Feature matching	No	Slow	Very high
Yottixel	Mosaic	Any feature extraction method	Barcoding	Barcode matching	Yes	Fast	Very low
SISH	Yottixel's mosaic	DenesNet / Custom autoencoder	Yottixel's barcoding	Barcode matching / Tree matching	No	Sluggish	Excessive
RetCCL	Yottixel's mosaic	Custom	None	Feature matching	No	Slow	Very high
HSDH	None	Custom	None	Feature matching	No	Slow	Very high
HSHR	None	Custom	Hypergraphs	Correlation fusion	Yes	Slow	Very high

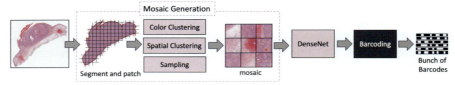

FIG. 11.4

The indexing structure of Yottixel [40] has sparked the inspiration for other search methodologies [45,49]. The mosaic and barcode stages contribute significantly to the rapid and efficient patch/WSI search process.

- Image search will be very slow if indexing is complicated.
- Image search will be infeasible/impractical if its indexing requires excessive storage.

Validation of image search methods

Lahr et al. recently conducted a thorough analysis and validation of BoVW, Yottixel, SISH, and RetCCL, utilizing both internal and external data [49]. Several other methods, namely SMILY, HSDH, and HBIR, were excluded due to a lack of a patching algorithm, while HSHR was excluded due to the unavailability of code (the absence of a patch selection could be an additional reason for ignoring a search scheme). The findings suggest that BoVW and Yottixel present comprehensive search solutions, combining high speed and efficient storage, offering valuable capabilities. However, to achieve high accuracy, integration with a well-trained backbone network and adjustments to the primary site are essential.

As Yottixel is a commercial product, BoVW provides researchers with more freedom to explore various improvement avenues (also, SISH provides limited access for research and usage because it borrows Yottixel's patented barcoding [50]. The manual settings in Yottixel's mosaic, particularly regarding cluster number and sampling percentage, could benefit from automation. Additionally, a logarithmic barcode comparison approach might enhance the speed of median-of-minimum Hamming distance calculations in Yottixel. On the other hand, Lahr et al. express concerns about SISH, which, as a Yottixel variant, deviates from Occam's Razor principle, introducing speed and scalability challenges due to its unnecessarily complex structure. SISH's reliance on vEB trees (an obsolete data structure mainly suitable for priority queues and not capable of dealing with hyperdimensional domains such as WSI analysis) with exponential space requirements renders it impractical for large datasets, impeding the loading and processing of terabytes of data. Regarding RetCCL, despite being labeled a search engine, it primarily focuses on the CCL network and lacks expressive embeddings for tissue morphology. Consequently, it struggles to qualify as an effective search engine.

In their conclusion, Lahr et al. provide, for the first time for histopathology, a ranking scheme for search methods, which can be found in Table 11.2 for investigated variations of four image search methods. As storage is a crucial factor in digital pathology, the storage overhead of search engines (additional storage they need to save indexing records) becomes a pivotal factor for search strategy selection (see Table 11.3).

Large deep models

The design and development of large deep models started with Large Language Models (LLMs) [51–53]. They are built on deep learning topologies, particularly transformer architectures [54], and are disrupting the way we deal with natural language processing. LLMs are trained with colossal amounts of diverse text documents to generate human-like language. LLMs like GPT-3 (Generative Pretrained Transformer 3) with an astronomical 175 billion parameters (i.e., weights similar to synapses in the human brain) are among the largest language models created to date [55] (see Fig. 11.5). LLMs are generally trained using many massive cohorts of text documents to learn general language patterns. Afterward they may be additionally fine-tuned to become specialized for specific downstream tasks [56].

The nuances and intricacies of natural language are learned first by an LLM to later adapt to different contexts. LLMs can summarize documents, translate a documents into different languages, and answer questions in an intuitive way [57,58]. They can create content, even generate ideas, and write different creative and professional text formats like poems, musical pieces, and computer programs. They can

Table 11.2 Lahr et al. ranked the performance of search engines based on their F1-score (top-1, majority of top-3, and majority of top-5), indexing time, searching time, failures, and storage (total performance ranking between 1 (best) and 6 (worst)) [49].

	Top-1	MV@3	MV@5	Indexing time	Searching time	Failures	Storage	Total ranking
Yottixel[a]	3	2	1	2	1	1	2	**1.71**
Yottixel-KR[b]	2	1	2	2	1	2	2	**1.71**
Yottixel-K[c]	1	3	4	2	1	2	2	**2.14**
BoVW	6	6	6	1	2	2	1	**3.43**
SISH[a]	4	4	3	3	3	4	4	**3.57**
SISH-N[d]	5	5	5	3	3	4	4	**4.14**
RetCCL-N[e]	8	8	7	4	4	3	3	**5.28**
RetCCL[a]	7	7	8	4	4	3	3	**5.43**

[a]As originally proposed.
[b]DenseNet replaced by KimiaNet, and using ranking after search.
[c]DenseNet replaced by KimiaNet.
[d]SISH with no ranking after search.
[e]RetCCL with no ranking after search.

Table 11.3 Index storage requirements for different search methods based on recent validation studios [49].

	Index size per WSI (kilobyte)	Index size for one million WSIs (gigabyte)
BoVW	0.03	10
Yottixel	0.38	119
RetCCL	5.76	1800
SISH	97.50	31,000

BoVW is the most storage-efficient approach, whereas SISH requires 31 terabytes of indexing storage for one million WSIs making it impractical for large datasets. At the present level of technology, loading terabytes of indexing data into memory is simply impossible.

FIG. 11.5

Size of CNNs like DenseNet and ResNet versus LLMs like BERT (one of the smallest LLMs) and ChatGPT. LLMs like LlaMa are with 65B parameters and PaLM with 540B parameters are much larger than ChatGPT.

also serve as conversation chatbots to perform all these tasks in a user-friendly manner [59]. Despite the fascination and obvious utilities, LLMs also exhibit major pitfalls. They may *hallucinate* and generate wrong text [60,61]. They may also—due to relying on the colossal data from public sphere—be biased [62]. Ethical considerations about data privacy is another issue of LLMs [63]. The pathology community is concerned about these challenges [64,65]. However, the potentials of LLMs to derive clinical factors for extraction of relevant information from huge medical records, among other pathology reports [66]. Other researchers and clinicians point to the expected future of LLMs and encourage pathologists to "embrace this technology" by identifying in what ways LLMs may support pathology for educational, clinical, and research purposes [67]. As well, the active involvement of pathologists is necessary in the improvement of AI and in helping to "design user-friendly interfaces" to integrate AI within the pathology workflow.

Perhaps more significant for histopathology, large vision-language models (LVLMs) are another type of large deep network constellation with the capabilities of LLMs but also fluent in computer vision, hence being able to process tissue images. The most notable model in this category is CLIP, short for Contrastive

Language-Image PreTraining [68]. Relying on bidirectional training between images and captions, LVLMs can generate both human language and digital images. Expectedly, such models are trained on massive datasets comprising images and corresponding textual descriptions, e.g., captions. The training on image-text pairs enables LVLMs like CLIP to build a two-way bridge between digital images and textual descriptions. In the case of CLIP, that was 400,000,000 image-text pairs. With such colossal training data, LVLMs become capable of answering questions about the content of images. Equipped with this additional capability compared to LLMs, LVLMs can perform image classification, object detection, and image retrieval based on textual descriptions. The *conventional* image search can be made tremendously more useful by LVLMs by offering the possibility to the user to utilize natural language queries for a more intuitive and efficient experience but retrieving content-based image matches.

Like LLMs, LVLMs also face challenges. Among others, they need a very large set of image-text pair of high quality, facing biases in training data, restricted understanding of complex visual concepts such as tissue images of rare diseases, and ethical concerns regarding generated content. Cross-modal topologies are the transition from conventional deep networks toward LVLMs. ***LILE*** *(Look In-Depth before Looking Elsewhere)* [69] is an example for such architectures, a dual attention network using transformers, that takes images and texts as inputs and extracts feature representations for each of those using a dedicated transformer. Then, a self-attention module [70] is applied. LILE is an architecture with a new loss term to help represent images and texts in a joint latent enabling bidirectional retrieval (Fig. 11.6). LILE was trained and tested with ARCH [71], a dataset of more than 7500 image-text histopathology images and showed better performance than foundation models like CLIP (see next section).

Cross-modal retrieval generally aims to identify a joint latent (feature) space where different modalities, such as image-text pairs, can be brought into close alignment. However, the main challenge often lies in the representation of expressive features for tissue morphology. LLMs can generally be trained relatively easily, whereas visual models often struggle due to the scarcity of labeled data [72]. To address this issue, a novel concept called *"harmonization"* has been introduced into processing histopathology images that can improve DINO (distillation without supervision) [73]. The harmonization of scale, a paramount factor in digital pathology, refines the DINO paradigm through a novel patching approach, overcoming the complexities posed by gigapixel whole slide images (see Fig. 11.7).

Foundation models

Foundation models (FMs) [74] mark a recent milestone in artificial intelligence that could drastically transform the role of machines within society and medicine. Generally, FMs are understood to be "base models" that offer adaptability to different domains. As such, one could understand FMs as very large models with unique

Foundation models   221

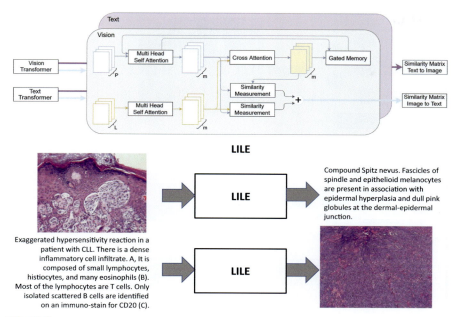

FIG. 11.6

Bidirectional information retrieval. LILE (Look In-Depth before Looking Elsewhere) [69] (top architects consisting of two transformers) uses a dual attention mechanism for cross-modal learning. This enables the model to auto-caption a query image (middle) as well as retrieve an image for a query description (bottom) (example inputs taken from ARCH [71] processed by LILE to retrieve the output) (Fig. 11.7).

capability to adjust to downstream tasks [75]. By training on colossal amounts of data (which understandably requires massive computation power), FMs can learn pervasive patterns and general relationships within an enormous amount of unlabeled data, and not for specific tasks. Of course, FMs must be trained with good quality, conflict-free, and diverse data types such as text and images (e.g., duplicate and new-duplicate data instances must be filtered out from the training data). As the data cannot be labeled at such massive scale that FMs require, the dominant training paradigm is unsupervised or self-supervised [52,53].

Major FMs include GPT-3 (multilingual translation, question answering, and creative text generation), Jurassic-1 Jumbo factual language processing, and LaMDA (generating realistic chat conversations). Generally, vision transformers can take large topologies to be used for object detection and image captioning. As a versatile base with vast knowledge learned from massive data, FMs can be fine-tuned for new downstream tasks and hence democratize AI usage by significantly reducing the development time of deep approaches for a large number of applications.

222 CHAPTER 11 Foundation models and information retrieval

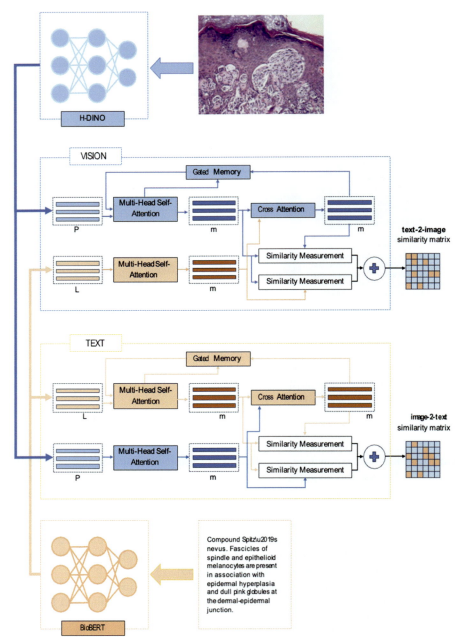

FIG. 11.7

The architecture of harmonizing LILE using DINO for image-text data is shown. It comprises H-DINO architecture and BioBERT to extract features for both vision and text modalities and LILE backbone to align feature representation of pair images and text.

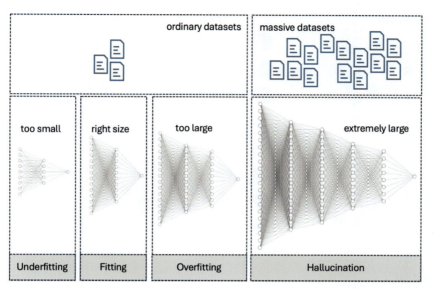

FIG. 11.8

The relationship between size of data and size of neural networks may be optimal for fitting, or may cause different issues like under- and overfitting for conventional networks, and hallucinations for larger models (network depictions taken from alexlenail.me).

Foundation models, as the superset of LLMs and LVLMs, may also be susceptible to biased responses and be unfair to specific social groups due to the inevitable societal preferences and predispositions in the training data [76]. The "black-box" concern about neural networks becomes more pressing as the lack of explainability and interpretability becomes even more pronounced for how FMs make decisions. One must emphasize that the massive computational resources required for training, maintaining, and running these models are certainly a major barrier for small clinics and community hospitals with limited resources. In general, the size of data and its relation to the size of a network can create different challenges. Whereas we have been mainly worried about overfitting, the advent of foundation models adds new concerns like "hallucination" to our terminology (see Fig. 11.8).

Generative AI

Most AI models are discriminative in nature, meaning that they draw lines (or planes) to separate different classes against each other. A form of deep models that, instead of classification, can rather produce new data such as text and images are called Generative AI [77] (see Fig. 11.9). Like any other AI approach, generative AI learns by being trained on available datasets, but rather on massive datasets in order to capture the data distribution and existing correlations among data instances. During such

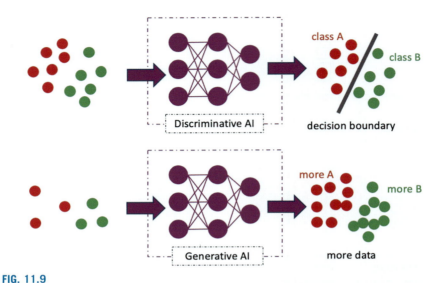

FIG. 11.9

The main task for discriminative AI is to classify data (top). Generative AI, in contrast, learns to generate more data of the same type/distribution.

training, the data patterns are identified and processed to grasp the meaning of a cluster of pixels in an image or a group of words in a sentence [78,79]. Furthermore, generative AI can "manufacture" novel data (or a combination of data that has not been seen before). Once trained, the model applies its learned knowledge to produce new text/images, mincing the distribution and recreating the correlations learned during training. Additionally, generative AI can be adjusted, i.e., fine-tuned, through "prompting," allowing users to guide the type of content they desire [80].

Large models, and generally foundation models, exhibit adaptability as a fundamental characteristic by handling various prompts and generating different new data formats (although they are supposed to be quite capable as they are as well, the co-called zero-shot learning). Generative models demonstrate versatile capabilities such as translating between many languages, generating digital images based on user descriptions, crafting various forms of text like poems, and answering questions at different levels of detail [81]. Table 11.4 summarizes the differences between generative versus discriminative models.

Various key topologies are employed to build generative AI models. One of the early and prominent approaches involves Generative Adversarial Networks (GANs) [82]. GANs mostly consist of two competitive parts: A *generator* produces (fake) data samples based on learned distribution, while a *discriminator* attempts to differentiate between real and fake samples. This "adversarial" interaction guides the generator to generate better and better outputs to fool the discriminator, a framework that can even be used for generating complex patterns like human tissue [83,84]. Another

Information retrieval and foundation models

Table 11.4 Discriminative versus generative AI.

	Discriminative AI	Generative AI
Model size	Small (<100M)	Large (>300M)
Dominant learning mode	Supervised	Self-supervised
Data size	Lots of data (e.g., 1M images)	Mass data (e.g., 400M images)
Task	Makes decisions	Creates content
Domain	Specific	General
Reliability	Consistent	May be distorted

generative scheme is implemented via Variational Autoencoders (VAEs) [85]. VAEs also operate based on two main parts although quite different from GANs: an encoder network compresses data into a "core representation," and a decoder network reconstructs new data based on that core. VAEs can capture the saliencies and substance of training data, i.e., the complex nonlinear data distributions, and use it to generate similar examples. Lastly, transformer architectures [70,86] have emerged as the most powerful generative engines. To accommodate data generation, they use many layers of purposefully arranged artificial neurons to analyze and connect different parts/fragments/segments of the input data. This exhaustive but soft correlation analysis, that can only be implemented in the form of a deep network, enables Transformers to understand correlations and complex relationships within the data and leverage this knowledge to generate new image/text content by activating trajectories of learned meaningful likelihoods.

Information retrieval and foundation models

Considering the success and dominance of deep learning models, one may think that information retrieval may be replaced by foundation models. LLMs and LVLMs, and more generally foundation models, can perform conversations with the pathologists and, through displaying immense knowledge, convince the expert to accept what they are putting forward. However, the trend in developments seems to go toward the combination and merging of these two distinct technologies. Among possibilities, the Retrieval-Augmented Generation (RAG) is one of the major ideas that synergizes the capabilities of information retrieval and generative AI [87]. Among others, information retrieval can help foundation models with "source attribution" [88] and point to the relevant information in small or large datasets solidifying the FM outputs with facts and details. This will reduce the "black box" disadvantage of models that persists in different forms also for FMs. On the other hand, generative AI excels at summarization, translation, and descriptions. However, FMs are limited in accessing the most accurate and up-to-date information and cannot point to the source.

226   CHAPTER 11 Foundation models and information retrieval

Recently, some researchers have collected online images (e.g., from Twitter and PubMed) to re-train CLIP for histopathology [89,90]. The general wisdom in computer science is that employing online images is not a suitable venue: *Garbage in, garage out*. Validations with high-quality clinical data have clearly shown these models won't provide any value in pathology compared to simpler models trained with high-quality data [91,92]. Cross-modal retrieval with vision transformers (Fig. 11.7), when trained with high-quality clinical data can indeed be much more reliable [72]. In case the research community designs and trains a few foundation models for pathology with high-quality data, one has to redefine the requirements for image search. In this case, the tasks will be intrinsically multimodal and the Divide & Conquer has been implemented within the pathology-aware foundation model. A new type of search will then perhaps be part of an FM-derived question and answer. Most crucial concern in this regard will be the "generative" aspect of FMs, which is undesired in medicine when it degenerates into hallucinations. It's essential to note that foundation models with conversational capabilities are closely tied to search, serving as implicit information retrieval. Table 11.5 offers a comparison between conventional information retrieval and foundation models.

Conclusions

With almost 400 years of history behind light microscopy, any innovation to bring about lasting change in histopathology may require some time. From today's perspective, deep learning's impact on digital (computational) pathology will be rather significant. The rise of large deep models is further amplifying the importance of deep networks in histopathology workflows. However, the future holds even greater promise when we regard the recent emergence of foundation models. These models have the potential to revolutionize computational pathology by offering valuable insights through "soft" information retrieval and fostering insightful conversations with researchers. While artificial intelligence through large foundational models, equipped with generative features of deep networks can excel at processing massive datasets and uncovering complex patterns, it cannot replace the need for traditional information retrieval platforms. As a matter of fact, AI is in dire need of information retrieval: Large or small, deep models struggle with transparency, simplicity, source attribution, and resource efficiency—all crucial aspects for their affordable and sustainable deployment in digital pathology. In contrast, conventional information retrieval methods like image search offer interpretable results, straightforward understanding, clear source attribution, and efficient resource utilization. Therefore, a comprehensive and balanced approach to information processing and knowledge extraction in digital pathology requires to focus on the synergy between large deep models and conventional information retrieval methods. This collaboration will leverage the strengths of both approaches to create more robust and informative systems of the future.

Table 11.5 Comparison of "search" versus "foundation models" as two technologies that may be used to assist pathologists in decision making (also see Ref. [39]).

	Information retrieval (IR)	Foundation models (FMs)
Model size	Small	Very large
Size of dataset needed	Small	Very large
Computational footprint	Small	Very large
Strength	IR convinces through retrieving evidence, i.e., the evidently diagnosed cases from the past	FMs convince through knowledgeable conversations
Disease type suitability	All diseases including rare cases with only a few examples	Mainly common diseases with a lot of data available
Information processing type	Explicit information retrieval	Implicit information retrieval
Source attribution for responses to query	Visible; accessible; explainable	Invisible; not accessible; not easily explainable
Maintenance	• Low dependency on hardware updates • Adding/deleting cases straightforward • Newer models can replace the old ones	• High dependency on hardware updates • High efforts for prompting to customize specific tasks • Expensive re-training cycles may be necessary

References

[1] Al-Janabi S, Huisman A, Van Diest PJ. Digital pathology: current status and future perspectives. Histopathology 2012;61(1):1–9.

[2] Hanna MG, Reuter VE, Ardon O, Kim D, Sirintrapun SJ, Schüffler PJ, Busam KJ, et al. Validation of a digital pathology system including remote review during the COVID-19 pandemic. Mod Pathol 2020;33(11):2115–27.

[3] Pantanowitz L. Digital images and the future of digital pathology. J Pathol Inf 2010;1:15. https://doi.org/10.4103/2153-3539.68332. PMID: 20922032. PMCID: PMC2941968.

[4] Evans AJ, Brown RW, Bui MM, Chlipala EA, Lacchetti C, Milner Jr DA, Pantanowitz L, et al. Validating whole slide imaging systems for diagnostic purposes in pathology: guideline update from the College of American Pathologists in collaboration with the American Society for Clinical Pathology and the Association for Pathology Informatics. Arch Pathol Lab Med 2022;146(4):440–50.

[5] Kumar N, Gupta R, Gupta S. Whole slide imaging (WSI) in pathology: current perspectives and future directions. J Digit Imaging 2020;33(4):1034–40.

[6] Griffin J, Treanor D. Digital pathology in clinical use: where are we now and what is holding us back? Histopathology 2017;70(1):134–45.

[7] Kiran N, Sapna FNU, Kiran FNU, Deepak Kumar FNU, Raja SS, Paladini A, et al. Digital pathology: transforming diagnosis in the digital age. Cureus 2023;15(9).

[8] Parwani AV, Amin MB. Convergence of digital pathology and artificial intelligence tools in anatomic pathology practice: current landscape and future directions. Adv Anat Pathol 2020;27(4):221–6.

[9] Bera K, Schalper KA, Rimm DL, Velcheti V, Madabhushi A. Artificial intelligence in digital pathology—new tools for diagnosis and precision oncology. Nat Rev Clin Oncol 2019;16(11):703–15.

[10] Klein C, Zeng Q, Arbaretaz F, Devêvre E, Calderaro J, Lomenie N, Maiuri MC. Artificial intelligence for solid tumour diagnosis in digital pathology. Br J Pharmacol 2021;178 (21):4291–315.

[11] Tizhoosh HR, Pantanowitz L. Artificial intelligence and digital pathology: challenges and opportunities. J Pathol Inf 2018;9(1):38.

[12] Browning L, Colling R, Rakha E, Rajpoot N, Rittscher J, James JA, Salto-Tellez M, Snead DRJ, Verrill C. Digital pathology and artificial intelligence will be key to supporting clinical and academic cellular pathology through COVID-19 and future crises: the PathLAKE consortium perspective. J Clin Pathol 2021;74(7):443–7.

[13] Zarella MD, McClintock DS, Batra H, Gullapalli RR, Valante M, Tan VO, Dayal S, et al. Artificial intelligence and digital pathology: clinical promise and deployment considerations. J Med Imaging 2023;10(5):051802.

[14] Niazi MK, Khan AV, Parwani, and Metin N. Gurcan. Digital pathology and artificial intelligence. Lancet Oncol 2019;20(5):e253–61.

[15] Hanna MG, Ardon O, Reuter VE, Sirintrapun SJ, England C, Klimstra DS, Hameed MR. Integrating digital pathology into clinical practice. Mod Pathol 2022;35(2):152–64.

[16] Tizhoosh HR, Diamandis P, Campbell CJV, Safarpoor A, Kalra S, Maleki D, Riasatian A, Babaie M. Searching images for consensus: can AI remove observer variability in pathology? Am J Pathol 2021;191(10):1702–8.

[17] Kalra S, Tizhoosh HR, Shah S, Choi C, Damaskinos S, Safarpoor A, Shafiei S, et al. Pancancer diagnostic consensus through searching archival histopathology images using artificial intelligence. npj Digital Med 2020;3(1):31.

[18] Manning CD. An introduction to information retrieval. Cambridge University Press; 2009.

[19] Singhal A. Modern information retrieval: a brief overview. IEEE Data Eng Bull 2001; 24(4):35–43.

[20] Guo J, Fan Y, Ai Q, Croft WB. Semantic matching by non-linear word transportation for information retrieval. In: Proceedings of the 25th ACM international on conference on information and knowledge management; 2016. p. 701–10.

[21] Sivarajkumar S, Mohammad HA, Oniani D, Roberts K, Hersh W, Liu H, He D, Visweswaran S, Wang Y. Clinical information retrieval: a literature review. J Healthcare Inf Res 2024;1–40.

[22] Hersh W. Information retrieval: a health and biomedical perspective. Springer Science & Business Media; 2008.

[23] Ting SL, See-To EWK, Tse YK. Web information retrieval for health professionals. J Med Syst 2013;37:1–14.

[24] Vanopstal K, Buysschaert J, Laureys G, Stichele RV. Lost in PubMed. Factors influencing the success of medical information retrieval. Expert Syst Appl 2013;40(10):4106–14.

[25] Reis D, Cesar J, Bonacin R, Perciani EM. Intention-based information retrieval of electronic health records. In: 2016 IEEE 25th international conference on enabling technologies: infrastructure for collaborative enterprises (WETICE). IEEE; 2016. p. 217–22.

[26] Wang Y, Wen A, Liu S, Hersh W, Bedrick S, Liu H. Test collections for electronic health record-based clinical information retrieval. JAMIA Open 2019;2(3):360–8.

[27] McInerney DJ, Dabiri B, Touret A-S, Young G, Meent J-W, Wallace BC. Query-focused ehr summarization to aid imaging diagnosis. In: Machine learning for healthcare conference. PMLR; 2020. p. 632–59.

[28] Kamath S, Mayya V, Priyadarshini R. A probabilistic precision information retrieval model for personalized clinical trial recommendation based on heterogeneous data. In: 2021 12th International Conference on Computing Communication and Networking Technologies (ICCCNT). IEEE; 2021. p. 1–5.

[29] Chaudhary KK, Mishra N. A review on molecular docking: novel tool for drug discovery. Databases 2016;3(4):1029.

[30] David L, Thakkar A, Mercado R, Engkvist O. Molecular representations in AI-driven drug discovery: a review and practical guide. J Cheminf 2020;12(1):1–22.

[31] Cornsweet T. Visual perception. Academic Press; 2012.

[32] Suetens P. Fundamentals of medical imaging. Cambridge University Press; 2017.

[33] Rosai J. Why microscopy will remain a cornerstone of surgical pathology. Lab Investig 2007;87(5):403–8.

[34] Tseng L-J, Matsuyama A, MacDonald-Dickinson V. Histology: the gold standard for diagnosis? Can Vet J 2023;64(4):389.

[35] Babaie M, Kalra S, Sriram A, Mitcheltree C, Zhu S, Khatami A, Rahnamayan S, Tizhoosh HR. Classification and retrieval of digital pathology scans: a new dataset. In: Proceedings of the IEEE conference on computer vision and pattern recognition workshops; 2017. p. 8–16.

[36] Pantanowitz L, Michelow P, Hazelhurst S, Kalra S, Choi C, Shah S, Babaie M, Tizhoosh HR. A digital pathology solution to resolve the tissue floater conundrum. Arch Pathol Lab Med 2021;145(3):359–64.

[37] Tayebi RM, Youqing M, Dehkharghanian T, Ross C, Sur M, Foley R, Tizhoosh HR, Campbell CJV. Automated bone marrow cytology using deep learning to generate a histogram of cell types. Commun Med 2022;2(1):45.

[38] Zhou XS, Zillner S, Moeller M, Sintek M, Zhan Y, Krishnan A, Gupta A. Semantics and CBIR: a medical imaging perspective. In: Proceedings of the 2008 international conference on Content-based image and video retrieval; 2008. p. 571–80.

[39] Tizhoosh HR, Pantanowitz L. On image search in histopathology. arXiv preprint arXiv:2401.08699; 2024.

[40] Kalra S, Tizhoosh HR, Choi C, Sultaan S, Diamandis Phedias JV, Clinton C, Liron P. Yottixel—an image search engine for large archives of histopathology whole slide images. Med Image Anal 2020;65, 101757.

[41] Zhang X, Liu W, Dundar M, Badve S, Zhang S. Towards large-scale histopathological image analysis: hashing-based image retrieval. IEEE Trans Med Imaging 2014;34(2):496–506.

[42] Zhu S, Li Y, Kalra S, Tizhoosh HR. Multiple disjoint dictionaries for representation of histopathology images. J Vis Commun Image Represent 2018;55:243–52.

CHAPTER 11 Foundation models and information retrieval

[43] Hegde N, Hipp JD, Liu Y, Emmert-Buck M, Reif E, Smilkov D, Terry M, Cai CJ, Amin MB, Mermel CH, et al. Similar image search for histopathology: Smily. npj Digital Med 2019;2(1):56.

[44] Chen C, Lu MY, Williamson DFK, Chen TY, Schaumberg AJ, Mahmood F. Fast and scalable search of whole-slide images via self-supervised deep learning. Nat Biomed Eng 2022;6(12):1420–34.

[45] Sikaroudi M, Afshari M, Shafique A, Kalra S, Tizhoosh HR. Comments on 'Fast and scalable search of whole-slide images via self-supervised deep learning'. arXiv preprint arXiv:2304.08297; 2023.

[46] Wang X, Yuexi D, Yang S, Zhang J, Wang M, Zhang J, Yang W, Huang J, Han X. Retccl: clustering-guided contrastive learning for whole-slide image retrieval. Med Image Anal 2023;83, 102645.

[47] Alizadeh M, Helfroush MS, Müller H. A novel siamese deep hashing model for histopathology image retrieval. Expert Syst Appl 2023;225, 120169.

[48] Li S, Zhao Y, Zhang J, Yu T, Zhang J, Gao Y. High-order correlation-guided slide-level histology retrieval with self-supervised hashing. IEEE Trans Pattern Anal Mach Intell 2023;45(9):11008–23.

[49] Lahr I, Alfasly S, Nejat P, Khan J, Kottom L, Kumbhar V, Alsaafin A, et al. Analysis and validation of image search engines in histopathology. arXiv preprint arXiv:2401.03271; 2024.

[50] Tizhoosh HR. Systems and methods for barcode annotations for digital images. U.S. Patent 11,270,204, issued March 8; 2022.

[51] Yang J, Jin H, Tang R, Han X, Feng Q, Jiang H, et al. Harnessing the power of llms in practice: a survey on chatgpt and beyond. ACM Trans Knowl Discov Data 2024; 18(6):1–32.

[52] Zhao T, Wei M, Preston JS, Poon H. Automatic calibration and error correction for large language models via pareto optimal self-supervision. arXiv preprint arXiv:2306.16564; 2023.

[53] Zhao WX, Zhou K, Li J, Tang T, Wang X, Hou Y, Min Y, et al. A survey of large language models. arXiv preprint arXiv:2303.18223; 2023.

[54] Liu Y, He H, Han T, Zhang X, Liu M, Tian J, Zhang Y, et al. Understanding llms: a comprehensive overview from training to inference. arXiv preprint arXiv:2401.02038; 2024.

[55] Wu T, He S, Liu J, Sun S, Liu K, Han Q-L, Tang Y. A brief overview of ChatGPT: the history, status quo and potential future development. IEEE/CAA J Autom Sin 2023;10 (5):1122–36.

[56] Zhang Z, Zheng C, Da Tang KS, Ma Y, Bu Y, Zhou X, Zhao L. Balancing specialized and general skills in llms: the impact of modern tuning and data strategy. arXiv preprint arXiv:2310.04945; 2023.

[57] Hadi MU, Qureshi R, Shah A, Irfan M, Zafar A, Shaikh MB, Akhtar N, Wu J, Mirjalili S. A survey on large language models: applications, challenges, limitations, and practical usage. Authorea Preprints; 2023.

[58] Laban P, Kryściński W, Agarwal D, Fabbri AR, Xiong C, Joty S, Wu C-S. SummEdits: measuring LLM ability at factual reasoning through the lens of summarization. In: Proceedings of the 2023 Conference on Empirical Methods in Natural Language Processing; 2023. p. 9662–76.

[59] Freire SK, Wang C, Niforatos E. Chatbots in knowledge-intensive contexts: comparing intent and LLM-based systems. arXiv preprint arXiv:2402.04955; 2024.

References   231

[60] Rawte V, Sheth A, Das A. A survey of hallucination in large foundation models. arXiv preprint arXiv:2309.05922; 2023.

[61] Xu Z, Jain S, Kankanhalli M. Hallucination is inevitable: an innate limitation of large language models. arXiv preprint arXiv:2401.11817; 2024.

[62] Salewski L, Alaniz S, Rio-Torto I, Schulz E, Akata Z. In-context impersonation reveals large language models' strengths and biases. Adv Neural Inf Proces Syst 2024;36.

[63] Peris C, Dupuy C, Majmudar J, Parikh R, Smaili S, Zemel R, Gupta R. Privacy in the time of language models. In: Proceedings of the sixteenth ACM international conference on web search and data mining; 2023. p. 1291–2.

[64] Hart SN, Hoffman NG, Gershkovich P, Christenson C, McClintock DS, Miller LJ, Jackups R, Azimi V, Spies N, Brodsky V. Organizational preparedness for the use of large language models in pathology informatics. J Pathol Inf 2023; 100338.

[65] Ullah E, Parwani A, Baig MM, Singh R. Challenges and barriers of using large language models (LLM) such as ChatGPT for diagnostic medicine with a focus on digital pathology—a recent scoping review. Diagn Pathol 2024;19(1):1–9.

[66] Choi HS, Song JY, Shin KH, Chang JH, Jang B-S. Developing prompts from large language model for extracting clinical information from pathology and ultrasound reports in breast cancer. Radiat Oncol J 2023;41(3):209.

[67] Arvisais-Anhalt S, Gonias SL, Murray SG. Establishing priorities for implementation of large language models in pathology and laboratory medicine. Acad Pathol 2024;11(1), 100101.

[68] Radford A, Kim JW, Hallacy C, Ramesh A, Goh G, Agarwal S, Sastry G, et al. Learning transferable visual models from natural language supervision. In: International conference on machine learning. PMLR; 2021. p. 8748–63.

[69] Maleki D, Tizhoosh HR. LILE: look in-depth before looking elsewhere—a dual attention network using transformers for cross-modal information retrieval in histopathology archives. In: International conference on medical imaging with deep learning. PMLR; 2022, December. p. 879–94.

[70] Vaswani A, Shazeer N, Parmar N, Uszkoreit J, Jones L, Gomez AN, Kaiser Ł, Polosukhin I. Attention is all you need. Adv Neural Inf Proces Syst 2017;30.

[71] Gamper J, Rajpoot N. Multiple instance captioning: learning representations from histopathology textbooks and articles. In: Proceedings of the IEEE/CVF conference on computer vision and pattern recognition; 2021. p. 16549–59.

[72] Maleki D, Rahnamayan S, Tizhoosh HR. A self-supervised framework for cross-modal search in histopathology archives using scale harmonization. Preprint on Springer's Research Square 2024.

[73] Caron M, Touvron H, Misra I, Jégou H, Mairal J, Bojanowski P, Joulin A. Emerging properties in self-supervised vision transformers. In: Proceedings of the IEEE/CVF international conference on computer vision; 2021. p. 9650–60.

[74] Bommasani R, Hudson DA, Adeli E, Altman R, Arora S, von Arx S, Bernstein MS, et al. On the opportunities and risks of foundation models. arXiv preprint arXiv:2108.07258; 2021.

[75] Touvron H, Martin L, Stone K, Albert P, Almahairi A, Babaei Y, Bashlykov N, et al. Llama 2: open foundation and fine-tuned chat models. arXiv preprint arXiv:2307.09288; 2023.

[76] Dehkharghanian T, Bidgoli AA, Riasatian A, Mazaheri P, Campbell CJV, Pantanowitz L, Tizhoosh HR, Rahnamayan S. Biased data, biased AI: deep networks predict the acquisition site of TCGA images. Diagn Pathol 2023;18(1):67.

[77] Feuerriegel S, Hartmann J, Janiesch C, Zschech P. Generative AI. Bus Inf Syst Eng 2024;66(1):111–26.

[78] Epstein Z, Hertzmann A, Investigators of Human Creativity, Akten M, Farid H, Fjeld J, Frank MR, et al. Art and the science of generative AI. Science 2023;380(6650):1110–1.

[79] Jo A. The promise and peril of generative AI. Nature 2023;614(1):214–6.

[80] Dang H, Mecke L, Lehmann F, Goller S, Buschek D. How to prompt? Opportunities and challenges of zero-and few-shot learning for human-AI interaction in creative applications of generative models. arXiv preprint arXiv:2209.01390; 2022.

[81] Gozalo-Brizuela R, Garrido-Merchán EC. A survey of generative AI applications. arXiv preprint arXiv:2306.02781; 2023.

[82] Pan Z, Weijie Y, Yi X, Khan A, Yuan F, Zheng Y. Recent progress on generative adversarial networks (GANs): a survey. IEEE Access 2019;7:36322–33.

[83] Afshari M, Yasir S, Keeney GL, Jimenez RE, Garcia JJ, Tizhoosh HR. Single patch super-resolution of histopathology whole slide images: a comparative study. J Med Imaging 2023;10(1):017501.

[84] Safarpoor A, Kalra S, Tizhoosh HR. Generative models in pathology: synthesis of diagnostic quality pathology images. J Pathol 2021;253(2):131–2.

[85] Kingma DP, Welling M. An introduction to variational autoencoders. Found Trends Mach Learn 2019;12(4):307–92.

[86] Hudson DA, Zitnick L. Generative adversarial transformers. In: International conference on machine learning. PMLR; 2021. p. 4487–99.

[87] Lewis P, Perez E, Piktus A, Petroni F, Karpukhin V, Goyal N, Küttler H, et al. Retrieval-augmented generation for knowledge-intensive NLP tasks. Adv Neural Inf Proces Syst 2020;33:9459–74.

[88] Kamalloo E, Jafari A, Zhang X, Thakur N, Lin J. Hagrid: a human-llm collaborative dataset for generative information-seeking with attribution. arXiv preprint arXiv:2307.16883; 2023.

[89] Huang Z, Bianchi F, Yuksekgonul M, Montine TJ, Zou J. A visual–language foundation model for pathology image analysis using medical Twitter. Nat Med 2023;1–10.

[90] Lu MY, Chen B, Williamson DFK, Chen RJ, Liang I, Ding T, Jaume G, et al. Towards a visual-language foundation model for computational pathology. arXiv preprint arXiv:2307.12914; 2023.

[91] Alfasly S, Nejat P, Hemati S, Khan J, Lahr I, Alsaafin A, Shafique A, et al. When is a foundation model a foundation model. arXiv preprint arXiv:2309.11510; 2023.

[92] Alfasly S, Nejat P, Hemati S, Khan J, Lahr I, Alsaafin A, Shafique A, et al. Foundation models for histopathology—fanfare or flair. Mayo Clin Proc Digital Health 2024;2 (1):165–74.

CHAPTER

Precision medicine in digital pathology via image analysis and machine learning

12

Peter D. Caie[a], Neofytos Dimitriou[b], and Ognjen Arandjelović[b]

[a]*School of Medicine, QUAD Pathology, University of St Andrews, St Andrews, United Kingdom,*
[b]*School of Computer Science, University of St Andrews, St Andrews, United Kingdom*

Introduction

Precision medicine

The field of medicine is currently striving toward more accurate and effective clinical decision-making for individual patients. This can be done through many forms of analysis and be put into effect at multiple stages of a patient's disease progression and treatment journey. However, the overarching goal is for higher treatment success rates with lower side effects from potentially ineffectual, but toxic, therapies and of course better patient well-being and overall survival. For example, understanding if a specific treatment for an individual patient's cancer may help their treatment or in fact be detrimental to their overall survival, as is the case with cetuximab treatment in colorectal cancer. This process of treating the patient as an individual and not as a member of a broader and heterogeneous population is commonly termed precision medicine and has traditionally been driven by advances in targeted drug discovery with accompanying translatable companion molecular tests. These tests report on biomarkers measured from patient samples and inform if specific drugs will be effective for patients, normally based on the molecular profile of their diagnostic tissue sample. The tests may derive from our knowledge of biological processes, like designing inhibitors against EGFR receptor pathways, or from machine learning-based mining of large multiomic datasets to identify novel drug targets or resistance mechanisms. In the current era of digital medicine, precision medicine is being applied throughout the clinical workflow from diagnosis to prognosis to prediction. Large flows of data can be tapped from multiple sources and no longer solely through molecular pathology. This is made possible by the digitization, and availability, of patient history and lifestyle records as well as clinical reports and through the adoption of digital pathology and image analysis in both the realm of research and the clinic. In fact, prior to the interrogation of digitized histopathological datasets by

Artificial Intelligence in Pathology. https://doi.org/10.1016/B978-0-323-95359-7.00012-1
Copyright © 2025 Elsevier Inc. All rights are reserved, including those for text and data mining, AI training, and similar technologies.

233

image analysis, in vitro high-content biology-based drug screens were being developed by the pharmaceutical industry [1]. These screens also applied image analysis, but to cultured cells exposed to molecular manipulation, to segment and classify cellular structures before capturing large multiparametric datasets that inform on novel targets or drug efficacy. A similar methodology taken from high-content biology and image analysis can be applied to digital pathology. This is the case for classical object threshold-based image analysis or for artificial intelligence. The overarching aim is to quantify and report on specific biomarkers or histological patterns of known importance or by capturing unbiasedly collected multiparametric data and pattern recognition across segmented objects and whole-slide images (WSI). The aim of distilling and reporting on the extracted data from digital pathology images is to allow for the stratification of patients into distinct groups that may inform the clinician on their optimal and personalized treatment regimen.

Digital pathology

Digital pathology, the high-resolution digitization of glass-mounted histopathological specimens, is becoming more commonplace in the clinic. This disruptive technology is on track to replace the reporting of glass slides down a microscope, as has been the tradition for centuries. There remain certain obstacles to overcome for widescale adoption of digital pathology, such as IT infrastructure, scanning workflow costs, and the willingness of the pathology community. However, as more institutes trend to full digitization, these obstacles diminish. The adoption of digital pathology holds advantages over traditional microscopy in teaching, remote reporting and image sharing, and, not least, the ability to perform image analysis on the resultant digitized specimens.

In essence, the technology is currently moving from the glass slide and microscope to the digital image and high-resolution screen, although manual viewing and diagnosis remain the same. Clinical applications using digital pathology and WSIs are currently restricted to the primary diagnosis of H&E stained slides. However, future applications, such as immunofluorescence, can bring advantages to digital pathology. Indeed, scanner vendors are frequently combining both brightfield and multiplexed immunofluorescence visualization into their platforms. Immunofluorescence allows the identification, classification, and quantification of multiple cell types or biomarkers colocalized at the single-cell resolution and on a single tissue section. The importance of this capability is becoming increasingly apparent as we realize that the complex intercellular and molecular interactions within the tumor microenvironment, on top of cellular morphology, play a vital role in a tumor's progression and aggressiveness.

Humans are as adept at reporting from digital pathology samples as they are from glass-mounted ones. They do this by identifying diagnostic and prognostic patterns in the tissue while at the same time disregarding artifact and nonessential histology. However, an inherent human setback in the field of pathology has been the standardization and consistency of inter- and intraobserver reporting. This is specifically true for the identification and semiquantification of more discrete and subtle morphologies as well as, for example, the counting of specific cell types, mitotic figures, or biomarker expression across WSIs [2–4]. The automated analysis of such features by computer algorithms can overcome these flaws and provide objective and standardized quantification. With ongoing research providing much-needed evidence of the capability of image analysis and artificial intelligence to accurately quantify and report on molecular and morphological features, it is only a matter of time before these too translate into the clinic.

The use of image analysis and artificial intelligence can be knowledge driven or data driven. The next section of this chapter will discuss how both methodologies can be applied to digital pathology before we expand on the theory and concepts behind the various artificial intelligence models commonly used in the field.

Applications of image analysis and machine learning
Knowledge-driven image analysis

The first applications of image analysis in digital pathology were the quantification of previously identified histopathological features or immunohistochemical results of known pathological significance. This image analysis typically relies on user-defined thresholds of either color or fluorescence intensity coupled with classifications based on object shape, extent, or texture. Examples of histopathological features quantified in such a manner include the quantification of tumor buds, lymphovascular density, and invasion, or immune infiltrate [5,6]. Similarly, the quantification of protein expression through immunolabeling, such as pHH3, PDL-1, or EGFR, whether studying predictive or prognostic outcomes, can also be quantified in such a manner [7–9]. This image analysis can be applied to user-defined regions of interest through manual annotations or across WSIs. A further advantage of the automatic reporting of such features, over and above reproducibility, is that they can be quantified across a continuum instead of traditional semiquantified categorical reporting. The analysis of continuous data is far more amenable to personalized pathology through machine learning than creating distinct categories to group patients.

Machine learning for image segmentation

Pathologists regularly quantify features from specific locations within the tissue section, for example, only within the tumor or stromal cells or at the cancer's invasive margin. However, to manually annotate images to differentiate these regions of interest prior to performing image analysis is a subjective and tedious task, specifically if needed to be applied to the large patient cohorts required for statistically relevant studies. One can therefore apply machine learning algorithms, such as random forest, directly to the images to overcome the need for manual annotations. This machine learning-based classification relies on the user teaching the algorithm what the specific regions are that they want to segment. They do so by selecting examples across a training subset of images. Once the algorithm has learned the difference between the regions to be differentiated, it can be applied to a larger sample set in order to automatically segment the image of the tissue. An example of this would be using an antibody against cytokeratin to label the tumor and use this marker, on top of the cancer cell morphology, to differentiate the tumor from the stroma (Fig. 12.1A). Once this is performed, one can employ the before-mentioned threshold-based image analysis to quantify, for example, Ki67 expression in only the tumor (Fig. 12.1B), tumor buds only at the invasive margin [10], or CD3 and CD8 positive lymphocytic infiltration within either the tumor core or invasive margin [6] (Fig. 12.2).

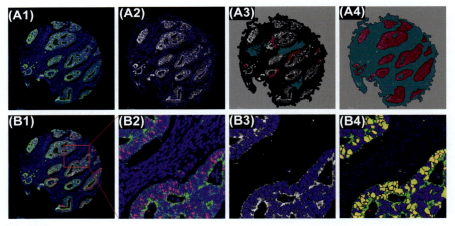

FIG. 12.1

Automated tumor to stroma segmentation and cellular classification. (A) A machine learning approach for automatic tissue segmentation using Definiens Tissue Studio software. In this example, segmenting the tumor from the stroma. (A1) A colorectal TMA core labeled by immunofluorescence for tumor (Pan Cytokeratin, *green*) and nuclei (Hoechst, *blue*). (A2) The software automatically segments the image based on pixel heterogeneity *(blue outline)*. (A3) The user tags segmented example objects of the features to be segmented, here turquoise for stroma and purple for tumor. The machine learning algorithm learns these features and automatically segments the rest of the images in the project, as demonstrated in Aiv. (B) The same TMA core was labeled for ki67, which is coregistered in (B1) in *red*. (B2) shows all nuclei automatically segmented *(blue outline)* by the Definiens software, prior to the exclusion of stroma nuclei *(gray)* and the retaining of tumor nuclei *(blue)* in (B3). (B4) shows the tumor nuclei being classified as positive for ki67 expression *(yellow)*.

Applications of image analysis and machine learning 237

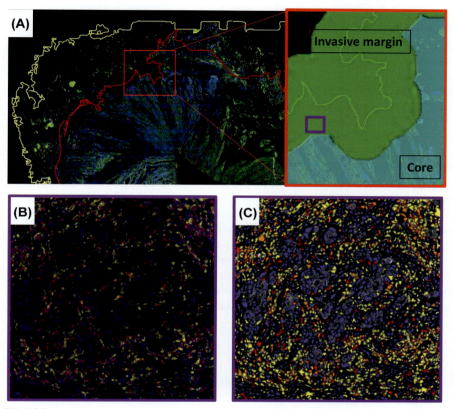

FIG. 12.2

Quantification of lymphocytic infiltration with the tumor's invasive margin or core. (A) A whole slide image of a colorectal cancer tissue section labeled for tumor cells *(green)* and nuclei *(blue)*. The image analysis in this figure is performed using Indica Labs HALO software. The red outline is the automatically detected deepest invasion of tumor cells, and the inset shows a zoomed-in example of the image segmented into an invasive margin *(green)* and the tumor core *(blue)*. The purple square denotes where figures (B) and (C) originate from in the invasive margin. (B) Multiplexed immunofluorescence visualized CD3 *(yellow)* and CD8 *(red)* positive lymphocytes. (C) Automated quantification of these lymphocytes within just the invasive margin of the tumor.

Deep learning for image segmentation

There exists great inter- and intrapatient heterogeneity, specifically in the molecular and phenotypic makeup of a patient's tumor. This also equates to great variation within the images, not only morphologically but also when applying a specific marker to differentiate regions of interest for image analysis. Furthermore, the tissue specimen is imperfect, and so, therefore, is the digitized WSI. These imperfections can originate from multiple stages of tissue preparation. They can result in folds, tears, or uneven thickness of the tissue to more subtle morphological differences due to ischemia or fixation times. If performing immunolabeling, nonspecific staining, edge-effect, or autofluorescence can further create confounding issues for automated image analysis. Examples of such artifacts can be seen in Fig. 12.3. All of the above may cause inaccurate reporting, such as false positives when applying image analysis and machine learning to segment tissue across large patient cohorts. However, many of these issues can be overcome by employing deep learning architecture to segment the digital WSI. To do this, trained experts can annotate the regions of interest selected for quantification while further identifying and training the algorithm to ignore the tissue artifact [11]. The human brain is extraordinary at pattern recognition, and the pathologist routinely and automatically ignores imperfect tissue specimens in order to hone in on the area containing the information needed to make their diagnosis or prognosis. The ever-developing sophistication of deep learning architectures now allows this level of analysis by automated algorithms.

For accurate tissue segmentation, using either machine learning or deep learning methodology, a strong and standardized signal to noise is required within the feature one is using to differentiate regions of interest. However, the inherently heterogeneous sample is also reflected in the marker used for segmentation, and this may vary in brightness or intensity between patient samples, and even within the same sample.

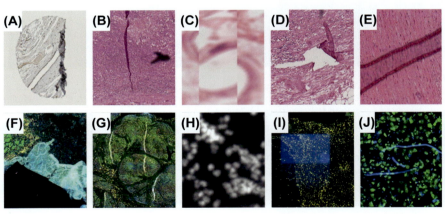

FIG. 12.3

Examples of image artifacts within digital pathology. (A–E) Examples of artifacts from brightfield image capture. (A) Folded over a section of a TMA core. (B) Cutting artifact *(dark pink line)* and piece of dust *(gray)* on an H&E-stained image. (C) Out of focus and incorrectly stitched image of an H&E-stained image. (D) Tear and fold in H&E-labeled tissue. (E) Out-of-focus section bordered by a foreign object on the H&E-labeled specimen. (F–J) Examples of artifacts from fluorescence image capture. (F) A tear in a tissue section resulting in nonspecific fluorescence. (G) Cutting artifact *(bright lines)*. (H) Out-of-focus nuclei. (I) Illumination artifact resulting in large blue squares. (J) Autofluorescence from hair in a urine cytology sample *(blue lines)*.

Image color and intensity standardization algorithms can be employed prior to the analysis of the image by artificial intelligence [12]. This can lead to a more accurate, fully automated tissue segmentation across diverse and large patient cohorts; however, there is controversy over whether this is the best method to achieve this goal and whether this methodology may reduce real diagnostic information.

Deep learning is not only being applied to segment tissue prior to biomarker or cellular quantification but also has the ability to recognize histopathological patterns in digitized H&E-stained tissue sections (Fig. 12.4). This has been demonstrated in studies where the expert pathologist has annotated features of significance in order to

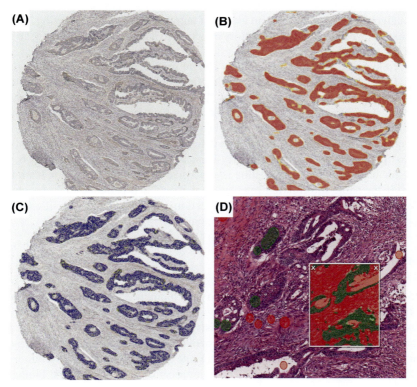

FIG. 12.4

Tissue segmentation performed by deep learning. When segmenting tumor from stroma without the aid of a tumor visualization marker, shallow learning-based image segmentation is usually inaccurate. Deep learning can be applied in this instance for accurate tissue segmentation. These examples utilize Indica Labs HALO AI software to segment tumor from stroma. (A) A TMA core stained with only hematoxylin. (B) HALO AI algorithm trained to recognize only tumor cells *(red)* within the hematoxylin-stained digital image. (C) The software next classifies each tumor nuclei *(dark blue)*. (D) Deep learning algorithm's ability to be trained to recognize tumor *(green overlay)* from stroma *(red overlay)* and from H&E-stained digital images.

240 CHAPTER 12 Precision medicine in digital pathology

train the deep learning algorithms to identify these in unseen test and validation sets [13,14]. The clinical application and methodology that allow the computer to visualize and recognize more subtle prognostic-associated patterns are covered in more detail elsewhere in this book. Algorithms such as these are being developed to aid the pathologist in their diagnosis, and it is only a matter of time until they are applied routinely in the clinical workflow of pathology departments. Currently, these algorithms are not being designed to automatically report on patient samples, without any human verification, but rather to act as an aid to the pathologist in order to increase the speed of their reporting, for example, by highlighting the areas of interest, as a method to triage urgent cases or as a second opinion. However, as deep learning becomes more sophisticated, there is a strong possibility that in the future, computer vision algorithms may perform an aspect of autonomous clinical reporting. Later in this chapter, we discuss what regulatory concerns to address when designing an algorithm for translation into the clinic.

Deep learning architectures are now able to predict patient molecular subgroups from H&E-labeled histology. They do this from only the morphology and histological pattern of the tissue sample [15,16]. This may be quite remarkable to imagine; however, the histopathologist would most likely have predicted this to be possible. They have been fully aware of the complex and important variations of morphology present in the tissue and how they affect patient outcome, even if they have not been able to link these to molecular subtypes. This, however, has real implications beyond academic research. Molecular testing is expensive and requires complex instrumentation. If the same information relevant to personalized pathology can be gleaned from the routine and cheap H&E-labeled section, this could have a significant monetary impact when calculating health economics.

Spatial resolution

We have briefly covered the importance of reporting the tissue architecture in the field of precision medicine when quantifying biomarkers of interest, as opposed to other molecular technology that destroys the tissue and thus the spatial resolution of its cellular components. Histopathologists know that context is key to an accurate diagnosis and prognosis. The tumor microenvironment is complex, with many cellular and molecular interactions that play a role in inhibiting or driving tumor progression and responses to therapy. The quantification of a stand-alone biomarker, even by image analysis, may not be enough to predict accurate prognosis or prediction for an individual patient. This is the case for PDL-1 immunohistochemical testing, where even patients with PDL-1-positive tumors may not respond to anti-PDL-1 therapy [17]. Similarly, there is an advantage to quantifying prognostic histopathological features such as lymphocytic infiltration or tumor budding in distinct regions within the tumor microenvironment. Traditionally, image analysis has quantified a single prognostic feature across a single tissue section with proven success at patient stratification. However, by applying multiplexed immunofluorescence, it is now possible to visualize multiple biomarkers and histological features within a single tissue section. Image analysis software can furthermore calculate and export the exact x and

Applications of image analysis and machine learning

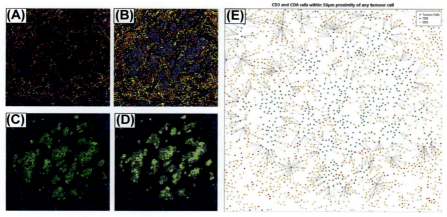

FIG. 12.5

Quantifying the spatial resolution of cellular subpopulations. This figure demonstrates how CD3 and CD8 positive lymphocyte densities can be mapped to tumor cells when multiplexed immunofluorescence is quantified using Indica Lab HALO software. Immunofluorescence labeled CD3 *(yellow)* and CD8 *(red)* positive lymphocytes (A) segmented and quantified (B). (C) Tumor cells *(green)* captured from the same image as (A) and segmented and quantified (D). The exported *x* and *y* spatial coordinates of the tumor cells *(green dots)* are plotted alongside the coordinates of the CD3 *(yellow dots)* and CD8 *(red dots)* positive lymphocytes (E). This allows for the spatial statistics of the lymphocytes within a 50-μm diameter of a tumor cell being calculated.

y spatial coordinates of each feature of interest across the WSI or recognize specific patterns within the interacting cellular milieu of the tumor microenvironment (Fig. 12.5). This not only brings the advantage of measuring more than one prognostic feature but also allows new insights into the understanding of disease progression based on quantifying novel interactions at the spatial resolution of the tissue. This was demonstrated by Nearchou et al., who showed that the density of tumor budding and immune infiltrate were significantly associated with stage II colorectal cancer survival, but furthermore, that their specific interaction added value to a combined prognostic model [6]. Studies such as this show that complex spatial analysis may be key to the success of accurate prognosis for the individual patient.

Machine learning on extracted data

The image analysis solutions described above allow accurate image segmentation and cellular and molecular classifications of known biomarkers or histopathological features. The quantification of these known features can be utilized and applied to deliver personalized pathology. However, image analysis can further extract a wealth of unbiasedly collected data across multiple classified objects and their spatial interactions. Such data may be the specific density of specific objects surrounding others or their distance from each other, as well as a wealth of extent, shape, and texture measurements of individual objects. The value of these data is often not known a priori and is difficult, if not impossible, to sort and analyze by eye. Machine learning can be applied to large datasets from both molecular and digital pathology in order to understand the optimal features that allow patient stratification and thus clinical decision-making in the field of personalized medicine. However, caution must be taken when deciding on which machine learning algorithm to apply to your data. If one model is superior at analyzing one dataset, based on, e.g., the search for the optimal area under the receiver operator, it does not mean that the same algorithm will be as successful at analyzing a second and distinct dataset under similar computational restrictions. This forms part of the "no free lunch theorem" [18], which we will touch upon again later in the chapter. In plain terms, different machine learning algorithms are superior to others when applied to specific datasets. For example, some algorithms excel at analyzing data with low dimensionality (such as k-nearest neighbor) and, however, become intractable or return poor results when the data dimensions increase. On the other hand, random forests excel at analyzing high-dimensional data. Similarly, some models are better than others at separating data in a linear fashion. As we rarely know a priori which model, or which settings of their hyperparameters, is optimal at separating a specific dataset, it is prudent to test multiple machine learning methodologies across a single dataset. Automated workflows can be designed that split data into balanced training and test sets, apply feature reduction algorithms (if needed), test multiple machine learning models, and set their hyperparameters before returning the model and features used to best separate ones data and answer the clinical question being asked [19].

An example that demonstrates the usefulness of combining many of the topics that we have discussed in this chapter is that of Schmidt et al., who designed an image analysis and deep learning workflow to better predict a patient's response to ipilimumab [20]. They combined basic image analysis thresholding to classify lymphocytic infiltrating cells and applied deep learning to negate artifacts and necrosis in the images. Furthermore, they used pathologist annotations to train for specific regions of interest, where they compartmentalized the tumor and stroma prior to quantification of the lymphocytic contexture. Finally, they applied multiple methods of analyzing their resultant data before reporting the final optimal model that predicts response to treatment.

Beyond augmentation

To make deep learning an effective method for patient diagnosis in the clinic, there must be a sufficient amount of labeled annotations and a wide variety of samples to be representative of the larger population. In the case of deep learning, this requires training and validation on thousands of patient samples obtained from multiple international institutes and prepared by multiple individuals. This is not an easy or fast task to perform. A major drawback to the field is a lack of such large, well-annotated, and curated datasets that are available to data scientists. However, there is a wealth of data held in each hospital located in their glass microscope slide archives that go back decades. These data can be traced back to each patient treated in that institute along with their clinical reports. A sample tissue section from each patient in the archive will be stained with H&E, and as each prospective patient's samples are also stained with H&E, it makes sense to concentrate deep learning efforts on digital pathology samples prepared with this stain. The bottleneck is therefore not access to patient samples but the expert's digitized annotations of regions of interest that pertain to diagnosis. To overcome this bottleneck, researchers are forgoing image-level annotations and developing weakly supervised deep learning architectures that rely only on slide-level annotations, namely the diagnosis of the patient. Simply put, the computer is not told where in the image the cancer is, but rather that somewhere in the image there are cancer cells. This methodology has been shown to be effective and relies on the data-driven analysis of patients, where the computer vision algorithm identifies subtle and complex morphologies that relate to diagnosis or prognosis. Thus, the machine can now inform the human on what is of pathological significance within the patient's individual tissue specimen. These patterns may have gone unnoticed to date or have been too complex to allow for a standardized reporting protocol to be produced by human effort. This type of methodology has been tested in colorectal cancer [21], prostate cancer, and lymph nodal metastasis of breast cancer [22]. Further information on these methodologies can be found in the following chapter.

Practical concepts and theory of machine learning
Machine learning and digital pathology

The existing work in the realm of digital pathology, which analyzes various types of clinical data, could be conceptually organized in a variety of ways, that is, according to different criteria. Considering the diversity of pathologies of possible interest, one approach would be to consider solutions developed with specific diseases in mind, or indeed tissues or populations. Another criterion would be based on the type of annotations that is used for training feedback, available for learning. Yet another criterion could distinguish between different types of input data and the modalities used to acquire them.

For reasons which will become apparent as we develop our narrative, herein we found it useful to start with a broad categorization of work in the area into two categories in terms of their technical approach to the problem. The first of these could be roughly described as involving "conventional" (or "traditional") computer vision, and the second one as deep learning based. This distinction is somewhat loose, but in that it reflects the current trends in the field, namely the shift in focus from the former to the latter, it will be useful for illustrating the driving challenges and opportunities in the field, as well as for highlighting potential pitfalls and limitations of the current thinking, and for explaining our thoughts and recommendations for future work.

CHAPTER 12 Precision medicine in digital pathology

Considering the variety of challenges that a pathologist has to deal with in the analysis of real-world cases, it should come as no great surprise to the reader that algorithmic solutions addressing a wide range of possibly useful tasks have been explored and proposed. Much, indeed probably most, of research has focused on diagnostic tasks or prognostic stratification (e.g., high vs. low risk). Others provide a more assistive role to a human expert by providing automatic or semiautomatic segmentation of images, classification of cells, detection of structures of interest (cells, tissue types, tumors, etc.) from images, and so on. Although future work may lead to the fully automatic diagnosis of patients, currently, the drive is for clinically transferable tools to aid the pathologist in their diagnosis.

Common techniques

The common clinical questions highlighted in the previous section can all be broadly seen as having the same form as the two most frequently encountered machine learning paradigms: classification (e.g., "is a disease present or not?" and "is the patient in the high-risk category or not"?) and regression (e.g., "what is the severity of disease?" and "what is the patient's life expectancy?"). Therefore it can hardly come as much of a surprise to observe that most of the work in the area to date involves the adoption and adaptation of well-known and well-understood existing techniques and their application on pathology data. Here we give the reader a flavor of the context and pros and cons of some of these that have been applied to analyzing the multiparametric data extracted from the image analysis of digital pathology specimens.

Supervised learning

Even the mere emergence of digitization in clinical pathology, that is, the use of computerized systems for storing, logging, and linking information, has led to the availability of vast amounts of labeled data as well as the associated meta-data. The reduced cost of computing power and storage, increased connectivity, and widespread adoption of technology have all contributed greatly to this trend, which has escalated yet further with the rising recognition of the potential of artificial intelligence. Consequently, supervised machine learning in its various forms has attracted a great amount of research attention and continues to be one of the key focal points of ongoing research efforts.

Both shallow and deep learning algorithms have been successfully applied to clinical problems. Deep learning strategies based on layer depth and architecture of neural network-based models have been discussed in detail in Chapters 2–4. This section will therefore focus on the mathematical underpinnings of shallow learning algorithms such as Naïve Bayes, logistic regression, support vector machines, and random forests. In contrast, the mathematics of deep learning models are highly complex, as they encompass multilayer perceptron models, probabilistic graphical models, residual and recurrent networks, reinforcement and evolutionary learning, and so on. A detailed overview of these is presented in Chapters 2–4, but the mathematics involved are beyond the scope of the book. For a rigorous mathematical treatment of deep learning, see *Deep Learning* by Goodfellow, Bengio, and Courville or any of the other modern texts in the field.

Naïve Bayes assumption-based methods

Naïve Bayes classification applies the Bayes theorem by making the "naïve" assumption of feature independence. Formally, given a set of n features $x_1, ..., x_n$, the associated pattern is deemed as belonging to the class y which satisfies the following condition:

$$y = \arg_j P(C_j) \prod_{i=1}^{n} p(x_i|C_j) \tag{12.1}$$

where $P(C_j)$ is the prior probability of the class Cj and $p(x_i|C_j)$ is the conditional probability of the feature x_i given class C_j (readily estimated from data using a supervised learning framework) [23].

The key potential weakness of naïve Bayes-based algorithms, be they regression or classification oriented, is easily spotted, and it lies in the unrealistic assumption of feature independence. Yet, somewhat surprisingly at first sight, these simple approaches often work remarkably well in practice and often outperform more complex and, as regards the fundamental assumptions of feature relatedness, more expressive and more flexible models [19].

There are a few reasons why this might be the case. One of these includes the structure of errors—if the conceptual structure of data relatedness under a given representation is in a sense symmetrical, errors in the direction of overestimating conditional probabilities and those in the direction of underestimating them can cancel out in the aggregate, leading to more accurate overall estimates [24]. Another equally important factor contributing to often surprisingly good performance of methods which make the naïve Bayes assumption emerges as a consequence of the relationship between the amount of available training data (given a problem of a specific complexity) and the number of free parameters of the adopted model. It is often the case, especially considering that in digital pathology class imbalance poses major practical issues, that more complex models cannot be sufficiently well trained; thus, even if, in principle, able to learn a more complex functional behavior, this theoretical superiority cannot be exploited.

While a good starting point and a sensible baseline, naïve Bayes-based methods are in the right circumstances outperformed by more elaborate models, some of which we summarize next.

Logistic regression-based methods

Logistic regression is another widely used, well-understood, and often well-performing supervised learning technique. In logistic regression, the conditional probability of the dependent variable (class) y is modeled as a logit-transformed multiple linear regression of the explanatory variables (input features) $x_1, ..., x_n$:

$$P_{LR}(y = \pm 1|x, w) = \frac{1}{1 + e^{-yw^Tx}}. \tag{12.2}$$

The model is trained (i.e., the weight parameter w learned) by maximizing the likelihood of the model on the training dataset, given by:

$$\prod_{i=1}^{2} \Pr(y_i | x_i, w) = \prod_{i=1}^{2} \frac{1}{1 + e^{-y_i w^T x_i}},$$

(12.3)

penalized by the complexity of the model:

$$\frac{1}{\sigma \sqrt{2\pi}} e^{\frac{-1}{2\sigma^2} w^T w},$$

(12.4)

which can be restated as the minimization of the following regularized negative log-likelihood:

$$L = C \sum_{i=1}^{2} \log\left(1 + e^{-y_i w^T x_i}\right) + w^T w.$$

(12.5)

A coordinate descent approach, such as the one described by Yu et al. [25], can be used to minimize L.

Support vector-based methods

Support vector machines perform classification by constructing a series of class separating hyperplanes in a high-dimensional (potentially infinitely dimensional) space into which the original input data are mapped [26]. For comprehensive detail of this regression technique, the reader is referred to the original work by Vapnik [27]; herein we present a summary of the key ideas.

In the context of support vector machines, the seemingly intractable task of mapping data into a very high-dimensional space is achieved efficiently by performing the aforesaid mapping implicitly rather than explicitly. This is done by employing the so-called *kernel trick*, which ensures that dot products in the high-dimensional space are readily computed using the variables in the original space. Given labeled training data (input vectors and the associated labels) in the form $\{(x_1, y_1), \ldots, (x_n, y_n)\}$, a support vector machine aims to find a mapping which minimizes the number of misclassified training instances, in a regularized fashion. As mentioned earlier, an implicit mapping of input data $x \rightarrow \Phi(x)$ is performed by employing a Mercer-admissible kernel [28] $k(x_i, x_j)$, which allows for the dot products between mapped data to be computed in the input space: $\Phi(x_i) \cdot \Phi(x_j) = k(x_i, x_j)$. The classification vector in the transformed, high-dimensional space of the form

$$w = \sum_{i=1}^{n} c_i y_i \Phi(x_i)$$

(12.6)

is sought by minimizing

$$\sum_{i=1}^{n} c_i - \frac{1}{2} \sum_{i=1}^{n} \sum_{j=1}^{n} y_i c_i k\left(x_i, x_j\right) y_j c_j$$

(12.7)

subject to the constraints $\sum_{i=1}^{n} c_i y_i = 0$ and $0 \leq c_i \leq 1/(2n\lambda)$. The regularizing parameter λ penalizes prediction errors. Support vector-based approaches usually

Practical concepts and theory of machine learning **247**

perform well even with relatively small training datasets and have the advantage of well-understood mathematical behavior (which is an important consideration in the context of regular compliance, among others).

Stepping back for a moment from the technical detail, intuitively what is happening here is that the algorithm is learning which class exemplars are the "most problematic" ones, i.e., which exemplars are nearest to the class boundaries and thus most likely to be misclassified. These are the support vectors that give the approach its name. Inspection of these is insightful. Firstly, a large number of support vectors (relative to the total amount of training data) should immediately raise eyebrows as it suggests overfitting. Secondly, by examining which exemplars end up as support vectors, an understanding of the nature of learning that took place can be gained along with that of the structure of the problem and data representation, which can lead to useful and novel clinical insight.

Nonparametric, k-nearest neighbor-based methods

The k-nearest neighbor classifier classifies a novel pattern comprising features $x_1, \ldots, x_n$ to the class dominant in the set of k-nearest neighbors to the input pattern (in the feature space) among the training patterns with known class memberships [29]. The usually employed similarity metric used is the Euclidean distance.

Random forests

Random forest classifiers fall under the broad umbrella of ensemble-based learning methods [30]. They are simple to implement, fast in operation, and have proven to be extremely successful in a variety of domains [31,32]. The key principle underlying the random forest approach comprises the construction of many "simple" decision trees in the training stage and the majority vote (mode) across them in the classification stage. Among other benefits, this voting strategy has the effect of correcting for the undesirable property of decision trees to overfit training data [33]. In the training stage, random forests apply the general technique known as bagging to individual trees in the ensemble. Bagging repeatedly selects a random sample with replacement from the training set and fits trees to these samples. Each tree is grown without any pruning. The number of trees in the ensemble is a free parameter which is readily learned automatically using the so-called out-of-bag error [29].

Much like in the case of naïve Bayes- and k-nearest neighbor-based algorithms, random forests are popular in part due to their simplicity on the one hand and generally good performance on the other. However, unlike the former two approaches, random forests exhibit a degree of unpredictability as regards the structure of the final trained model. This is an inherent consequence of the stochastic nature of tree building. As we will explore in more detail shortly, one of the key reasons why this characteristic of random forests can be a problem is regulatory—clinical adoption often demands a high degree of repeatability not only in terms of the ultimate performance of an algorithm but also in terms of the mechanics as to how a specific decision is made.

Unsupervised learning

We have already mentioned the task of patient stratification. Indeed, the need for stratification emerges frequently in digital pathology, for example, due to the heterogeneity of many diseases or differential response of different populations to treatment or the disease itself [19].

Given that the relevant strata are often unknown a priori, often because of the limitations imposed by previously exclusively manual interpretation of data and the scale at which the data would need to be examined to draw reliable conclusions, it is frequently desirable to stratify automatically. A common way of doing this is by means of unsupervised learning, by applying a clustering algorithm such as a Gaussian mixture model or, more frequently, due to its simplicity and fewer free parameters, the k-means algorithm [21]. Then any subsequent learning can be performed in a more targeted fashion by learning separate models for each of the clusters individually.

Let $X = \{x_1, x_2, ..., x_n\}$ be a set of d-dimensional feature vectors. The k-means algorithm partitions the points into K clusters, $X_1, ..., X_K$, so that each datum belongs to one and only one cluster. In addition, an attempt is made to minimize the sum of squared distances between each data point and the empirical mean of the corresponding cluster. In other words, the k-means algorithm attempts to minimize the following objective function:

$$J(X_1, ..., X_k) = \sum_{i=1}^{k} \sum_{x \in X_i} (c_i - x)^2,$$ (12.8)

where the empirical cluster means are calculated as:

$$c_i = \sum_{x \in X_i} x / |X_i|.$$ (12.9)

The exact minimization of the objective function in Ref. [1] is an NP-hard problem [34]. Instead, the k-means algorithm only guarantees convergence to a local minimum. Starting from an initial guess, the algorithm iteratively updates cluster centers and data-cluster assignments until (1) a local minimum is attained or (2) an alternative stopping criterion is met (e.g., the maximal desired number of iterations or a sufficiently small sum of squared distances). The k-means algorithm starts from an initial guess of cluster centers.

Often, this is achieved simply by choosing k data points at random as the centers of the initial clusters, although more sophisticated initialization methods have been proposed [35,36]. Then, at each iteration $t = 0, ...$ the new datum-cluster assignment is computed:

$$X_i^{(t)} = \left\{ x : x \in X \wedge \arg_j \left(x - c_j^{(t)} \right)^2 = i \right\}.$$ (12.10)

In other words, each datum is assigned to the cluster with the nearest (in the Euclidean sense) empirical mean. Lastly, the locations of cluster centers are recomputed from the new assignments by finding the mean of the data assigned to each cluster:

$$c_j^{(t+1)} = \sum_{x \in X_i^{(t)}} x/|X_i^{(t)}|. \tag{12.11}$$

The algorithm is guaranteed to converge because neither of the updates in Ref. [3] nor [4] can ever increase the objective function of Ref. [1]. Various extensions of the original k-means algorithm include fuzzy c-means [37], k-medoid [38], kernel k-means [39], as well as discriminative k-means when partial labeling is available [40]. The interested reader is referred to the cited publications for further information.

Image-based digital pathology

Probably one of the most interesting areas of research in digital pathology, in terms of both practical potential and the nature of intellectual challenges, concerns the use of images.

Due to the inherent complexity of images of interest in pathology, the aforementioned distinction between what we term "conventional" machine learning-based and deep learning-based methods is particularly pronounced in this realm of digital pathology. In an outline, the former usually takes on a modular pipeline form whereby explicitly engineered features (sometimes referred to as "handcrafted" features) are extracted from input images first and then fed into a machine learning algorithm of the sort described in the previous section. In contrast, in deep learning-based approaches, the process is not atomized [41] in this manner—good features, such as complex functions of directly sensed data (e.g., pixel intensities), are learned automatically from labeled training data. This distinction has often led to deep learning-based methods being characterized as "data agnostic" [42]; colloquially put, the idea is that no matter what the input is, the algorithm will learn what good features are. However, to describe deep learning in this manner would be a misnomer and highly misleading. For any learning to take place, the model has to be constrained and, by virtue of these constraints, contain implicit information about the problem at hand; succinctly put, a tabular rasa cannot learn. Indeed, one of the theorems popularly known as "no free lunch theorem" implicitly says the very same:

> *Any two optimization algorithms are equivalent when their performance is averaged across all possible problems. [18]*

Thus, deep learning algorithms too are not endowed with a quasi-magical ability to overcome this fundamental limitation but are also constrained by some prior knowledge (and are hence not agnostic). In particular, a deep neural network contains constraints, which emerge from the type of its layers, their order, the number of neurons in each layer, the connectedness of layers, and other architectural aspects. This is important to keep in mind, and deep learning should not be seen as necessarily superior to "conventional" approaches or as the universal solution to any problem.

Conventional approaches to image analysis

As mentioned already, the first broad category of algorithms that have been proposed as a means of analyzing images used in clinical pathology is one which uses what we term conventional, or traditional, computer vision. By this, we primarily refer to the first stage in the processing, which concerns the extraction of a priori, explicitly defined features. The key premise is appealing and intuitively sensible to use human expertise to identify what elementary, low-level elements of an image are salient in that they capture important discriminative information on the one hand and are yet compact enough to facilitate computational efficiency and reliable learning from available data.

Having demonstrated success on a wide variety of images of so-called natural scenes (to say the least, they have rather revolutionized the field) [43], local appearance descriptors originally proposed for more day-to-day applications of computer vision (e.g., location recognition, synthetic panoramic image generation, object localization, etc.) have been adopted first and applied on a diverse range of pathology image types [44,45]. Popular and widely used examples include local binary pattern (LBP) [46], scale invariant feature transform (SIFT) [47], and histogram of oriented gradients (HOG)-based descriptors [48].

While also "handcrafted" in contrast to the above general purpose descriptor-based features are features which directly and explicitly exploit domain-specific information, that is, human expert knowledge of pathology. Thus, body of interest counts (e.g., tumor and immune cells, etc.), their morphology quantifiers (e.g., size, eccentricity), or spatial statistics are some of the widely used ones. The appeal of these is twofold. First, by their very nature they can be reasonably expected to exhibit saliency in the context of pathology slide analysis. Moreover, they are usually rather effortlessly measured (extracted) from images: different staining and imaging modalities or biomarkers can be used to highlight the targets of interest, and simple image processing or computer vision algorithms (such as flood fill algorithms [49], morphological operators [50], blob detectors [45], etc.) usually suffice as regard to the computational side.

Deep learning on images

One of the reasons why deep learning has become such a popular area of research lies in its success when applied to many computer vision problems, including a number of those which have been prohibitively challenging using more traditional approaches [51]. Its application to image analysis in digital pathology has not disappointed either, with deep learning quickly establishing state-of-the-art performance in the context of a number of diverse tasks. Considering that the technical detail specifically of deep learning is addressed by several other chapters in the present volume, herein we focus our attention on the bigger picture with the aim of highlighting the key trends in the field, the main challenges, and inherent advantages and disadvantages of deep learning.

Rather than joining the choir singing praises to deep learning, a more useful, insightful, and instructive question to ask concerns the outstanding limitations. If the key obstacle were merely the availability of sufficient data (undoubtedly a difficulty encountered by virtually any single research group, perhaps save for the few tech giants), the solution would be merely a matter of time and concerted effort. Unfortunately, from the practical point of view, or fortunately, from the point of view of intellectual challenge, the answer is more nuanced. In particular, the major problem with current methods is posed by the sheer size of individual images, which are often in excess of 10 GB each. Training deep neural network images of this size (and bear in mind that deep learning methods are very training data hungry) is computationally impractical.

A survey of existing work in the field shows two main ways of dealing with the problem posed by the size of WSIs. The first of these involves severe downsampling—a multiple gigapixel image is scaled down to an image with mere tens of thousands of pixels. The second approach consists of dividing a WSI into small patches (usually nonoverlapping), performing learning on these as individual images, and then aggregating patch-level decisions to arrive at a slide-level decision. Both approaches can be seen to lead to information loss, albeit in different ways. As formally demonstrated by Nyquist in the 1930s, but equally easily understood intuitively, the type of information loss incurred by the downsampling approach comes in the form of loss of fine detail (i.e., high spatial frequencies). Even to a human expert, this information is important in the interpretation of a tissue slide and can be expected to be even more so to computer-based methods, which do not suffer from many of the imperfections of the human visual system. The type of information loss that takes place in patch-based algorithms is rather different and more nuanced, so it is useful to highlight a few influential works before continuing the discussion.

In the simplest form, patches are treated entirely separately from one another. This approach has yielded promising results in patch-level distinction between tumorous and normal tissues [52] and the segmentation of precursor lesions [53], but on prognostic tasks, the success has been much more limited [54]. Moving away from patch level decision as the ultimate aim, Hou et al. [55] proposed a method whereby a deep neural network is trained on individual patches, and patch-level decisions are aggregated by virtue of voting to make slide-level predictions. Recently, a few methods which address the challenges of intraslide heterogeneity have been introduced, broadly comprising adaptive selection of patches, phenotype-based patch clustering, automatic clustering selection, and learned aggregation of cluster-level predictions. Initial attempts were disappointing [56], but more subsequent refinements of the methodology have led to much more promising results [21].

What the above fly-through the field brings out to the fore are some of the key limitations of patch-based methods, which dominate the existing literature. First, by decomposing a WSI into patches, it is virtually unavoidable but to resort to a form of ad hoc labeling of patches, usually attaching the whole slide label to all of the patches extracted from it. This is clearly a rather unattractive proposition as not every region in a slide containing a pathology is equally informative; indeed, some may contain perfectly healthy tissue. Hence, the learning algorithm is effectively fed incorrect data. Moreover, problems arise in the aggregation stage too. Lastly, there is a loss of spatial information effected by the separation of patches and the arbitrariness of the loci of patch boundaries with respect to the imaged physical tissue.

Regulatory concerns and considerations

Thus far, our focus has been on the fundamental challenges associated with the use of images for tasks in digital pathology, the manner in which these have shaped the field to date, and the trends most likely to yield advances in the future. Related to these, equally important in practice though somewhat different in nature are the requirements and demands associated with the adoption of artificial intelligence in everyday clinical processes. Oftentimes, these are imposed by various regulatory bodies, such as the Food and Drug Administration (FDA) in the United States or, more loosely, the end users themselves. While a comprehensive examination of this topic is beyond the scope of this chapter, it is important to draw the reader's attention to some of the key issues within this area. In particular, we highlight three issues:

- Explainability
- Repeatability
- Generalizability

Explainability in the context of relevance here concerns the ability for a decision made by an algorithmic, artificial intelligence-based solution to be communicated to a human on a semantically appropriate level. In other words, the question being asked is not just what the decision is but rather how or why the algorithm made that decision. As alluded to before, for many of the algorithms covered under the umbrella of conventional approaches to image understanding, this is reasonably straightforward. For example, in the k-nearest neighbor-based solutions, the nearest neighbors, that is, previously seen and human-labeled examples which exhibit the greatest similarity to a new example under consideration can be brought up. With support vector–based methods, the closest examples (in a kernel sense) or the support vectors which define class boundaries can be similarly used to gain insight into what was learned and how a decision was made. With random forests, a well-known method of substituting features with dummy features can be used to quantify which features are the most important ones in decision-making and "sanity checked" by an expert. This process can be not only confirmatory but can also lead to novel clinical insight. While explainability has for a long time been seen as a potential disadvantage (cf. with human memory and brain: can one localize where the concept of "cake" is stored?) of deep learning-based approaches, which used to be seen as proverbial "black boxes," recent years have seen huge strides of progress in this area [57]. For example, looking at which neurons in a network fire together (cf. the mantra of biological reinforcement learning "what fires together wires together") can be insightful. A higher (semantic) level of insight can be gained by looking at different layers and visualizing features learned—typically layers closer to input tend to learn simple, low-level appearance elements, which are then combined to compose more complex visual patterns downstream [58]. Another ingenious technique involves so-called "occlusion," whereby parts of an input image are occluded by a uniform pattern (thus effecting localized information loss) and quantifying the impact such occlusion has on the decision—the greater the impact, the greater the importance of a particular image locus [59].

The issues of repeatability and reproducibility have gained much prominence in academic circles, and wider, in recent years. The difference between the two concepts has been convincingly highlighted and discussed by Drummond [60], but this nuance does not appear to have penetrated regulatory processes as of yet. In particular, the practical impossibility of perfect reproducibility of experimental outcomes for some machine learning algorithms poses a major obstacle to their adoption in clinical practice due to the stochastic nature of their operation. We have already alluded to this in the previous section in our overview of random forests. In particular, even if the same training data are used and the same manner of training is employed, the parameters of a trained random forest or neural network will differ from instance to instance. This is an inherent consequence of the stochastic elements of their training processes and is something that does not sit comfortably with some. While this sentiment is not difficult to relate to, it does illustrate what can be argued to be an inconsistency in how human and machine expertise are treated. In particular, the former has been shown to exhibit major interpersonal variability (two different, competent, and experienced pathologists arriving at different conclusions from the same data), as well as intrapersonal one (e.g., depending on the degree of fatigue, time of day, whether the decision is made in the anteprandial or postprandial period, etc.). This kind of double standard is consistent with a broad range of studies involving human-human and human-machine interaction and is neurologically well understood, with the two types of engagement differently engaging important brain circuitry such as the ventromedial prefrontal cortex and the amygdala [61]. Generalizability is a concept that pervades machine learning. It refers to the ability of an algorithm to learn from training data some information contained within it, which would allow it to make good decisions on previously unseen input, often seemingly rather different from anything seen in the training stage. The issue of generalizability underlies the tasks of data representation, problem abstraction, mathematical modeling of learning, etc. Herein we are referring to generalizability in a very specific context. Namely, a major challenge in the practice of pathology concerns different protocols and conditions in which data are acquired. Put simply, the question being asked is what performance can I expect to see from an algorithm evaluated using data a technician acquired using particular equipment on one cohort of patients in a specific lab when it is applied to data acquired by a different technician in a different lab from a different cohort. It can be readily seen that slight changes in the data acquisition (such as the duration of exposure to a dye), different physical characteristics of lab instruments, or indeed different demographics of patients all pose reasonable

Acknowledgments

The authors would like to acknowledge Inés Nearchou for kindly providing images for the figures within this chapter.

References

[1] Caie PD, Walls RE, Ingleston-Orme A, Daya S, Houslay T, Eagle R, et al. High-content phenotypic profiling of drug response signatures across distinct cancer cells. Mol Cancer Ther 2010;9(6):1913–26.

[2] Deans GT, Heatley M, Anderson N, Patterson CC, Rowlands BJ, Parks TG, et al. Jass' classification revisited. J Am Coll Surg 1994;179(1):11–7.

[3] Lim D, Alvarez T, Nucci MR, Gilks B, Longacre T, Soslow RA, et al. Interobserver variability in the interpretation of tumor cell necrosis in uterine leiomyosarcoma. Am J Surg Pathol 2013;37(5):650–8.

[4] Chandler I, Houlston RS. Interobserver agreement in grading of colorectal cancers— findings from a nationwide web-based survey of histopathologists. Histopathology 2008;52(4):494–9.

[5] Caie PD, Zhou Y, Turnbull AK, Oniscu A, Harrison DJ. Novel histopathologic feature identified through image analysis augments stage II colorectal cancer clinical reporting. Oncotarget 2016;7(28):44381–94.

[6] Nearchou IP, Lillard K, Gavriel CG, Ueno H, Harrison DJ, Caie PD. Automated analysis of lymphocytic infiltration, tumor budding, and their spatial relationship improves prognostic accuracy in colorectal cancer. Cancer Immunol Res 2019;7(4):609–20.

[7] Khameneh FD, Razavi S, Kamasak M. Automated segmentation of cell membranes to evaluate HER2 status in whole slide images using a modified deep learning network. Comput Biol Med 2019;110:164–74.

[8] Widmaier M, Wiestler T, Walker J, Barker C, Scott ML, Sekhavati F, et al. Comparison of continuous measures across diagnostic PD-L1 assays in non-small cell lung cancer using automated image analysis. Mod Pathol 2019;33.

[9] Puri M, Hoover SB, Hewitt SM, Wei BR, Adissu HA, Halsey CHC, et al. Automated computational detection, quantitation, and mapping of mitosis in whole-slide images for clinically actionable surgical pathology decision support. J Pathol Inform 2019;10:4.

[10] Brieu N, Gavriel CG, Nearchou IP, Harrison DJ, Schmidt G, Caie PD. Automated tumour budding quantification by machine learning augments TNM staging in muscle-invasive bladder cancer prognosis. Sci Rep 2019;9(1):5174.

[11] Brieu N, Gavriel CG, Harrison DJ, Caie PD, Schmidt G. Context-based interpolation of coarse deep learning prediction maps for the segmentation of fine structures in immunofluorescence images. SPIE 2018.

[12] Roy S, Kumar Jain A, Lal S, Kini J. A study about color normalization methods for histopathology images. Micron 2018;114:42–61.

[13] Cruz-Roa A, Gilmore H, Basavanhally A, Feldman M, Ganesan S, Shih NNC, et al. Accurate and reproducible invasive breast cancer detection in whole-slide images: a deep learning approach for quantifying tumor extent. Sci Rep 2017;7:46450.

[14] Litjens G, Bandi P, Ehteshami Bejnordi B, Geessink O, Balkenhol M, Bult P, et al. 1399 H&E-stained sentinel lymph node sections of breast cancer patients: the CAMELYON dataset. GigaScience 2018;7(6).

[15] Coudray N, Ocampo PS, Sakellaropoulos T, Narula N, Snuderl M, Fenyö D, et al. Classification and mutation prediction from non–small cell lung cancer histopathology images using deep learning. Nat Med 2018;24(10):1559–67.

[16] Sirinukunwattana K, Domingo E, Richman S, Redmond KL, Blake A, Verrill C, et al. Image-based consensus molecular subtype classification (imCMS) of colorectal cancer using deep learning. bioRxiv 2019; 645143.

[17] Havel JJ, Chowell D, Chan TA. The evolving landscape of biomarkers for checkpoint inhibitor immunotherapy. Nat Rev Cancer 2019;19(3):133–50.

[18] Wolpert DH, Macready WG. No free lunch theorems for optimization. Trans Evolut Comput 1997;1(1):67–82.

[19] Dimitriou N, Arandjelović O, Harrison DJ, Caie PD. A principled machine learning framework improves accuracy of stage II colorectal cancer prognosis. NPJ Digit Med 2018;1(1):52.

[20] Harder N, Schonmeyer R, Nekolla K, Meier A, Brieu N, Vanegas C, et al. Automatic discovery of image-based signatures for ipilimumab response prediction in malignant melanoma. Sci Rep 2019;9(1):7449.

[21] Yue X, Dimitriou N, Arandjelovic O. Colorectal cancer outcome prediction from H&E whole slide images using machine learning and automatically inferred phenotype profiles, 2019. 1 February. Available from: https:/ui.adsabs.harvard.edu/abs/2019arXiv190203582Y.

[22] Campanella G, Hanna MG, Geneslaw L, Miraflor A, Werneck Krauss Silva V, Busam KJ, et al. Clinical-grade computational pathology using weakly supervised deep learning on whole slide images. Nat Med 2019;25(8):1301–9.

[23] Bishop CM. Pattern recognition and machine learning. New York, USA: Springer-Verlag; 2007.

[24] Vente D, Arandjelović O, Baron V, Dombay E, Gillespie S. Using machine learning for automatic counting of lipid-rich tuberculosis cells influorescence microscopy images. In: Proc. AAAI conference on artificial intelligence workshop on health intelligence; 2019.

[25] Yu H-F, Huang F-L, Lin C-J. Dual coordinate descent methods for logistic regression and maximum entropy models. Mach Learn 2011;85(1–2):41–75.

[26] Schölkopf B, Smola A, Müller K. Advances in kernel methods – SV learning, chapter kernel principal component analysis. Cambridge, MA: MIT Press; 1999. p. 327–52.

[27] Vapnik V. The nature of statistical learning theory. Springer-Verlag; 1995.

[28] Mercer J. Functions of positive and negative type and their connection with the theory of integral equations. Philos Trans Roy Soc A 1909;209:415–46.

[29] Cunningham P, Delany SJ. K-nearest neighbour classifiers. In: Multiple classifier systems; 2007. p. 1–17.

[30] Breiman L. Random forests. Mach Learn 2001;45(1):5–32.

[31] Cutler DR, Edwards TC, Beard KH, Cutler A, Hess KT, Gibson J, Lawler JJ. Random forests for classification in ecology. Ecology 2007;88(11):2783–92.

[32] Ghosh P, Manjunath B. Robust simultaneous registration and segmentation with sparse error reconstruction. IEEE Trans Pattern Anal Mach Intell 2013;35(2):425–36.

[33] Zadrozny B, Elkan C. Obtaining calibrated probability estimates from decision trees and naive Bayesian classifiers. In: Proceedings IMLS international conference on machine learning, 1; 2001. p. 609–16.

[34] Dasgupta S. The hardness of k-means clustering technical report CS2007-0890. San Diego: University of California; 2007.

[35] Khan SS, Ahmadb A. Cluster center initialization algorithm for k- means clustering. Pattern Recogn Lett 2004;25(11):1293–302.

[36] Peña JM, Lozano JA, Larrañaga P. An empirical comparison of four initialization methods for the k-means algorithm. Pattern Recogn Lett 1999;20(10):1027–40.

[37] Dunn JC. A fuzzy relative of the ISODATA process and its use in detecting compact well-separated clusters. J Cybern 1973;3:32–57.

[38] Kaufman L, Rousseeuw PJ. Finding groups in data: an introduction to cluster analysis. Wiley; 2005.

[39] Schölkopf B, Smola A, Müller K-R. Nonlinear component analysis as a kernel eigenvalue problem. Neural Comput 1998;10(5):1299–319.

[40] Arandjelović O. Discriminative k-means clustering. In: Proc. IEEE international joint conference on neural networks; 2013. p. 2374–80.

[41] Arandjelović O. A more principled use of the p-value? Not so fast: a critique of Colquhoun's argument. Royal Society Open Science; 2019.

[42] Kooi T, Litjens G, Van Ginneken B, Gubern-Mérida A, Sánchez I, Mann R, den Heeten A, Karssemeijer N. Large-scale deep learning for computer aided detection of mammographic lesions. Med Image Anal 2017;35:303–12.

[43] Niemeyer M, Arandjelović O. Automatic semantic labelling of images by their content using non-parametric Bayesian machine learning and image search using synthetically generated image collages. In: Proc. IEEE international conference on data science and advanced analytics; 2018. p. 160–8.

[44] Alhindi TJ, Kalra S, Ng KH, Afrin A, Tizhoosh HR. Comparing lbp, hog and deep features for classification of histopathology images. In: 2018 International joint conference on neural networks (IJCNN). IEEE; 2018. p. 1–7.

[45] Xing F, Yang L. Robust nucleus/cell detection and segmentation in digital pathology and microscopy images: a comprehensive review. IEEE Rev Biomed Eng 2016;9:234–63.

[46] Fan J, Arandjelović O. Employing domain specific discriminative information to address inherent limitations of the LBP descriptor in face recognition. In: Proc. IEEE international joint conference on neural networks; 2018. p. 3766–72.

[47] Lowe DG. Distinctive image features from scale-invariant keypoints. Int J Comput Vis 2003;60(2):91–110.

[48] Arandjelović O. Object matching using boundary descriptors. In: Proc British machine vision conference; 2012. https://doi.org/10.5244/C.26.85.

[49] Mehta N, Rajas A, Chaudhary V. Content based sub-image retrieval system for high-resolution pathology images using salient interest points. In: 2009 Annual international conference of the IEEE engineering in medicine and biology society. IEEE; 2009. p. 3719–22.

[50] Karunakar Y, Kuwadekar A. An unparagoned application for red blood cell counting using marker controlled watershed algorithm for android mobile. In: 2011 Fifth international conference on next generation mobile applications, services and technologies. IEEE; 2011. p. 100–4.

[51] Zhu J-Y, Park T, Isola P, Efros AA. Unpaired image-to-image translation using cycle-consistent adversarial networks. In: Proceedings of the IEEE international conference on computer vision; 2017. p. 2223–32.

[52] Jamaluddin MF, Fauzi MFA, Abas FS. Tumor detection and whole slide classification of h&e lymph node images using convolutional neural network. In: Proc. IEEE International conference on signal and image processing applications; 2017. p. 90–5.

[53] Albayrak A, Ünlü A, Çalık N, Bilgin G, Türkmen İ, Çakır A, Çapar A, Töreyin BU, Ata LD. Segmentation of precursor lesions in cervical cancer using convolutional neural networks. In: Proc. Signal processing and communications applications conference; 2017. p. 1–4.

[54] Mackillop WJ. The importance of prognosis in cancer medicine. TNM Online; 2003.

[55] Hou L, Samaras D, Kurc TM, Gao Y, Davis JE, Saltz JH. Patch-based convolutional neural network for whole slide tissue image classification. In: Proc. IEEE conference on computer vision and pattern recognition; 2016. p. 2424–33.

[56] Zhu X, Yao J, Zhu F, Huang J. Wsisa: Making survival prediction from whole slide histopathological images. In: IEEE conference on computer vision and pattern recognition; 2017. p. 7234–42.

[57] Erhan D, Bengio Y, Courville A, Vincent P. Visualizing higher-layer features of a deep network. Univ Montreal 2009;1341(3):1.

[58] Cooper J, Arandjelović O. Visually understanding rather than merely matching ancient coin images. In: Proc. INNS conference on big data and deep learning; 2019.

[59] Schlag I, Arandjelović O. Ancient Roman coin recognition in the wild using deep learning based recognition of artistically depicted face profiles. In: Proc. IEEE international conference on computer vision; 2017. p. 2898–906.

[60] Drummond C. Replicability is not reproducibility: NOr is it good science; 2009.

[61] Kätsyri J, Hari R, Ravaja N, Nummenmaa L. The opponent matters: elevated fmri reward responses to winning against a human versus a computer opponent during interactive video game playing. Cereb Cortex 2013;23(12):2829–39.

CHAPTER

Generative deep learning in digital pathology

13

David Morrison[a,b,c], David Harris-Birtill[b], and Peter D. Caie[d]

[a]*School of Medicine, University of St Andrews, St Andrews, United Kingdom,* [b]*School of Computer Science, University of St Andrews, St Andrews, United Kingdom,* [c]*Sir James Mackenzie Institute for Early Diagnosis, University of St Andrews, St Andrews, United Kingdom,* [d]*Indica Labs, Albuquerque, NM, United States*

Introduction

Clinical histopathology is at an exciting paradigm shift, with many laboratories replacing traditional microscopy with high-resolution scanners and large digital displays. Unlike traditional slides, digital images can be shared electronically, marked up simultaneously by multiple pathologists, and assessed automatically [1]. The deployment into clinical practice of systems that automate and augment diagnostic reporting is expected to lead to a significant increase in assessment capacity alongside quicker reporting times. This chapter provides a brief introduction to deep generative models, reviews their current use in digital pathology, and envisions their future applications within the field. To contextualize this work, deep generative models are discussed in relation to the current state-of-the-art deep learning techniques for pathology and the problems that generative techniques can solve within a conventional pipeline.

Before discussing the place that generative models could take in the field of automated histopathology, it is necessary to describe the current typical workflow of machine learning in digital pathology and some of the common issues that can hinder downstream reporting tasks. A taxonomy of data science tasks, independent of pathology, organized into five categories, undertaken sequentially: obtain, scrub, explore, model, and interpret (dataists, http:/www.dataists.com/2010/09/a-taxonomy-of-data-science, accessed 18 September 2020) are shown in Fig. 13.1. This model can be used to understand the process of applying machine learning in digital pathology. Data are obtained through the fixing, staining, and scanning of tissue to transform into a set of whole slide images. These images are then scrubbed, or preprocessed, to remove artifacts and prepared to be used in the modeling phase. Tasks such as stain normalization, data augmentation, and patch generation fall into this category. In the exploration phase, resulting scrubbed data are analyzed, either

Artificial Intelligence in Pathology. https://doi.org/10.1016/B978-0-323-95359-7.00013-3
Copyright © 2025 Elsevier Inc. All rights are reserved, including those for text and data mining, AI training, and similar technologies.

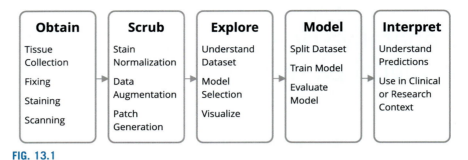

FIG. 13.1

A taxonomy of data science tasks applied to automated whole slide image analysis.

automatically or by a human, to determine an appropriate modeling technique such as a specific neural network architecture. A large number of different pathologies and tissue types may be of interest in digital pathology. This makes it impractical to iteratively try every possible modeling technique and, in the case of ensemble learning, every combination of techniques. The machine learning system is trained and evaluated during the modeling phase. In the interpretation phase, human pathologists are presented with the predictions of the model, which can be used for clinical or research work.

The automation of whole slide image (WSI) analysis and diagnosis presents several significant challenges [2]. First and foremost is the issue of data size; whole slide images are multigigabyte images in the range of approximately $100{,}000 \times 100{,}000$ pixels. This makes a direct application of modern computer vision algorithms on nonspecialist computing hardware impractical. Typical solutions to overcome this include downsampling the image and breaking the image up into smaller subimages called patches.

Second, data availability is problematic for most researchers. Supervised machine learning requires labels for each sample. In WSI analysis, this may mean assigning a category to each slide as a whole, identifying a set of points of interest on the tissue, or drawing around areas to segment tissue types or pathologies. For each of these, a trained specialist in histopathology is required. The process is time consuming and expensive, and there is often a lot of interobserver and intraobserver variability between the labels provided by pathologists. As a result, data sets used to train automated digital pathology models tend to be small compared with those available in other computer vision subfields, such as ImageNet [3], where nonspecialists can straightforwardly provide labels (e.g., labeling a cat vs. a dog). This situation, however, has been improved by the release of tissue-annotated open data sets, such as Camelyon16 [4] and Camelyon17 [5]. Furthermore, initiatives such as iCAIRD (iCAIRD, https://icaird.com) and PathLAKE (PathLAKE, https://www.pathlake.org, both accessed 18 September 2020) provide large, well-annotated, and curated WSI data sets linked to clinicopathologic data. These make rich digital pathology training material widely available, albeit within narrowly defined clinical reporting and specific tissue types.

Third, WSI analysis experiences several domain-specific image artifacts caused by the process of surgical removal, fixing, cutting, staining, and scanning the tissue. These can include folds in the tissue, retraction artifacts, variations in the application of chemicals in the staining process, small cracks and imperfections in the glass slide and coverslip, partial blurring of the image caused by focusing errors, and image resolution and compression differences between different scanners and file formats.

Despite these challenges, computer vision techniques based on supervised and weakly supervised learning have been used to successfully automate some common assessment tasks in histopathology. These include, for example, cell nucleus identification, pathology classification, and cancer segmentation [6]. Unsurprisingly, state-of-the-art results on slide classification tasks, such as the work by Campanella et al. [7] on prostate cancer, basal cell carcinoma, and breast cancer nodal metastases, rely on large data sets.

Deep generative models

This section briefly introduces required terminology from computer vision, deep learning, and generative modeling before describing their uses in a digital pathology workflow. First, an image filter or kernel is a rectangular matrix that can be applied to parts of a digital image to extract information, called features, from it. To apply a filter, a dot-product is performed (component-wise multiplication followed by a sum) between the filter and a section of the image with the same dimensions. In computer vision, this operation is referred to as a convolution. By sliding the filter across the image and performing the convolution at each point, this operation can produce a new matrix known as a feature map. Filters that recognize primitive features, such as horizontal or vertical lines, can be hand crafted; however, more complex features must be learned by the model. Neural networks are the most commonly used machine learning approaches [8]. A convolutional neural network [9] is a machine learning approach that enables image filters to be learned from data rather than programmed explicitly.

Generative models are an approach to machine learning in which systems attempt to estimate the probability of a specific sample being picked at random based on training data [10]. Once there is an estimate for the probability density function over the training set, the model can be used to generate new examples. For example, a model can be trained to generate new images of cats by training it on a large number of images of cats. Generative models are contrasted with discriminative models, which estimate the probability of an output value given an input value (this includes classification and regression problems). Recently, generative models based on deep learning have shown promise in generating novel data across a range of domains and tasks.

The most effective techniques, such as generative adversarial networks (GANs) [10] and variational autoencoders [11], come from a class of models known as latent

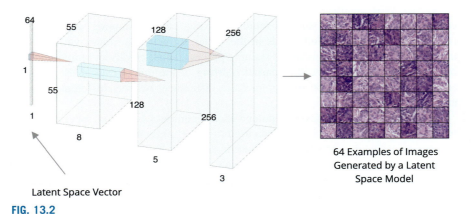

FIG. 13.2

In latent space models, a low-dimensional vector is used to generate data. In this example, a series of convolutions are used to achieve this transformation. The example output of the network shows 64 images, each 64 pixels wide and 64 pixels high.

variable generative models. In such systems, a model is trained that takes the lower-dimensional representation of data, called the latent space vector, and generates high-dimensional data from it. GANs and variational autoencoders differ in the way they are trained, but both conceptualize generation as decoding. By changing what data are passed in, as the latent space vector, model parameters can be learned that enable the model to perform data translation tasks. Fig. 13.2 shows an example of a latent space vector and generated images. In their recent review of GANs in pathology, Tschuchnig et al. [12] split the GANs up based on what kind of translation task the model is training for. This puts the emphasis on task (e.g., image-to-image translation vs. label-to-image translation). The rest of this review describes how different generative models have been trained to perform different translation tasks and how these could be usefully applied to the automated reporting of a clinical task within a digital pathology pipeline.

Generative adversarial networks [10] are a class of generative models in which a network, known as the generator, is trained by having it attempt to trick a second model, known as the discriminator. The discriminator and the generator are trained simultaneously. During training, the generator is sampled by having to translate noise into fake data. The discriminator is then trained on a combination of the fake data, labeled as fake, and the real data, labeled as real. The generator is then trained by having it generate fake data and asking the discriminator to predict labels for it. The training loss for the generator is based on how well the discriminator can tell them apart (i.e., how well the generator can fool it [generating fake data that the discriminator classes as real]). This simultaneous training procedure can cause GANs to both be computationally expensive and experience difficulty in converging to an accurate solution.

Generative models in the digital pathology pipeline

Generative models have the potential to overcome several issues that come up when developing computer vision systems for digital diagnosis and reporting. For example, data sets stained at different institutions can often have a lot of variation in color and intensity. It can be expensive and time-consuming to acquire high-quality labeled training data. Generative models can generate synthetic data sets to overcome this. They can also be used to virtually stain tissue, reducing the tissue preparation overhead.

Color and intensity normalization

During tissue preparation, particularly staining, variations in color and intensity can be introduced between different whole slide images. These artifacts can complicate the interpretation of the slide by pathologists and computers. When this occurs, similar tissue features can present differently or different ones similarly. Such artifacts are introduced from several sources, such as differences between scanners, the thickness of the cut tissue samples, and the amounts and concentrations of chemicals used in varying staining protocols. These issues can be mitigated in three ways: ignoring color information, training models to learn features insensitive to the artifacts [13], or normalizing images to account for differences.

By converting the image to grayscale, much of the information provided by the staining process is lost. Analysis techniques for grayscale pathology images have to rely on other features (e.g., texture and morphology [14,15]), leading to lower performance on downstream tasks. In other situations, artifacts can be compensated for by applying a large number of color perturbations to the training data so that a wide range of variations are presented to the model during training [13]. This technique requires the perturbations to be statistically similar to the color and intensity variations across the data to be assessed, information that is not always available and requires increased computational and memory overheads because of the large amount of data augmentation.

Ruifrok and Johnston [16] proposed a novel method based on color deconvolution that depends on user-determined color information to reconstruct images for each stain. This method provides state-of-the-art results for stain normalization but is limited in its applicability to extensive studies because the user needs to estimate the values used in the deconvolution manually. Magee et al. [17] presented a method for estimating the required color deconvolution parameters from the image data, eliminating the need for user input. This work was extended by Khan et al. [18] to account for image-specific color variations and to improve the training data used to separate the different stains.

A limitation of color-deconvolution techniques is their failure to take into account information outside of the image color (e.g., tissue structure or texture). Generative models are able to address this limitation. Stain normalization can be thought of as an image generation problem. Generative models have proved useful for image generation and recently have been applied to generate normalized pathology slides. Three different approaches have been applied to this task: stain-style transfer [19], CycleGAN-based [20] image-to-image translation, and Pix2Pix-based translation [21].

Stain-style transfer

Neural style transfer [22] is an image translation technique that transfers the style of one source image onto the content of another to generate a target image. The terms style and content can be a little misleading at first; content refers to aspects of the image, like the shape and arrangement of nuclei and cells and the tissue architecture that they comprise; and style refers to aspects such as color, like the hematoxylin and eosin (HE) shades, and texture (e.g., the nuclear chromatin). Style representations are derived from correlations between the same location in different activation maps of the same layer of a neural network. For example, there might be a filter that recognizes blue pixels and another that recognizes a curve. If they consistently activate together, then this would represent that curves are generally blue. Stain normalization can be thought of as a kind of style transfer from the source to the target; however, it is important that only the color distribution is transformed, not other histopathologic features.

Stain-style transfer [19] uses a modification on GANs to perform color normalization, as indicated by its application on patches extracted from the Camelyon16 data set [4]. The normalized patches improve tumor classification. In this technique, the input into the GAN generator is changed from noise to the unnormalized image. A conditional GAN [23] is then used in which both the generator and discriminator are trained to generate and discriminate class labels for each patch, in this case tumor or nontumor, in addition to fake or real labels. On its own, this produces distortion in the patches' noncolor histopathologic features. To address these issues, two other loss functions were added to the system: reconstruction loss, to minimize the difference between the source and generated images, and feature-preserving loss, which derives a loss by comparing the activations of the final layer of the discriminator when the source and generated images are passed through the network. This approach improves the classification accuracy of a convolutional neural network-based model trained on image patches extracted from the Camelyon16 data set. Ben-Taieb and Hamarneh [24] propose a similar approach in which the generator architecture is replaced with a U-Net encoder-decoder style network, called the stain transfer network, and the discriminator is given an additional classification task. This approach was assessed on both classification and segmentation tasks across three separate datasets, showing it can be used to improve the identification of a wide range of tissue and pathology types.

Pix2Pix-base image-to-image translation

Pix2Pix [21] is an extension of conditional GANs, which, like other image-to-image translation models, learns the mapping from one image domain to another. The difference with Pix2Pix is that it also learns a loss function to train the translation model. This means that models based on Pix2Pix can be trained to translate between different domains without the need to specify a specific loss function for that translation, something that is hard to do. Like conditional GANs for image-to-image translation, Pix2Pix requires image pairs, one from each domain, as example translations. Salehi and Chalechale [25] applied this approach successfully to stain normalization using five different HE data sets. The method involves destaining the patches by reducing them to grayscale before synthetically restaining them in a way that ensures that the color is consistent. This is similar to the artificial staining proposed by Rana et al. [26], discussed under data adaptation, and has been shown to perform well across a range of statistical measurements comparing ground-truth stained images against those restained using the GAN. This indicates that they may improve downstream assessment tasks, such as tumor classification and segmentation, in a similar way to the stain-style transfer techniques [19,24].

CycleGAN-based image-to-image translation

One of the key disadvantages of Pix2Pix is the need for paired images from the source and target domain (e.g., coregistered images before and after staining). CycleGAN [20] bypasses this requirement, allowing models to be trained to translate from a source to a target domain without the need for paired examples. This is done by training an inverse mapping from the source to target domain at the same time as training the translation. By comparing the original image with one that has had the forward and inverse transformation applied to it, a loss called cycle-consistency loss is derived. When the generator is trained, cycle-consistency loss is minimized, as is the conventional adversarial loss derived from trying to fool the discriminator.

de Bel et al. [27] showed that modifying the original CycleGAN [20] to use a U-Net [28] style architecture made it more suitable for use with pathology images. This system can be used to artificially stain images to a high quality. The technique was applied to two data sets of renal tissue sections stained with periodic acid-Schiff from different staining centers. Models trained using the normalized data had increased accuracy when segmenting various objects of interest within the renal slides, such as arteries, tubuli, and glomeruli. However, the system was able to generate changes in texture, something that breaks the constraint that the transform should preserve noncolor tissue features and potentially introduces unwanted bias into the generated datasets.

Data adaptation

Data adaptation is the task of taking the data in one domain, such as HE WSIs, and translating them into images that resemble those in a different domain, such as immunofluorescence WSIs. This can be useful as a data augmentation technique, allowing for images labeled in one domain to be used effectively for learning in another domain. Doing this relies on the image translation process retaining the correct labels. For example, if something is labeled as a cell nucleus, it still has to look like a cell nucleus once it has been translated.

One possible use of this data adaptation is to enrich patches with additional channels showing different fluorescence labels that highlight different kinds of information. This is called multiplexing and has traditionally been achieved through relabeling the same tissue multiple times and scanning in each fluorophore separately. There are two issues with this: after multiple relabeling, the tissue quality begins to degrade, and scanning requires the slides to be precisely aligned to allow the tissue to be coregistered. By doing virtual staining, the tissue is not degraded, and because a single scan is used, there are no issues related to alignment.

A histopathologic-to-immunofluorescence translation model that uses Pix2Pix [21] has been introduced by Burlingame et al. [29]. They adapt Pix2Pix by adding an adaptive regularization term during training that changes based on the prevalence of stained tissue in the patch. Patches with a low amount of stained tissue are penalized. This composites for the relative ease of translating patches with low amounts of tissue. The system's ability to generate realistic immunofluorescence stains from HE stains opens up the possibility of quickly providing information about cellular complexity when only a standard HE stain is available.

Another possible application of image translation is artificial staining, in which the source domain is an unstained image and the target is stained ones. If it is possible to do this in a consistent way, it can remove the need for laboratory-based staining with its associated variations, requiring stain normalization, and for the potential for human error. Rana et al. [26] apply a modified Pix2Pix [21] model in which the generator made use of a U-Net architecture [28] to translate between an unstained WSI taken from a prostrate core biopsy and virtual HE stains of the same image. Examination of the virtually stained images by pathologists showed that the system correctly stained many different histologic structures, including glands, stroma, nerve, and vascular spaces.

Data adaptation can also be used as a form of data augmentation. In domain adaptation and segmentation GAN (DASGAN), Kapil et al. [30] use a CycleGAN [20] to generate virtual programmed death ligand 1 stains from existing cytokeratin stain that has been marked up with a costly segmentation label. These data were then used to train an image segmentation model for the tumor epithelium that outperformed the same model without the additional data.

Data synthesis

Data synthesis is perhaps the most exciting prospect for generative models, especially in the field of artificial intelligence-based reporting of histopathology. In digital pathology, generating a ground truth is expensive and time-consuming. If accurately labeled synthetic samples could be generated, then this problem would be alleviated. The amount of data would only be limited by the resources available to run the generative model. However, a conundrum exists here. If there is enough data to train a generative model to generate new labeled data, then it is likely there is enough data to train an accurate classifier. Useful data synthesis requires one or both of two things: that the generative model is able to learn different representations, more useful in generation than a possible classifier, or that extra information is somehow added to the generative process (e.g., using a guide image).

The first of these approaches has been tested by a number of studies. Pathology-GAN [31] combines a variation of BigGAN [32] with a novel orthogonal regularization scheme for the generator and a relativistic average discriminator [33] to produce synthetic images of breast and colorectal cancer at a size of 224×224 pixels. Levine et al. [34] utilize a Progressive GAN [35] that upscales the generator and discriminator during training. This both increases the stability of the training process and allows for the generation of larger images, in this case, 1024×1024 pixels. They assess this technique across five different cancer types and five different histotypes of ovarian carcinoma and find that the synthetic images are classified with similar accuracies and cannot be disguised visually by pathologists. Similar results are shown by Krause et al. [36], who train a Conditional GAN [23] to produce colorectal cancer images at a size of 512×512 pixels, and by Falahkheirkhah et al. [37], whose system generates 1024×1024-pixel images of prostate and colon tissue. The Style-Guided Instance-Adaptive Normalization (SIAN) system [38] uses a novel architecture inspired by StyleGAN [39] to generate 224×224-pixel patches that are shown to have a lower Fréchet Inception Distance (FID) [40] score than similar approaches. FID is a common distance metric that evaluates the distribution of the generated images against the distribution of the synthetic ones.

The generation of new images can also be achieved by reformulating the problem as image translation, as shown by Wei et al. [41]. They take normal colonic mucosa images and generate synthetic colorectal polyp images on them. As with many other image-to-image translation models, their system is a variation on CycleGAN [20]. The Stitching Across the FROntier Network (SAFRON) framework [42] uses masks showing the locations of different tissue components as the input to a U-Net-based [28] GAN to generate overlapping patches of 256×256 pixels that are then stitched back together to synthetic images of any desired size. It is shown to score well both in terms of the FID and when assessed visually.

Future directions

High-quality synthetic data sets with labels generated using GANs [31,43] improve the performance of discriminative models trained on their data. Currently, these techniques are applied to the synthesis of patches rather than complete whole slide images. When diagnosing, a human pathologist mostly works at low magnification (e.g., $\times 10$) and relies on architectural features that are lost when the image is broken down into patches. There is potential to train on similar low-magnification images to exploit these features. A single WSI can be split into many thousands of patches, meaning that the training sets for patch classifiers are many times larger when the image is patched at high magnifications. At a lower magnification, the number of images available for training reduces dramatically, making such approaches less feasible. This is where generative approaches, such as those above, could be used to generate a large number of low-magnification synthetic images containing architectural features. Using other kinds of image synthesis, such as traditional computer graphics techniques, in combination with generative models [44] may provide a useful method in this domain and is an exciting future direction.

Generative models have the potential to enable medical data of all kinds, including pathology slides, to be used to train machine learning models without them needing access to the original patient-identifiable data set. Training a model on nonanonymized data and then using the model to generate a new artificial anonymous data set may provide a way to overcome the clinical firewalls that, because of patient confidentiality, prohibit many researchers from accessing the original data. This topic is the subject of a large amount of research within the deep learning community [45–47]. Sufficiently deep generative models are capable of memorizing their training data in a way that can cause potentially confidential information to leak into any synthetic data. This has important implications for data governance going forward. To make generative models public, it is critical to ensure they are trained in such a way that removes the possibility of confidential data being leaked into any synthetic data set, and therefore, guidelines for doing this while maintaining privacy are required and will need to be adhered to.

Conclusion

This chapter reviews recent advances in the application of generative models to digital pathology. Work in this domain seeks to address issues of color and intensity artifacts, data adaptation, and data synthesis, as well as how generative models can address these. Generative models can assist with several open challenges in the digital pathology workflow. Multiresolution WSI synthesis may provide a way to train deep models that exploit architectural tissue features in a way that is currently unpractical because of a lack of data. Additionally, differential privacy for WSI data sets may allow for a much larger amount of useful data to be released publicly. The application of generative models has proved useful in improving digital pathology workflows, and this fast-developing technology holds much promise in this field.

Acknowledgments

Supported by the Sir James Mackenzie Institute for Early Diagnosis, University of St. Andrews, United Kingdom (DM) and Industrial Centre for Artificial Intelligence Research in Digital Diagnostics, United Kingdom, grant TS/S013121/1 (D.M., D.H.-B., P.D.C.).

References

[1] Dimitriou N, Arandjelovic O, Caie PD. Deep learning for whole slide image analysis: an overview. Front Med 2019;6.

[2] Komura D, Ishikawa S. Machine learning methods for histopathological image analysis. Comput Struct Biotechnol J 2018;16:34–42.

References **269**

[3] Deng J, Dong W, Socher R, Li L-J, Li K, Fei-Fei L. Imagenet: a large-scale hierarchical image database. In: 2009 IEEE conference on computer vision and pattern recognition. IEEE; 2009. p. 248–55.

[4] Bejnordi BE, Veta M, Van Diest PJ, Van Ginneken B, Karssemeijer N, Litjens G, Van Der Laak JAWM, Hermsen M, Manson QF, Balkenhol M, et al. Diagnostic assessment of deep learning algorithms for detection of lymph node metastases in women with breast cancer. JAMA 2017;318(22):2199–210.

[5] Bandi P, Geessink O, Manson Q, Van Dijk M, Balkenhol M, Hermsen M, Bejnordi BE, Lee B, Paeng K, Zhong A, et al. From detection of individual metastases to classification of lymph node status at the patient level: the camelyon17 challenge. IEEE Trans Med Imaging 2018;38(2):550–60.

[6] Janowczyk A, Madabhushi A. Deep learning for digital pathology image analysis: a comprehensive tutorial with selected use cases. J Pathol Inform 2016;7.

[7] Campanella G, Hanna MG, Geneslaw L, Miraflor A, Silva VWK, Busam KJ, Brogi E, Reuter VE, Klimstra DS, Fuchs TJ. Clinical-grade computational pathology using weakly supervised deep learning on whole slide images. Nat Med 2019;25(8):1301–9.

[8] Goodfellow I, Bengio Y, Courville A, Bengio Y. Deep learning. vol. 1. Cambridge: MIT Press; 2016.

[9] LeCun Y, Boser B, Denker JS, Henderson D, Howard RE, Hubbard W, Jackel LD. Backpropagation applied to handwritten zip code recognition. Neural Comput 1989;1 (4):541–51.

[10] Goodfellow I, Pouget-Abadie J, Mirza M, Xu B, Warde-Farley D, Ozair S, Courville A, Bengio Y. Generative adversarial nets. In: Advances in neural information processing systems; 2014. p. 2672–80.

[11] Hinton GE, Zemel RS. Autoencoders, minimum description length and helmholtz free energy. In: Advances in neural information processing systems; 1994. p. 3–10.

[12] Tschuchnig ME, Oostingh GJ, Gadermayr M. Generative adversarial networks in digital pathology: a survey on trends and future potential. arXiv 2020. arXiv-2004.

[13] Liu Y, Gadepalli K, Norouzi M, Dahl GE, Kohlberger T, Boyko A, Venugopalan S, Timofeev A, Nelson PQ, Corrado GS, et al. Detecting cancer metastases on gigapixel pathology images. arXiv 2017. preprint arXiv:1703.02442.

[14] Mosquera-Lopez C, Agaian S, Velez-Hoyos A, Thompson I. Computer-aided prostate cancer diagnosis from digitized histopathology: a review on texture-based systems. IEEE Rev Biomed Eng 2014;8:98–113.

[15] Liu Y-Y, Chen M, Ishikawa H, Wollstein G, Schuman JS, Rehg JM. Automated macular pathology diagnosis in retinal oct images using multi-scale spatial pyramid and local binary patterns in texture and shape encoding. Med Image Anal 2011;15(5):748–59.

[16] Ruifrok AC, Johnston DA, et al. Quantification of histochemical staining by color deconvolution. Anal Quant Cytol Histol 2001;23(4):291–9.

[17] Magee D, Treanor D, Crellin D, Shires M, Smith K, Mohee K, Quirke P. Colour normalisation in digital histopathology images. In: Proc optical tissue image analysis in microscopy, histopathology and endoscopy (MICCAI workshop), vol. 100. Citeseer; 2009. p. 100–11.

[18] Khan AM, Rajpoot N, Treanor D, Magee D. A nonlinear mapping approach to stain normalization in digital histopathology images using image-specific color deconvolution. IEEE Trans Biomed Eng 2014;61(6):1729–38.

[19] Cho H, Lim S, Choi G, Min H. Neural stain-style transfer learning using gan for histopathological images. arXiv 2017. preprint arXiv:1710.08543.

270 CHAPTER 13 Generative deep learning in digital pathology

[20] Zhu J-Y, Park T, Isola P, Efros AA. Unpaired image-to-image translation using cycle-consistent adversarial networks. In: Proceedings of the IEEE international conference on computer vision; 2017. p. 2223–32.

[21] Isola P, Zhu J-Y, Zhou T, Efros AA. Image-to-image translation with conditional adversarial networks. In: Proceedings of the IEEE conference on computer vision and pattern recognition; 2017. p. 1125–34.

[22] Gatys LA, Ecker AS, Bethge M. A neural algorithm of artistic style. arXiv 2015. preprint arXiv:1508.06576.

[23] Mirza M, Osindero S. Conditional generative adversarial nets. arXiv 2014. preprint arXiv:1411.1784.

[24] BenTaieb A, Hamarneh G. Adversarial stain transfer for histopathology image analysis. IEEE Trans Med Imaging 2017;37(3):792–802.

[25] Salehi P, Chalechale A. Pix2pix-based stain-to-stain translation: a solution for robust stain normalization in histopathology images analysis. arXiv 2020. preprint arXiv:2002.00647.

[26] Rana A, Yauney G, Lowe A, Shah P. Computational histological staining and destaining of prostate core biopsy rgb images with generative adversarial neural networks. In: 2018 17th IEEE international conference on machine learning and applications (ICMLA). IEEE; 2018. p. 828–34.

[27] de Bel T, Hermsen M, Kers J, van der Laak J, Litjens GJS, et al. Stain-transforming cycle-consistent generative adversarial networks for improved segmentation of renal histopathology. In: MIDL; 2019. p. 151–63.

[28] Ronneberger O, Fischer P, Brox T. U-net: convolutional networks for biomedical image segmentation. In: International conference on medical image computing and computer-assisted intervention. Springer; 2015. p. 234–41.

[29] Burlingame EA, Margolin AA, Gray JW, Chang YH. Shift: speedy histopathologicalto-immunofluorescent translation of whole slide images using conditional generative adversarial networks. In: Medical imaging 2018: Digital pathology, vol. 10581. International Society for Optics and Photonics; 2018. p. 1058105.

[30] Kapil A, Wiestler T, Lanzmich S, Silva A, Steele K, Rebelatto M, Schmidt G, Brieu N. Dasgan—joint domain adaptation and segmentation for the analysis of epithelial regions in histopathology pd-l1 images. arXiv 2019. preprint arXiv:1906.11118.

[31] Quiros AC, Murray-Smith R, Yuan K. Pathology gan: Learning deep representations of cancer tissue. arXiv 2019. preprint arXiv:1907.02644.

[32] Brock A, Donahue J, Simonyan K. Large scale gan training for high fidelity natural image synthesis. arXiv 2018. preprint arXiv:1809.11096.

[33] Jolicoeur-Martineau A. The relativistic discriminator: a key element missing from standard gan. arXiv 2018. preprint arXiv:1807.00734.

[34] Levine AB, Peng J, Farnell D, Nursey M, Wang Y, Naso JR, Ren H, Farahani H, Chen C, Chiu D, et al. Synthesis of diagnostic quality cancer pathology images by generative adversarial networks. J Pathol 2020;252(2):178–88.

[35] Karras T, Aila T, Laine S, Lehtinen J. Progressive growing of gans for improved quality, stability, and variation. arXiv 2017. preprint arXiv:1710.10196.

[36] Krause J, Grabsch HI, Kloor M, Jendrusch M, Echle A, Buelow RD, Boor P, Luedde T, Brinker TJ, Trautwein C, et al. Deep learning detects genetic alterations in cancer histology generated by adversarial networks. J Pathol 2021;254(1):70–9.

[37] Falahkheirkhah K, Tiwari S, Yeh K, Gupta S, Herrera-Hernandez L, McCarthy MR, Jimenez RE, Cheville JC, Bhargava R. Deepfake histological images for enhancing digital pathology. arXiv 2022. preprint arXiv:2206.08308.

References

[38] Wang H, Xian M, Vakanski A, Shareef B. Sian: style-guided instance-adaptive normalization for multi-organ histopathology image synthesis. arXiv 2022. preprint arXiv: 2209.02412.

[39] Karras T, Laine S, Aila T. A style-based generator architecture for generative adversarial networks. In: 2019 IEEE/CVF conference on computer vision and pattern recognition (CVPR); 2019. p. 4396–405.

[40] Heusel M, Ramsauer H, Unterthiner T, Nessler B, Hochreiter S. Gans trained by a two time-scale update rule converge to a local nash equilibrium. Adv Neural Inf Process Syst 2017;30.

[41] Wei J, Suriawinata A, Vaickus L, Ren B, Liu X, Wei J, Hassanpour S. Generative image translation for data augmentation in colorectal histopathology images. arXiv 2019. preprint arXiv:1910.05827.

[42] He K, Zhang X, Ren S, Sun J. Deep residual learning for image recognition. In: Proceedings of the IEEE conference on computer vision and pattern recognition; 2016. p. 770–8.

[43] Hou L, Agarwal A, Samaras D, Kurc TM, Gupta RR, Saltz JH. Unsupervised histopathology image synthesis. arXiv 2017. preprint arXiv:1712.05021.

[44] Shrivastava A, Pfister T, Tuzel O, Susskind J, Wang W, Webb R. Learning from simulated and unsupervised images through adversarial training. In: Proceedings of the IEEE conference on computer vision and pattern recognition; 2017. p. 2107–16.

[45] Beaulieu-Jones BK, Zhiwei Steven W, Williams C, Lee R, Bhavnani SP, Byrd JB, Greene CS. Privacy-preserving generative deep neural networks support clinical data sharing. Circ Cardiovasc Qual Outcomes 2019;12(7), e005122.

[46] Fan L. A survey of differentially private generative adversarial networks. In: The AAAI Workshop on Privacy-Preserving Artificial Intelligence; 2020.

[47] Chugui X, Ren J, Zhang D, Zhang Y, Qin Z, Ren K. Ganobfuscator: mitigating information leakage under gan via differential privacy. IEEE Trans Inf Forensics Secur 2019;14 (9):2358–71.

CHAPTER

Artificial intelligence methods for predictive image-based grading of human cancers

14

Gerardo Fernandez, Abishek Sainath Madduri, Bahram Marami, Marcel Prastawa, Richard Scott, Jack Zeineh, and Michael Donovan
Department of Pathology, Icahn School of Medicine at Mount Sinai, New York, NY, United States

Introduction

Surgical pathology and the practice of general medicine are often described as having legitimate components of art and science. That perception is perhaps less relevant in the current information age and with the adoption of genomic medicine, largely due to the growth in disease-specific causal knowledge. As a surgical pathologist, gestalt and intuition are words you may not use to describe how you arrive at a specific patient's diagnosis, but it is certainly understood that with experience, in particular with a deep immersion in a subspecialty, comes an understanding of histopathology that cannot easily be transmitted or formalized. This is not in any way to delegitimize pathology or relegate it to the unscientific, but rather to illustrate the strength of the human brain. For well over 100 years using a microscope, pathologists have been applying this strength to group and separate disease entities into diagnoses and their subtypes. The brain seems to be good at pattern recognition, filtering noise, and dealing with variability, and from a histopathology perspective, this has complemented the medical need quite well for over a century. Although the concept of tumor differentiation and its relationship to clinical outcome dates back to the early days of histopathology until fairly recently, limited treatment options diminished the necessity for more precise and standardized grading.

Over the past several decades, however, treating physicians and patients have seen increasingly sophisticated therapeutic options, directed at narrower oncologic disease states and have therefore become increasingly reliant on accurate tumor classification to make informed decisions on the overall best treatment choice. The relevant information in this current era of personalized medicine ranges from robust prognostic information about specific diagnoses to the presence or absence of biomarkers that may dictate a targeted therapy. Although accurate classification clearly starts with tumor diagnosis, relevant subtypes, and even target biomarkers, it is not

Artificial Intelligence in Pathology. https://doi.org/10.1016/B978-0-323-95359-7.00014-5
Copyright © 2025 Elsevier Inc. All rights reserved, including those for text and data mining, AI training, and similar technologies.

complete without robust and reproducible grading schemes providing reliable prognostic information. Although grade is emphasized in this review, it is important to understand that clinically relevant prognostic clinical models rely on both the AJCC cancer staging system and histologic grade to guide treatment decisions and project outcomes. In addition, these clinical models have further evolved with the advent of molecular phenotyping and the interplay with these same clinical features.

If we restrict ourselves to one organ system or further to one cell of origin, there are many examples where tumor type and subtype provide the prognostic information required to guide treatment and therefore grading schemes are either less important or not applicable, as is the case, for example, with prognostically favorable tubular or medullary carcinomas of the breast and prognostically unfavorable small cell lung cancer. For many cancers, however, it is the degree of differentiation that determines the aggressiveness of the tumor and therefore the grade. This makes intuitive sense as the degree of differentiation is a measure of a tumor's morphologic departure from the normal histologic pattern of the organ. The greater the tumor's departure from normal histology, the more malignant, unchecked, and aggressive it is.

Grading schemes over the years represent attempts to create a histopathology-based prognostic model, capturing the degree of differentiation and providing treating clinicians with the most reliable insight into the potential aggressive behavior of a cancer and therefore risk to the patient. These models incorporate a range of histologic features that are quantitatively or semiquantitatively determined microscopically by the pathologist. Some grading schemes incorporate very few features or focus on a narrow group of features, as is the case in the Fuhrman grading system for renal clear-cell carcinoma which only uses nuclear features (size, shape, chromatin pattern, and nucleolar prominence) [1]. Other grading schemes incorporate a combination of features such as architectural or proliferative activities. Architectural features in carcinomas generally reflect the degree to which the tumor's epithelial component differs from the organ's normal epithelial component. Grading schemes for adenocarcinoma of the colon, adenoid cystic carcinoma of the salivary gland, and adenocarcinoma of prostate all primarily rely on architectural features [2–4]. Proliferative activity, namely mitotic figure counts or proliferative (Ki67) indices, is occasionally primarily relied on, such as in gastrointestinal stromal tumors and some neuroendocrine tumors of the gastrointestinal tract, whereas it is frequently a part of multiple-feature grading schemes, including breast's Nottingham modification of the Bloom-Richardson grade which additionally incorporates nuclear morphology and glandular architecture [5–8].

The battleground in histopathologic grading has been between complexity and reproducibility, where many schemes capture nuance at the expense of reproducibility. Numerous studies in different subspecialties have documented less than optimal reproducibility of grading systems, be it intra- or interobserver or whether the comparison is against a consensus of experts or general pathologists [9–12]. Recent efforts by international organizations representing subspecialties have tried consensus to guide grading recommendations in the direction of increased reproducibility

Tissue preparation and staining and clinical utility, perhaps none more than in the case of prostate cancer. The Gleason grading scheme has undergone a number of modifications to simplify ambiguity in the pattern categories. Most recently, the International Society of Urological Pathology (ISUP), focusing on clinical utility, modified how the score is reported to match pathologically defined subgroups to clinically relevant risk and treatment groups [13].

Prognostic models to establish risk for patients with cancer have evolved beyond grade and stage. In their simplest form, they have combined or incorporated the two to better assess prognosis. The Nottingham Prognostic Index, for example, combined grade with tumor size and lymph node status to arrive at clinically relevant prognostic groupings [14]. There are also numerous nomogram variations that combine grade with stage and clinical variables such as the Kattan preoperative and postoperative nomograms for prostate cancer [15,16].

Despite strategies for improving the reproducibility of grade by simplifying schemes or combining it with other predictive features such as stage or clinical variables for improved prognostic value, the central issue has not been addressed until recently. As long as humans are visually quantifying grading scheme features, there will always be a significant reproducibility problem. Not all feature quantitation suffers equally; in breast cancer, for example, we are significantly better at reproducibly assessing tubule formation than nuclear pleomorphism or mitotic counts [11,12]. In all cases, however, a tool for quantitative assessment has been missing. Although image analysis with traditional machine vision (MV) approaches has been slowly inching its way into this role, the recent feasibility of artificial intelligence (AI) approaches such as deep learning promises to change the field in the next half decade.

To date, there are no clinically validated let alone clinically accepted AI-based grading schemes, although there have been a few traditional MV-based models that combine morphometric assessment and biomarker quantitation to risk stratify patients [17–20]. Much more common in the literature are MV- and AI-based algorithms that focus on quantitating histologic features, and some have created models that correlate features to grade [21–23]. Although the latter, probably best described as computer-aided diagnosis (CAD), is a useful first step, the ultimate goal is to generate models that use these features to create a better prognostic tool, in terms of patient clinical outcome, than current grading schemes. The sections below will roughly follow a pipeline (Fig. 14.1) for developing an image analysis-based multivariate system, with attention drawn in each subsection to inherent obstacles and published state of the art, as well as to our own experience with solutions.

Tissue preparation and staining

A few basic principles and common factors related to tissue preparation and staining are worth mentioning that impact the design of an algorithm development cohort and the feature engineering strategy. Preanalytical factors known to affect the

276 CHAPTER 14 Artificial intelligence methods for predictive

> ⟶ Tissue preparation and staining (H&E, IF, IHC)
> ⟶ Image acquisition
> ⟶ Stain normalization or unmixing of IF spectral images
> ⟶ Automated region detection in whole-slide images
> ⟶ Image segmentation
>   ⟶ Nuclear and epithelial segmentation in IF and H&E images
>   ⟶ Mitotic Figures
>   ⟶ Ring segmentation
> ⟶ Protein Biomarker features
> ⟶ Morphological features
> ⟶ Modeling with patient outcome
> ⟶ Ground truth data for AI-based features

FIG. 14.1

Pathology image analysis pipeline.

antigenicity of tissue [24,25], such as length of fixation, tissue processing, and age of cut slides, can introduce significant variability into the set of biomarker-stained images used for algorithm development. This variability/noise particularly affects absolute biomarker intensity features that look for a correlation between raw intensity values and a clinical endpoint. H&E or IHC stain reagents, protocol variations, and cut tissue thickness can further affect color intensities and distribution. This variability is not only important when evaluating and optimizing feature performance but also when scaling any algorithm for broader use outside of the training environment where preanalytics and staining protocols are less controlled. Creating tissue microarrays when blocks are available can be an effective way to mitigate some of this variability in training by combining many patient samples on a single slide, although it does significantly limit the area of tumor analyzed and does not address preanalytical issues with fixation or the scaling for a broader population [26]. Some other strategies to mitigate the effect of these variables will be discussed below.

Image acquisition

There are many different instruments available for digital image capture. It is perhaps the relative abundance and accessibility of digital image data that, along with advances in computing technology, account for the recent interest and progress made in the field of computational pathology. Regulatory approval for the use of these systems as primary diagnostics tools will additionally increase the accessibility of digital images for development purposes [27]. A detailed discussion of each image capture technology is beyond the scope of this chapter, but we should mention some general categories. Whole-slide digital scanners have been around for over a decade; improved color fidelity, focal plane algorithms, and scanning speed have made these high-throughput devices the cornerstone of generating large digital image databases. Multispectral fluorescent scopes, including confocal image capture, have been around for even longer but until recently have been limited to static image capture.

Whole-slide fluorescent scanners are available; however, true multispectral whole-slide scanning is variably limited by speed.

Stain normalization

Color and intensity variations in histology images can not only cause inconsistencies in interpretation between pathologists but can also significantly impact the performance of CAD systems. These variations are in part due to variability in preparation, staining, and digitization of histological samples. These include preanalytic factors such as fixation and tissue preservation, paraffin embedding, deparaffinization, slide thickness, stain reagents, stain protocols, and mounting medium, as well as color fidelity algorithms of the digitization device. In order to compensate for this type of input variability, stain normalization strategies are often employed, in which the color and intensity range of an image is transformed to match those of another image. A review of the major techniques for stain normalization of digitized histologic images is discussed here.

An early method, originally developed for nonmedical images [28] and later applied to stain normalization of histology images [29], transforms the typical RGB image into an alternate color space ($l\alpha\beta$) which allows a statistical color distribution in each channel to match the reference image. This method, however, does not take into account the independent contribution of stain dyes to the final color and can result in improper color mapping [30]. Macenko et al. [31] proposed to convert the image to an optical density space and find the optimal stain vectors by means of singular value decomposition. This method does not preserve all information from the source image in the processed image and performs poorly in the presence of strong staining variations. The algorithm proposed by Vahadane et al. [32] preserves all the structure of the source image by using a sparse nonnegative matrix factorization method. Although this method outperforms all previous supervised stain separation methods, it is computationally costly and the algorithm might reach a nonrepresentative local optimal solution [33]. A whole-slide color standardization algorithm was proposed in Ref. [34] in which two chromatic distributions and one density distribution are computed for hematoxylin, eosin, and background in the image after applying an RGB-to-HSD model. The source image is then transformed to the space of the reference image by a weighted sum of three nonlinear transformations between the distributions. This algorithm has a high processing time and does not preserve all the source image information such as background and red blood cells color.

More recently, deep generative models have been used as style transfer and generative learning for stain normalization of histopathology images. BenTaieb et al. [35] used a generative adversarial network (GAN) for learning stain normalization across different datasets and integrated it with a task-specific discriminative model for supervised classification. Moreover, Shaban et al. [30] proposed StainGAN, using cycle-consistent adversarial networks (CycleGAN) for image-to-image

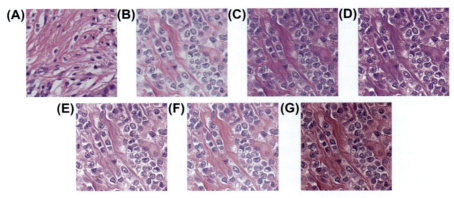

FIG. 14.2

Comparison of stain normalization methods: the source and the reference images are extracted from TCGA [36] and Mount Sinai Hospital whole-slide images, respectively. (A) Reference; (B) Shaban-StainGAN; (C) Bejnordi; (D) Vahadane; (E) Macenko; (F) Reinhard; (G) Source.

unpaired translation between two datasets. This method does not require picking a reference image for stain normalization, and the processed image has a good visual similarity to the reference domain. We have compared the performance of five different methods in Fig. 14.2.

Unmixing of immunofluorescence spectral images

The multispectral immunofluorescent microscope (e.g., Vectra imaging system [37]) acquires a spectral stack, a set of grayscale images taken at a series of wavelengths separated by 10 or 20 nm intervals. These images show proteins of interest tagged by fluorescent biomarkers, mixed with autofluorescence and other unwanted signals.

The unmixing process extracts the desired signals from the spectral stack, either for structural biomarkers, designed to respond to anatomical compartments, for example, nuclei (DAPI) and glands (CK18/pan-Cytokeratin), or for functional biomarkers such as cancer-related proteins, for example, androgen receptor (AR) in the prostate or estrogen receptor (ER) and progesterone receptor (PR) in the breast (Fig. 14.3). The unmixing algorithm for functional biomarkers is designed to produce a linear response proportional to the protein present in the tissue, typically using a deconvolution-type algorithm [38]. Careful calibration of staining and microscope is required to ensure that biomarker spectral profiles are stable over different images and tissues. For structural biomarkers, where the goal is to obtain an image that can be correctly segmented, the unmixing algorithm has more leeway to compensate for weak staining or other variations in spectral profiles. For prostate samples, we use a semisupervised approach based on nonnegative matrix factorization [39]. Fig. 14.3A shows a typical unmixed image output for CK18.

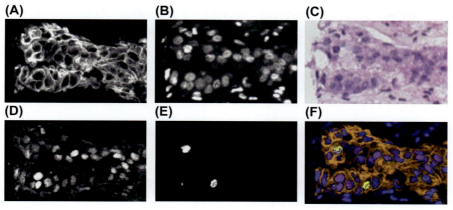

FIG. 14.3

Structural (A and B) and functional (D and E) immunofluorescent biomarkers in high-grade prostate cancer (20× objective lens), Gleason pattern 4, after unmixing. Panel (F) combines four biomarkers by showing the maximum value biomarker at each pixel, with AR *(purple)*, Ki67 *(yellow)*, DAPI *(blue)*, and CK18 *(gold)*. Compare with H&E stain of the same tissue in panel (C). (A), CK18; (B), DAPI; (C), H&E; (D), AR; (E), Ki67; (F), Max biomarker.

Automated detection of tumor regions in whole-slide images
Localization of diagnostically relevant regions of interest in whole-slide images

A digital whole-slide image (WSI) obtained from scanning glass slides often has more than a billion pixels in the highest resolution (40×). However, only small regions within the slide contain diagnostic information. The relevant region in the image is usually less than 50% of the digitized area. Additionally, artifacts such as ink, mounting medium air bubbles, tissue folds, crushed nuclei, and out-of-focus regions could potentially mislead feature extraction, segmentation, and detection algorithms.

A multiscale framework using high- and low-resolution imaging features was proposed in Ref. [40] to rapidly identify the high-power fields of interest in the WSIs. Bahlmann et al. [41] developed a classification method for discriminating between statistics of relevant versus irrelevant regions. They used visual descriptors that captured the distribution of color intensities observed for nuclei and cytoplasm. Moreover, Mercan et al. in Ref. [42] extracted regions of interest (ROIs) by tracking records of pathologists' actions, such as zooming, panning, and fixating. They developed a visual bag-of-words model based on the color and texture features of the extracted regions to predict diagnostically relevant ROI in the WSIs. We developed an automatic technique for detecting diagnostically important ROIs within the WSIs using convolutional neural networks (CNN). We trained the CNN to identify multiple

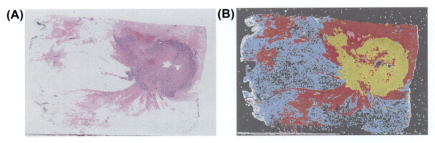

FIG. 14.4

Localization of region of interest in the H&E whole-slide image. Colors in the region of interest (ROI) detector represent the different regions of the slide. (A) H&E whole-slide image; (B) ROI detector output.

tissue types in the images excluding abovementioned regions from analysis. Using this tool, only diagnostically relevant regions are localized (orange color in Fig. 14.4B) and used for further processing and analysis. Localization of these ROIs not only significantly reduces the computational load for other processes but also improves the accuracy of these downstream processes by focusing on relevant areas and decreasing the possibility of artifact interference.

Tumor detection

Automated tumor detection in digital histopathology images is by itself quite a broad subject, with numerous techniques, from traditional features-based MV to deep learning approaches, with many different intended uses. Automated deep learning approaches to tumor detection have flourished over the past several years over traditional MV approaches as deep learning methods aim to automatically learn relevant characteristics and complex relationships directly from the data, and therefore require less supervision and actually perform better [43,44]. Although there is certainly overlap, the intended use and functional requirements tend to drive the technical design of the algorithm. These fall into a few categories which include screening applications, tumor classification or diagnosis, and tumor volume assessment.

Screening tools tend to be less supervised as the classification problem can be done at the ROI level and does not require manual annotations or object segmentation. For instance, Campanella et al. [45] employed a deep learning approach under the multiple instance learning assumption for prostate cancer diagnosis using over 12,000 needle biopsy slides, where only the overall slide diagnosis was available. Tumor classification and volume assessment can similarly require little input supervision if the requirement does not demand, for example, to differentiate distinct architectural patterns as would be required to separate in situ carcinoma or scattered normal structures mixed in with invasive carcinoma of breast. Cruz-Roa et al. [46] developed a CNN-based classifier for automatic detection of tumor (including both invasive and in situ carcinoma) on breast WSIs. Annotated tumor on the WSIs were used for patch-based training of the CNN classifier. They further developed their work [43] by creating a fast, high-throughput CNN-based method for automatic detection of invasive breast tumor on WSIs. For this, they used an efficient adaptive and iterative sampling method based on probability gradients to efficiently identify the precise extent of the invasive breast cancer, obtaining comparative results to those using dense sampling. Bejnordi et al. [47] assessed and compared the performance of deep learning algorithms submitted to a public challenge to detect metastases in H&E tissue sections of lymph nodes in women with breast cancer. Seven out of thirty-two deep learning algorithms submitted to this competition showed greater discriminating performance than a panel of eleven pathologists in a simulated time-constrained diagnostic setting.

The task becomes more complicated when attempting to both detect tumor and classify architectural heterogeneity on WSIs. This can be as simple as separating benign from malignant to distinguishing invasive disease, in situ carcinoma, proliferative entities, reactive patterns, and normal structures. Machine learning and more specifically deep learning techniques have provided state-of-the-art results in automated detection and morphologic classification in histopathology images. Bardou et al. [44] developed and compared two approaches for the classification of breast cancer histology images into benign and malignant classes, as well as their subclasses. In the first method, they extracted a set of handcrafted features encoded by bag-of-words and locality-constrained linear coding descriptors, and used support vector machines for the classification of features. In the second approach, they trained a CNN for the classification of a large-labeled image dataset. Although the deep learning approach has many advantages, including outperforming the handcrafted approach and automated feature generation, it is worth mentioning that one of the drawbacks of this approach is that the features are not transparent as they are embedded in the neural network architecture. Using a context-aware stacked CNN approach, Bejnordi et al. [48] classified whole-slide breast images into normal/benign, in situ, and invasive classes. They first trained a deep CNN with high pixel resolution, but small image size, to classify the tissue into different classes. Then, they fed much larger images to the model and used the extracted feature responses generated by the initial model as input to a second CNN. This architecture allowed them to use a larger input image to learn the global interdependence of various structures or context in different lesion categories.

We have developed a robust ensemble CNN-based technique for the classification of clinically significant architectural and morphologic patterns into four types: normal tissue, benign (proliferative) entities, in situ carcinoma, and invasive cancer. We incorporated image perturbation models specifically designed for digital pathology images for training the classifiers, which increased the robustness and generalizability of the classifier to real-world image data variability [49]. A review of the state-of-the-art methods with their performance and collective results on a public

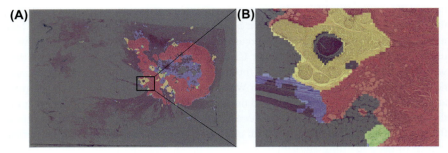

FIG. 14.5

Representation of colors in the tumor detector—invasive: *red*, DCIS: *orange*, benign: *green*, normal: *blue*, nontissue: *gray*. Tumor detection is only applied on the tissue region of Fig. 14.4. (A) Tumor detector output; (B) zoomed region.

challenge is given in Ref. [50]. Fig. 14.5 shows the results of our developed technique applied specifically to the histologically relevant areas of the WSI given in Fig. 14.4.

Image segmentation

Deep learning, as described above, is capable of learning complex relationships in histopathology images and has successfully been applied to learning differences in large sets of images to separate or classify them as either containing or not containing tumor. This black box or top-down approach certainly has its merits, particularly as a screening tool when the requirement is to detect areas of interest for a pathologist to pay special attention to. This approach, although arguably less clinically useful, has also been used to classify tumors into distinct patterns or grades [22]. Deep learning approaches to classify images using outcome though are a whole different endeavor, which will be discussed in more detail below, but at the highest level, it is difficult because large enough development cohorts with outcome are not generally available at the necessary scale.

This is where traditional machine vision handcrafted features and deep learning have converged with machine learning modeling to create the real possibility of automated histopathology image–based grading. We have shown that a bottom-up approach—segmenting and characterizing histologic compartments morphologically combined with protein expression on multiplexed immunofluorescent images—yields handcrafted feature-based models that provide a significant advantage over conventional grading for risk-stratifying patients [17]. This has never been done with H&E image-based models in large part due to the difficulties handcrafted features have in handling color variability and the lack of contrast between compartments. Leveraging deep learning's ability to distinguish subtlety in large datasets, the next step in automated grading of H&E images is to apply a bottom-up approach, segmenting histologic compartments for direct characterization and feature engineering.

Nuclear and epithelial segmentation in IF images

The goal of nuclear and epithelial segmentation is to delineate their borders in unmixed images (Fig. 14.6), accurately separating touching or overlapping objects and identifying each gland or nucleus with a unique label. A classic algorithm for IF nuclear segmentation is Ref. [51], which operates according to a popular nuclear detection pattern, by first finding seed points, one per nucleus, and then growing the seeds according to proximity, size, and shape constraints to find the nuclear boundaries. Many variations of this algorithm have been proposed for H&E and IHC images also, for example, Ref. [52], but they all share a common problem which is that the irregular texture and holes, the result of chromatin clumping and margination in high-grade nuclei, lead to the detection of multiple seed points per nucleus, producing fragmented segmentations. A second (inverse) problem is that closely packed nuclei may be merged together when too few seed points are detected or the nuclear borders are missed. A comprehensive review of nuclei segmentation algorithms [53] shows the successful application of deep neural networks to these problems.

A distinct problem, common in epithelium segmentation of IF images, is the inconsistency of cytokeratin intensity throughout the cytoplasm, often magnified by the negative impression where the nucleus sits or the DAPI holes. This problem has been addressed by combining epithelial with nuclear segmentation, as in Ref. [54], by filling in nuclear holes in the cytokeratin (the inverse of the DAPI image) along the gland border where the nuclei are clustered and the staining is often weaker. In order to compensate for unmixing errors, which may result in strong

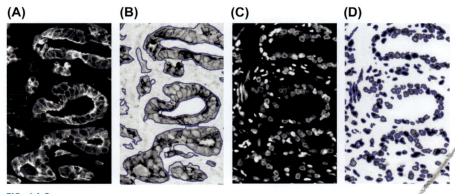

FIG. 14.6

Structural IF biomarkers (A-gland, C-nuclei) and their segmentations (B and D respectively) in low-grade prostate cancer, Gleason pattern 3 (20× objective lens). See Fig. 14.8 for segmentation of H&E images of the same tissue by deep NN. (A), CK18; (B), CK18 segmentation; (C), DAPI; (D), DAPI segmentation.

background signal or in weak response along gland borders, we have developed a combined unmixing/segmentation algorithm, which is robust against typical under- and over-segmentation errors.

Nuclei detection and segmentation in H&E images

Although not part of the Gleason grading system, manually characterizing nuclei is a common thread throughout most grading schemes and often one of the least reproducible [11,12]. As described above in IF images, H&E segmentation of nuclei suffers from similar challenges, including variability of size, shape, and chromatin textures, as well as touching and overlapping. Accounting for this variability is particularly challenging in H&E with a handcrafted approach. By leveraging large datasets, deep learning has the potential to generate robust and reproducible algorithms for both detecting the presence of and determining the boundaries of nuclei. These two tasks have somewhat different complexities and requirements for annotated data. Nuclei detection specifically is a task where an algorithm counts nuclei, often identifying them with bounding boxes. Segmentation algorithms, on the other hand, determine the precise pixels that form each nucleus with boundaries defined explicitly. Detecting and counting nuclei is comparatively simpler than segmenting nuclei, where the exact boundary is defined, distinguishing overlapping and touching nuclei. Fig. 14.7 shows the difference between these two tasks.

Nuclei detection with deep learning is typically performed by generating probability maps derived from training a CNN that is applied to image patches in a sliding window fashion or a fully convolutional neural network (FCN), and then finding local maximas within the nuclei probability map [55]. Naylor et al. proposed the use of an FCN for nuclei segmentation [56], combined with a watershed algorithm

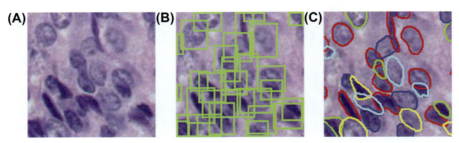

FIG. 14.7

Annotation for nuclei detection compared with segmentation. From left to right: (A) H&E image, (B) image overlaid with boxes indicating detected nuclei, and (C) image overlaid with nuclei boundary segmentation. In a detection task, the objective is to determine the presence of nuclei without exact boundaries. In a segmentation task, the objective is to determine nuclei membership for each pixel, accounting for overlaps and touching boundaries. The required annotation for a segmentation task is more demanding than detection.

for splitting touching nuclei. Nuclei segmentation provides more details and more flexibility in developing a cancer grading feature. Bounding boxes from a nuclei detector can only generate an approximation of sizes, while boundaries from a nuclei segmenter can provide measures such as elongation, orientation angle, and smoothness of chromatin boundaries.

Epithelial segmentation in H&E images

The task of epithelial segmentation in H&E images is considerably harder due to color variations and low contrast between cytoplasm and stroma, particularly in activated tumor-associated stroma. A good solution to these problems is to use deep neural networks, as seen in Fig. 14.8.

In colon cancer, the GlaS (gland segmentation) challenge [57] has inspired an impressive series of improvements to this gland-instance segmentation problem [58–60], using deep neural networks trained to define boundaries and separate, distinct epithelial regions. Chen et al. [58] used a modified version of the FCN with an explicit representation of gland contours. Ronnenberger et al. proposed a method called U-Net [61], which is also fully convolutional, with a penalty function for misclassifying pixels in the gaps between glands. Al-Milaji et al. [62] proposed a method that combines a CNN with an unsupervised classifier, where deep learning provides image features and unsupervised region segmentation refines the boundary of the epithelial segmentation. Bulten et al. [63] developed a segmentation approach for

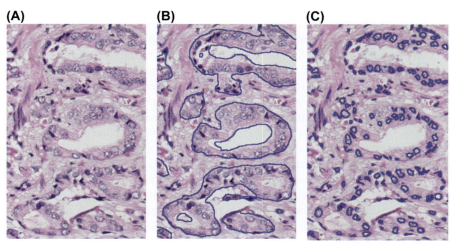

FIG. 14.8

H&E with deep NN segmentation on the same tissue as Fig. 14.6 (40× objective lens). Note the low contrast of glands, and nonuniform nuclei in panel (A), well segmented in panels (B) and (C). (A), H&E; (B), H&E epithelial segmentation; (C), H&E nuclei segmentation.

Mitotic figure detection

Establishing proliferative activity is central to determining the aggressive potential of many, if not most, tumors. Grading schemes call for manual mitotic counts or occasionally incorporate the use of immunohistochemical stains such as Ki67. The importance of proliferation in the prognostic evaluation of tumors is reinforced by a number of prognostic gene panels that include or are dominated by proliferation-associated genes [64]. As with the evaluation of nuclear features, mitotic counts are generally among the less reproducible components of grading schemes [11,12].

The same issues that make mitotic figures difficult to reproducibly count manually are the same that make them challenging to detect automatically, namely that they are relatively rare events, heterogeneously distributed, morphologically diverse, and with a number of nonmitotic objects that share similar properties. Cells undergoing cell division can be visually recognized at a number of morphologically different stages some, in the middle and later stages of the cycle, easier to identify than the earlier stages. Even in the later stages the morphologic variability is quite pronounced and the difference between mitotic objects and nonmitotic mimics can be as subtle as boundary fuzziness. Deep learning with large datasets is well suited for these types of subtle morphologic distinctions. One aspect of mitotic figure detection that is challenging for deep learning is that of class imbalance between mitotic figures, nonmitotic background objects such as regular nuclei and lymphocytes, and mimics such as apoptotic cells. This is generally handled by oversampling the mitotic figure class or by adjusting the loss function for training to account for the lower sample count. Ciresan et al. [65] developed a CNN approach for mitotic figure detection that is trained on image patches. Training of the neural network included enrichment of the limited mitotic training data by generating rotated and flipped versions of the images. These additional training examples helped alleviate the class imbalance issue and resulted in a robust network that had the best performance in the Assessment of Mitosis Detection Algorithms 2013 (AMIDA13) challenge [66]. Chen et al. [67] proposed the use of a cascade of two CNNs, with the first network generating a coarse classification of objects that are likely to be mitosis and may include other similar structures and a second network performing finer discrimination and verification to determine the final mitotic figure classification. Their approach utilized a specialized network that analyzed subtle patterns in images to reduce false-positive identifications. Albarqouni et al. [68] utilized data obtained from untrained annotators (crowds) in a CNN that contains an aggregation layer. Crowdsourcing provides enrichment of training data for mitosis, with noise handled using an explicit data aggregation step in training.

Ring segmentation

Gland structures, in general, are three-dimensional objects composed of tubules of varying width and length, with rounded ends, connected in a branching structure at lower grades, or forming an interconnected network at higher grades. Confocal microscopy of prostate samples in Ref. [69] and stacked histology slices in Ref. [70] show how the 3D geometry of glands relates to patterns in 2D histology. In the fragmented pattern of prostate cancer, illustrated with increasing grade on the right descending side of the Gleason diagram [71], tubules wrapped in stroma shrink down to small clusters and single cells, and central lumens disappear. In the cribriform and solid patterns, on the left descending side of the diagram, tubules are fused together without intervening stroma, as lumen size and frequency decrease. The morphogenesis from low grade to high grade, suggested by the Gleason diagram, is confirmed by 3D studies of gland structure to be a process of continuous growth along these two axes of variation.

There are a number of patterns that lumen, epithelium, and stroma can take in relation to one another. These patterns can be seen in many different kinds of cancer, with the frequency of occurrence and prognostic significance varying by organ. The cribriform pattern in breast, prostate, colon, lung, and other organs is discussed in Ref. [72]. From an engineering point of view, it is useful to identify organ-specific gland morphologic tendencies and their prognostic implications, to design segmentation strategies and features that correlate with the underlying biological processes, and so lead to accurate characterization of tumor differentiation.

This motivates the segmentation of gland rings, which we define as complete or fragmented tubules, appearing either as a ring of epithelial nuclei surrounding a gland lumen, or as a gland fragment with a few epithelial nuclei surrounded by stroma [73]. By looking at the adjacency between rings, stromal, and lumen regions, an algorithm can detect and evaluate these patterns across all grades of cancer.

Our ring segmentation algorithm works by forming a compact triangulation of epithelial nuclei centers, known as the Delaunay triangulation (characterized by the property that the interior of each triangle's circumcircle contains no other vertices). The algorithm then clusters the triangles across epithelial and lumen regions, to form broad areas separated by relatively short edges between ring nuclei, as shown in Fig. 14.9. This algorithm was originally developed based on IF epithelial and nuclear segmentation and has been validated as part of a clinical test [17] for risk stratification of prostate cancer patients. With the advent of accurate deep learning segmentations, rings and ring features can be applied to H&E images also with similar results (Fig. 14.14).

Few existing algorithms make the distinction between epithelial and gland ring segmentation, focusing instead on lower grade presentations with one ring of closely spaced nuclei per epithelial unit, surrounding a clear lumen, for example, Ref. [74]. These algorithms use nuclei as a guide for region-growing, starting from the lumen, an approach which can work well for the colon, less so for prostate and breast glands

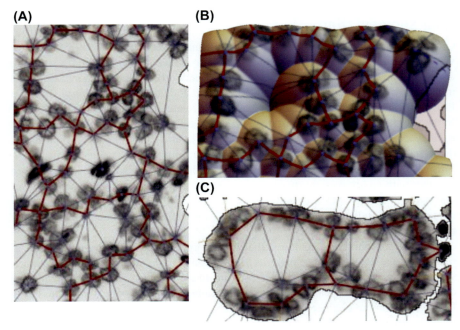

FIG. 14.9

Ring segmentation by watershed graph algorithm. (A) Rings and triangulation overlaid on nuclei (prostate, inverted DAPI, 20× objective lens) in fused gland. (B) Illustration of watershed graph segmentation algorithm applied to the top of image (A) showing nuclei centers as hilltops, joined by *red* rings which surround regions where simulated rainfall would flow down to the same valley. The algorithm effectively clusters triangular areas (*dark lines*, Delaunay triangulation, see text) to form broad regions separated by relatively short *(red)* edges [73]. (C) Geometry of two fused glands (or one U-tube-shaped tubule, or a branching tubule—the 2D histology image can be derived from several different 3D geometries). Tubule size in 3D = gland width in image = the longest triangle edge crossing the ring.

which tend to have a greater frequency of fragmentation and fusion. A notable exception is Ref. [75], which searches for polygonal gland rings using a randomized Markov Chain Monte Carlo approach, based upon approximate gland and nuclei maps. But this algorithm too does not generalize to handle the continuum of presentations between low- and high-grade cancers.

Protein biomarker features

Protein biomarkers provide valuable diagnostic and prognostic information not available in standard H&E-stained tissue. This section provides an overview of a few of the many ways of quantifying biomarkers in an image.

Immunohistochemistry (IHC), in its most popular form, uses a brown chromogenic detection system (3,3-diaminobenzidine or DAB) to tag a protein of interest against a nuclear hematoxylin counterstain, viewed on a brightfield microscope and manually quantified on a scale of 0–3. Multiplexing and automation extend this process to multiple, overlapping biomarkers. One technique stains serial sections with different markers, then coregisters the images with correlation software, for example, putting Ki67 on one slide and a cytokeratin epithelial marker on the next, to count epithelial Ki67 positive nuclei. This method was used in Ref. [76] with Visiopharm software [77] to analyze Ki67, ER, PR, and HER2 in breast cancer.

Alternatively, multiplex immunofluorescence puts a set of up to eight biomarkers on one slide [78,79], for example, analyzing tumor-infiltrating lymphocytes (TILs) in melanoma [80] or comparing TIL profiles in multiple tissues [81]. In prostate cancer, the androgen receptor (AR) signaling pathway drives progression of the disease, lowering AR levels in the stroma, while increasing AR in the epithelium [82], and thus the ratio of the two is a powerful prognostic feature [17].

Beyond simple counting and ratio features, the distance between different cell types can be used as a filter. In Ref. [83], for instance, the HALO package [84] analyzes colon cancer TIL immunoscore within 50 μm of invasive tumor buds only. Instead of averaging over the image, biomarker features may be calculated for clusters of cells, or for gland rings as in Fig. 14.10, or over other kinds of structural or morphological graphs. For example, in Ref. [85], breast cancer extracellular matrix (ECM) biomarkers integrin-α 3 and 6 are used to define the connectivity of cell graphs (covered in the next section) from which a set of features are derived.

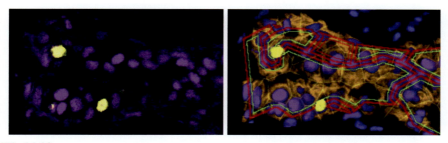

FIG. 14.10

Biomarker features combined with morphology. *Left*: Biomarker intensity image (20× objective lens). *Right*: Combination of morphology, AR and Ki67 features. Outer rings represent the SL-touch value (see next section) and inner rings represent the AR epistromal ratio (shown in *green* for high values, *red* for low values, i.e., worse outcome/grade). Nuclear AR strength is shown on a *blue* (low) to *purple* (high) scale, with Ki67 *(yellow)* on CK18 *(gold)* background.

Morphological features for cancer grading and prognosis

In the process of evaluating a slide, the pathologist takes into consideration features from several levels of image resolution. Although all resolutions are used to evaluate a slide, generally speaking, starting at the lowest resolution, composition (e.g., mucin, necrosis, stroma) and architectural patterns are noticed. At an intermediate resolution, size, shape, and the relationship (e.g., density and streaming) of nuclei to one another are prominent. And finally, nuclear texture, including chromatin pattern, margination, and nucleoli, depend on fine details seen at the highest resolution. Capturing and characterizing these relationships within and between scales is one of the creative sides of developing automated grading systems, as many of these potentially prognostic relationships are based on well-accepted histopathological principals, but in and of themselves are difficult to quantitate without computation. This is feature engineering.

Semiautomated attempts to quantify the relative importance of these features, for example, grading of invasive breast cancer in Ref. [86], or nuclear radius variance for prostate prognosis [87], show good results and encourage the search for fully automated methods.

As discussed throughout this section, layering is a common theme of feature engineering. Starting with the segmentation outputs discussed above, a variety of different techniques can be layered to quantitate or characterize relationships as well as to normalize features for robustness. Nuclei segmentation, for example, followed by ellipsoid fitting, can provide a high-level view of the shape and organization of nuclei within a region of interest (Fig. 14.11). Additionally derived features from the ellipsoidal representations could include density, shape, orientation, and regularity, among others (Fig. 14.12). Nuclear shape, texture, and orientation have recently

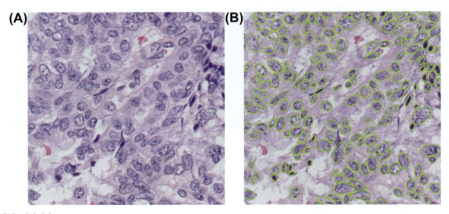

FIG. 14.11

Example nuclear segmentation and ellipse fitting. (A) Input H&E image. (B) Segmentation results with fitted ellipsoids.

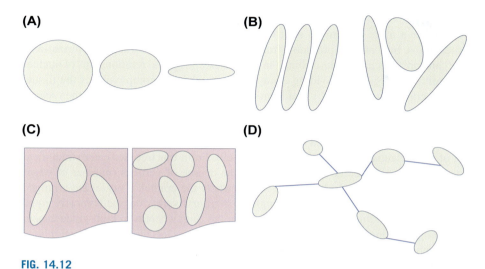

FIG. 14.12

Example of nuclear features. (A) Elongation of an individual nucleus can be computed from the ratio of major and minor axes of the ellipsoid. (B) The variability of the orientation of the major axis provides a measure of the level of organization in a tissue. (C) Density of nuclei within a given compartment is measured as the count of nuclei (or area of nuclei) divided by the area of the compartment. (D) Graphs can be used to provide a summary of the morphological relationships and structure of nuclei clusters.

been shown to be predictive of outcome in estrogen positive, lymph node negative breast cancer [21]. At the highest resolution, color and texture of nuclei successfully distinguish benign and malignant regions in colon cancer, as shown by Jørgensen et al. [88].

Although raw mitotic count may be predictive in TMA cohorts, normalizing to account for variability in the sample volume will stabilize the feature when developing in larger samples and when scaling outside of the test environment. Manual grading schemes generally require counting mitotic figures within a specified area or per a specified number of tumor cells to account for volume variability from patient to patient [89]. This provides a standard mechanism for comparing mitotic activity in different subjects. Applying that concept to an automated scheme, an algorithm applies the same specified area to each slide either randomly or looking for the area of highest proliferation, analogous to the mitotic count requirement of the modified Bloom-Richardson grading scheme for breast cancer. Another similar relative measure could be achieved by combining mitotic count and nuclear count, which would provide a density index that reflects a tumor's proliferative activity.

Characterizing the spatial distribution of epithelial cells and stromal cells in cancer is an evaluation primarily in the domain of computational pathology. Although glandular differentiation or the degree to which tumor glands deviate from normal is part of a number of grading schemes, computational pathology and graphs specifically offer a more bottom-up approach which can support gland segmentation as noted above but can also provide a framework for more free-form characterizations. We can construct graphs describing the adjacency and spatial relationships of nuclear and epithelial regions, providing a way to summarize the complexity of the tissue. Higher-grade cancer typically yields more irregular, complex graphs. We then search for graph features or statistics that predict grade and patient outcome.

The simplest predictive graph is perhaps the minimum spanning tree (MST) based on epithelial nuclei as vertices (Fig. 14.13). The MST is the minimum connected subset of the Delaunay triangulation (see Fig. 14.9). From this tree, we can extract measures such as the average edge length and the degree of vertices (number of edges connected to a vertex). In low-grade cancer where the epithelial nuclei are arranged in tightly spaced rings, MST edge length is short and vertex degree is mostly equal to two, corresponding to the two neighbors in the ring, whereas in high-grade cancer edge length is longer and a greater proportion of vertices has three or more neighbors.

Graphs based on clustering or segmentation of nuclei are better able to represent the higher-level structure of glands and different cell types. The cell cluster graph (CCG) [90], a popular hierarchical graph which contains clusters of nuclei, which in turn are probabilistically connected to nearby clusters, has been used to define

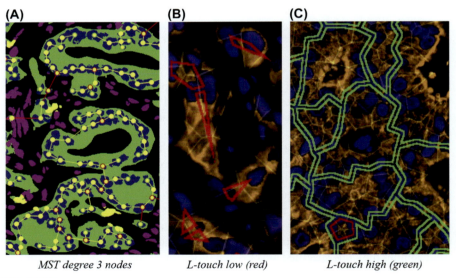

MST degree 3 nodes　　*L-touch low (red)*　　*L-touch high (green)*

FIG. 14.13

Graph features for prostate cancer. (A) Minimum spanning tree (MST) showing three-way vertices with *red* halos. The proportion of degree three vertices is a strong feature for grade and prognosis. (B) Gland rings in fragmented pattern, adjacent to stroma without lumens, having low value of L-touch (shown in *red*). (C) Gland rings in fused (cribriform) pattern not touching stroma, with luminal clearings, having high value of L-touch (shown in *green*).

numerous features such as gland angularity co-occurrence in the prostate [91] and infiltrating lymphocytes in the lung [92]. A challenge in setting up the CCG is that clusters may not correspond to biologically relevant structures and may not adapt to the dynamically changing scales present in the tissue. Good reviews covering pathology graphs and other texture and nuclear features are provided in Refs. [93, 94].

The gland ring or fragment, introduced in Ring segmentation section, is a natural clustering of nuclei that we have found to produce good features in prostate cancer across all grades, especially when combined with protein biomarker measurements. Since the growth and functioning of the prostate is mediated by signaling across the epithelial-stromal and epithelial-lumen boundaries, we have developed a set of simple ring-stroma-lumen adjacency features which account for the proportion of the ring boundary that is adjacent to stroma, S-touch, and the proportion adjacent to lumen, L-touch (Fig. 14.13). The product of these two, the SL-touch feature, covering gland fusion and fragmentation patterns, is highly prognostic in prostate cancer (Fig. 14.14).

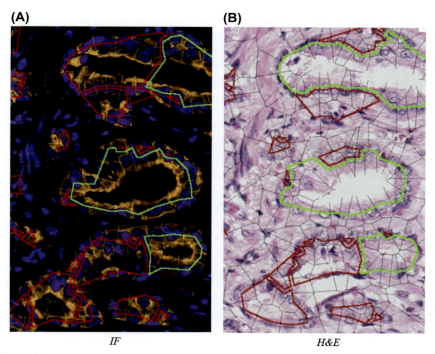

FIG. 14.14

SL-touch product feature. The product of stromal adjacency (S-touch) and lumen adjacency (L-touch) scores, with low value in *red* and high value in *green* (better prognosis) (compare segmentations in Figs. 14.6 and 14.8), showing similar results for (A) IF and (B) H&E images of the same tissue. *Gray lines* show the Voronoi diagram, the zone of influence around each epithelial nucleus.

294 **CHAPTER 14** Artificial intelligence methods for predictive

Given features characterizing nuclei clusters or rings, the features are aggregated over the graph representing the image or whole-slide by taking the mean, the median, or some other aggregation function. More sophisticated aggregation can use support vector machines (SVMs) learning (see *Modeling* section) with graph kernels [95] or deep graph neural networks [96].

Modeling

Modeling, as the final computational step in creating improved grading schemes, combines the output of image-derived features and clinically derived features to predict the probability of survival or the expected time until significant progression of the disease in a cohort of patients with known clinical outcome. Standard clinical features that go into models include age, race, sex, etc. as well as pathologic stage, laboratory values, and features from radiology and genetic testing [97]. Machine learning algorithms learn the most relevant feature combinations (imaging and clinically derived) and quantify the relative importance of each feature in a model that can then be applied to new patients outside of the training and test environment. Development cohorts for survival analysis are generally split into a set for training the model and a separate set for testing or validating the model. There are a number of disease- and cohort-specific considerations to be taken into account when creating these two groups, but in general, they should look very similar in terms of disease types, subtypes, grades, and outcome. This is to create a balanced testing environment.

Survival analysis involves methods which analyze and predict the time to a medical event such as death, disease recurrence, or disease progression [98]. The survival function, denoted by $S(t)$, is the probability that the patient has not experienced the event at time t and survives past this time. As time goes by, the probability of survival decreases, as shown in the Kaplan-Meier plot in Fig. 14.15 [99].

Concordance index (C-Index), a widely used measure of algorithm performance, is a form of rank correlation between the model predictions and the actual survival times, which takes into account missing or censored data (see below) [100]. The C-index has some elegant properties: it is the probability that the model correctly predicts which of a random pair of patients has the longer survival time, which in the binary outcome case is the area under the ROC curve—the curve that plots the model true-positive rate (sensitivity) against false-positive rate (1-specificity). The C-Index lies in the range 0–1, with a value of 0.5, meaning that the model results are no better than chance. Although a useful measure, some caveats are important to remember. C-Index indicates the discrimination power of an algorithm but not the predictive power. Since the index only looks at the proportion of correctly ordered pairs, it is agnostic to the actual predicted times.

Survival analysis can be treated as a regression problem of predicting survival times of which there are many methods [98], most of which, however, do not handle the problem of missing data caused by patients dropping out of the study over time.

Cox proportional hazards model

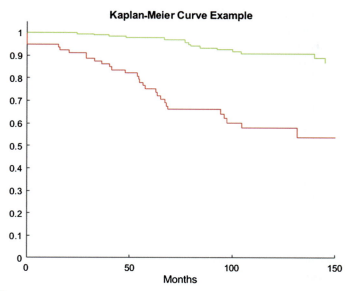

FIG. 14.15

An example, Kaplan-Meier (K-M) curve, plotting survival probability decreasing with time, in the range 0–1 on the *y*-axis against survival months on the *x*-axis, for two treatments. Top treatment *(green)* has better survival rates than bottom treatment *(red)*.

Missing information from these patients is known as censored data [101]. Patients who have experienced the event of interest (e.g., PSA recurrence, death, or metastasis) are considered events. Right-censored patients are those being followed at the end of the study who have not experienced the event or have dropped out of the study, and left-censored patients are patients who had an event but the timing is not clear.

Traditional ways of dealing with censored data in statistics include removing the sample points, treating them as nonevents, or more elaborate ways such as repeating the data points once as an event and once as a nonevent. There are also imputing methods which estimate the missing values before training on a survival model. Inevitably, however, all these methods introduce a bias to the resulting model. To rectify this, there are a variety of methods that can handle censored information, and these are outlined in the paragraphs below.

Cox proportional hazards model

One of the first and most popular methods of handling censored information is the Cox proportional hazards model [102]. In this model, the hazard function or risk (the slope of the survival function, $h_i(t)$) is proportional for any two individuals.

296 CHAPTER 14 Artificial intelligence methods for predictive

The proportionality is set up by taking the product of a baseline hazard function ($h_0(t)$) with baseline features, and an exponential function of weighted feature values, with the weights indicating the importance of each feature. The biggest caveat when using this model is to detect and account for time-varying effects where the proportionality assumption may break down [103], as is the case in many biological applications. The reliability of the Cox model (and other models) is not sound if the number of features is greater than the number of events.

Neural networks

In their earliest and simplest form, neural networks for survival analysis [104] learned the relationship between features and the hazard functions. Unfortunately, this method performed about the same as a linear Cox model. DeepSurv, a relatively new method which is rooted in the Cox model, uses deep learning to predict the risk of a patient experiencing an event [105]. It is a simple application of a multilayer neural network with one output node that predicts patient risk. The main advantage of DeepSurv is that, while the Cox model evaluates efficacy as a whole for the group of patients, DeepSurv produces a risk score tailored to each patient. DeepHit, a higher-performing improvement that does not make assumptions about the structure of the relationship between features and risk [106], learns the distribution of survival times directly. The structure of the network is more complicated than DeepSurv, with fully connected layers and a single softmax layer as the output layer. A disadvantage of using deep neural networks for survival analysis is that, in order to build robust models with reliable predictions, these algorithms tend to require large quantities of clinically annotated patient data (i.e., outcome), which is often not available at the necessary scale.

Decision trees and random forests

Tree-based methods can be used to solve problems in classification and regression by growing conceptual trees [107]. A simple traversal algorithm walks from the root to one of the leaves of the tree, choosing each branch limb depending on the value of a feature relative to a threshold the algorithm selects for that branching point. A regression value is therefore computed at each branch and leaf, creating a regression tree. Creating too many branching points overfits the data, and conversely, not creating a deep-enough tree risks not learning the appropriate structure of the data. Decision trees are unfortunately sensitive to noise, and small perturbation in the data can result in vastly different models due to the recursive identification of branch points.

Random forest [108] is a conceptual modification of the decision tree approach that introduces two main random components to generate multiple trees whose outputs are then aggregated. Random modifications include bootstrap aggregation of the data and random selection of features for each node. Using this modification to the decision-tree learning algorithm, multiple trees are generated and combined by averaging and arriving at a solution which reduces the bias and variance in the decision-tree models.

Random survival forest (RSF) is an extension of the random forest method [109], which includes the computation of an ensemble cumulative hazards function (EHF). The central idea of RSF is that the survival difference between low- and high-risk patients is maximized in the tree. For a given set of features, for every tree, a path is followed to a unique terminal node and EHF value. This is averaged over all survival trees to compute the survival estimate for the patient.

SVM-based methods: Survival-SVM, SVCR, and SVRc

Support vector machine, introduced in the early 1990s [110], is a classification algorithm that learns to distinguish between binary labels of given data. SVM is a linear method, extended to encapsulate nonlinear problems through projecting the data to a higher dimensional space. This method has enjoyed particular success in its ease of handling high-dimensional data and the interpretability of its linear models. This method was also extended to solving regression problems, known as support vector regression. Essentially, the SVM algorithm is an optimization algorithm that works by maximizing the margin of a data set and finding the separating hyperplane that neatly divides the data. The margin is the smallest distance between a data point and the separating hyperplane. Implicitly, we have assumed that the data we are trying to classify and learn are perfectly separable by a hyperplane, but if this is not the case, we can still find an approximation by optimizing the distance to the separating hyperplane (pink box; Fig. 14.16), with error variables (blue box), subject to constraints (red box).

SVCR, a simple extension of SVMs for survival analysis that accounts for censoring [112], ignores constraints where the prediction time is later than the censor time (right censoring, second line in red box). In this model, errors in events and nonevents and in early and late predictions are penalized identically, which creates a problem in that late prediction of an event is generally a far more serious error than early prediction, having consequences for patient treatment decisions.

SVRc, an effective and efficient solution to this problem, improves the handling of censored data by applying separate penalty factors [113]. The algorithm applies the most severe penalty to late prediction of events, followed by a lesser penalty to early prediction of nonevents, then early prediction of events, and finally a negligible penalty is applied to late prediction of nonevents (censored data). Optimal values for these parameters may be found using a hyper-parameter tuning algorithm to learn the values over multiple data sets. We have successfully used particle swarm optimization [114] to set up these parameters.

298 CHAPTER 14 Artificial intelligence methods for predictive

$$\min_{W,b} \quad \frac{1}{2}\|W\|^2 + C\sum_{n}^{i=1}(\xi_i + \xi_i^*)$$

given the constraints

$$y_i - (W \cdot \phi(x_i) + b) \le \epsilon + \xi_i$$

$$(W \cdot \phi(x_i) + b) - y_i \le \epsilon + \xi_i^*$$

$$\xi_i, \xi_i^* \ge 0 \quad i = 1, ..., n$$

FIG. 14.16

Support vector machine (SVM) equations. SVM is an optimization algorithm to find the optimal hyperplane that separates two classes of data in a given higher dimensional feature space. The operation of the algorithm is explained in three boxes. *Pink box*: minimize the weight of all features. The distance between a point of one class and the optimal hyperplane can be represented as 1. We want to maximize this distance or equivalently, minimize the denominator W, $\|W\|$. *Blue box*: penalize imperfectly separated data using error (slack) variables, with penalty factor C. *Red box*: constrain the slack variables to separate the data. Larger slack variables, with greater penalty, are needed for points whose projection to higher dimensional feature space by $\varphi(x_i)$ is further than ε from the target value [111]. The two constraint lines deal with points on either side of the hyperplane, or when extended to survival regression, to early predictions in the first line, and late predictions in the second line. The kernel trick is a way of speeding up the computation by calculating the matrix multiplication $W.\varphi$ once, and reusing the result for every data point. In SVCR (censored regression), the two hyper-parameter penalty values, C and ε, are used for early and late predictions and for censored data and events—an overly restrictive model. SVRc effectively extends the model by using separate penalty parameters for early/late and censored/event data. These eight hyper-parameters were optimized in advance over a set of cohorts by particle swarm optimization (see text).

Feature selection tools

The feature development sections above can generate many features, often anywhere between several hundreds to multiple thousands per application. Considering that for modeling purposes, there must be a limit and an optimal balance between features and events, methods for reducing the number of features must be a part of any modeling strategy. There are four possible classes of features—strongly relevant, weakly relevant, redundant, and irrelevant [115]. As the names suggest, relevant features are important for achieving good results [116]. Irrelevant features do not affect the result but may affect algorithm efficiency. Redundant features are those that contain the same information as another or a subset of features. Removing irrelevant and redundant features is the function of feature selection algorithms, with the goal of improving results and reducing computation time. Let us assume there are m features associated with a problem. To find the best subset of features, one needs to evaluate all $2^m - 1$ subsets, a computationally infeasible task. Hence, there is a need for approximation algorithms that can perform feature selection. Broadly there are three classes of feature selection methods: Filter, Wrapper, and Embedded methods [117].

Filter methods filter out irrelevant features by some mechanism that computes a score of importance. For example, Pearson's correlation measures the linear dependency of two random variables. This could be used to compute the correlation coefficients of features with respect to the target variable and then select the top features as input to the algorithm. Another coefficient, mutual information (MI), better captures nonlinear relationships between features and target variables.

Wrapper methods for feature selection are those methods that use the classification algorithm itself as a black box predictor and use the performance of the resulting model as an indicator of how good the feature subset is. Obviously, as explained above, not all subsets can be evaluated and thus, wrapper methods use clever algorithms to prune and search the space of feature subsets to arrive at an optimal subset. Broadly these methods can be classified as sequential search algorithms and heuristic search algorithms. Forward selection and backward elimination are examples of sequential search methods. In forward selection, we start with an empty set and add the feature that results in the best performance. Sequentially, we keep adding features and stop when there is no further improvement. Backward elimination works in a similar fashion where we start with the entire set of features and progressively remove features. These are both examples of greedy methods, where one hopes that sequentially taking the best step will result in a globally optimal solution. However, this is not usually the case when there are complicated interactions between features. Alternatively, heuristic-based wrapper algorithms, like particle swarm optimization and genetic algorithms, tend to find better locally optimal solutions. Given that feature selection is a problem for which no efficient algorithms exist, these solutions are the best we can do in a reasonable time.

Embedded methods perform feature selection as an intrinsic part of the classification algorithm, for example, decision trees only consider splits that result in best performance. The features that do not contribute to the model are not considered. Regularization is an important, optional addendum to linear feature selection models that falls under the class of embedded methods. When learning a linear model (linear regression, linear SVMs), lasso (least absolute shrinkage and selection operator) regularization is invoked by adding a term to the cost function. Linear models operate by learning the coefficients of the model that best fit the data, often including many small weights which do not meaningfully improve the model, and which do not generalize well to unseen data. To get over this limitation, we need to force some of the coefficients to zero and adjust other coefficients. This is done by adding an extra term to the cost function which penalizes the nonzero weights. In practice, this extra term limits the sum of the absolute value of the coefficients.

Ground truth data for AI-based features

Deep learning algorithms are heavily data dependent, and to ensure scalability outside of the development environment, these large datasets must conform to clinical reality both in terms of absolute numbers and in variety of representation. It is critical to not only represent different grades and morphologic subtypes but also different staining conditions and tissue quality. In addition to large datasets in a supervised or semisupervised setting, deep learning also demands ground truth annotations. This can be as simple as assigning a grade or binary label of tumor or nontumor to an input image, all the way to digitally drawing outlines around target objects for detection or segmentation. The choice of annotation strategy is dependent on the algorithms functional output. For example, if the output is boundary segmentation of an object, then precise boundaries must be part of the ground truth. If, on the other hand, the task is detection, then the annotation may involve putting a bounding box around or a seed in the target object. Clearly, segmentation provides much greater flexibility downstream in creating features, but it comes at a much higher upfront cost in terms of annotation.

Rapid outlining using semiautomatic methods can reduce the work demand for expert annotators, using minimal gestures and mouse clicks to initialize algorithms that create complete boundaries for structures of interest. Example algorithms include segmentation based on active contour evolution [118,119] where users provide an initial boundary and the algorithm evolves it to match an image boundary. This type of algorithm has been applied successfully for a variety of outlining tasks in medicine [120]. Random walker [121] is another algorithm that automatically generates contours, where the user provides a rough initial labeling of the different objects. This approach has less burden compared with active contour approaches as it does not require a complete definition of an outline. An alternative approach to contour initialization uses the watershed algorithm [122], which simulates the flooding of water in image valleys (low-intensity values surrounded by high values) to detect objects separated by boundaries. Annotators can interactively specify thresholds for the watershed algorithm and select pixel clusters that form the object of interest.

Due to the dependence of deep learning on large datasets, virtually all deep learning-based approaches conduct training of the neural network using data augmentation where training data are transformed to expand the existing datasets. Transformations are performed geometrically via rotations or flips, and they can also be performed based on color (e.g., stain transformation, contrast adjustment, etc.). Crowdsourcing also has great potential for generating larger datasets. Consistency and detection of low-quality annotations is a critical issue for such an approach, necessitating a data aggregation step such as the one proposed by Albarqouni et al. [68].

Conclusion

The importance of accurate histopathologic diagnosis in patient care is often underestimated. This perception is changing among treating clinicians as our molecular understanding of disease has exploded and targeted therapies and new specific treatment protocols continue to emerge. Having precise and reproducibly character-ized disease remains an important component in developing these specific treatment protocols. To this end, the use of deep learning methods for interrogating images and artificial intelligence to analyze complex, interrelated datasets are changing this landscape. The promise of this technological and knowledge revolution ultimately is to improve patient outcomes by creating a more robust basis for linking patients with the most effective treatments.

References

[1] Fuhrman SA, Lasky LC, Limas C. Prognostic significance of morphologic parameters in renal cell carcinoma. Am J Surg Pathol 1982;6(7):655–63.

[2] Bosman FT, Carneiro F, Hruban RH, Theise ND, et al. WHO classification of tumours of the digestive system. vol. 4. World Health Organization; 2010.

[3] Seethala RR. Histologic grading and prognostic biomarkers in salivary gland carcino-mas. Adv Anat Pathol 2011;18(1):29–45.

[4] Gleason DF, Mellinger GT. Prediction of prognosis for prostatic adenocarcinoma by combined histological grading and clinical staging. J Urol 1974;111(1):58–64.

[5] Mutch DG. The new figo staging system for cancers of the vulva, cervix, endometrium and sarcomas. Gynecol Oncol 2009;115(3):325–8.

[6] Miettinen M, Lasota J. Gastrointestinal stromal tumors: pathology and prognosis at dif-ferent sites. In: Seminars in diagnostic pathology, vol. 23. Elsevier; 2006. p. 70–83.

[7] Klimstra DS, Modlin IR, Coppola D, Lloyd RV, Suster S. The pathologic classification of neuroendocrine tumors: a review of nomenclature, grading, and staging systems. Pancreas 2010;39(6):707–12.

[8] Elston C, Ellis I. Assessment of histological grade. Breast 1998;13:356–84.

[9] Gleason DF. Histologic grading of prostate cancer: a perspective. Hum Pathol 1992; 23(3):273–9.

[10] Allsbrook Jr WC, Mangold KA, Johnson MH, Lane RB, Lane CG, Epstein JI. Interob-server reproducibility of Gleason grading of prostatic carcinoma: general pathologist. Hum Pathol 2001;32(1):81–8.

[11] Boiesen P, Bendahl P-O, Anagnostaki L, Domanski H, Holm E, Idvall I, Johansson S, Ljung- berg O, Ringberg A. Histologic grading in breast cancer: reproducibility between seven pathologic departments. Acta Oncol 2000;39(1):41–5.

[12] Meyer JS, Alvarez C, Milikowski C, Olson N, Russo I, Russo J, Glass A, Zehnbauer BA, Lister K, Parwaresch R. Breast carcinoma malignancy grading by Bloom–Richardson system vs proliferation index: reproducibility of grade and advan-tages of proliferation index. Mod Pathol 2005;18(8):1067.

[13] Epstein JI, Egevad L, Amin MB, Delahunt B, Srigley JR, Humphrey PA. The 2014 inter- national society of urological pathology (isup) consensus conference on Gleason grading of prostatic carcinoma. Am J Surg Pathol 2016;40(2):244–52.

[14] Galea MH, Blamey RW, Elston CE, Ellis IO. The Nottingham prognostic index in pri-mary breast cancer. Breast Cancer Res Treat 1992;22(3):207–19.

[15] Greene KL, Meng MV, Elkin EP, Cooperberg MR, Pasta DJ, Kattan MW, Wallace K, Carroll PR. Validation of the kattan preoperative nomogram for prostate cancer recur-rence using a community based cohort: results from cancer of the prostate strategic uro-logical research endeavor (capsure). J Urol 2004;171(6 Part 1):2255–9.

302 CHAPTER 14 Artificial intelligence methods for predictive

[16] Stephenson AJ, Scardino PT, Eastham JA, Bianco Jr FJ, Dotan ZA, DiBlasio CJ, Reuther A, Klein EA, Kattan MW. Postoperative nomogram predicting the 10-year probability of prostate cancer recurrence after radical prostatectomy. J Clin Oncol Off J Am Soc Clin Oncol 2005;23(28):7005.

[17] Donovan MJ, Fernandez G, Scott R, Khan FM, Zeineh J, Koll G, Gladoun N, Charytonowicz E, Tewari A, Cordon-Cardo C. Development and validation of a novel automated Gleason grade and molecular profile that define a highly predictive prostate cancer progression algorithm-based test. Prostate Cancer Prostatic Dis 2018;21(4):594.

[18] Beck AH, Sangoi AR, Leung S, Marinelli RJ, Nielsen TO, Van De Vijver MJ, West RB, Van De Rijn M, Koller D. Systematic analysis of breast cancer morphology uncovers stromal features associated with survival. Sci Transl Med 2011;3(108):108ra113.

[19] Donovan MJ, Khan FM, Fernandez G, Mesa-Tejada R, Sapir M, Zubek VB, Powell D, Fogarasi S, Vengrenyuk Y, Teverovskiy M, et al. Personalized prediction of tumor response and cancer progression on prostate needle biopsy. J Urol 2009;182(1):125–32.

[20] Blume-Jensen P, Berman DM, Rimm DL, Shipitsin M, Putzi M, Nifong TP, Small C, Choudhury S, Capela T, Coupal L, et al. Development and clinical validation of an in situ biopsy- based multimarker assay for risk stratification in prostate cancer. Clin Cancer Res 2015;21(11):2591–600.

[21] Lu C, Romo-Bucheli D, Wang X, Janowczyk A, Ganesan S, Gilmore H, Rimm D, Madab- hushi A. Nuclear shape and orientation features from H&E images predict survival in early-stage estrogen receptor-positive breast cancers. Lab Invest 2018;98(11):1438.

[22] Arvaniti E, Fricker KS, Moret M, Rupp N, Hermanns T, Fankhauser C, Wey N, Wild PJ, Rueschoff JH, Claassen M. Automated Gleason grading of prostate cancer tissue microarrays via deep learning. Sci Rep 2018;8.

[23] Couture HD, Williams LA, Geradts J, Nyante SJ, Butler EN, Marron J, Perou CM, Troester MA, Niethammer M. Image analysis with deep learning to predict breast cancer grade, er status, histologic subtype, and intrinsic subtype. NPJ Breast Cancer 2018; 4(1):30.

[24] Mirlacher M, Kasper M, Storz M, Knecht Y, Dürmüller U, Simon R, Mihatsch MJ, Sauter G. Influence of slide aging on results of translational research studies using immunohistochemistry. Mod Pathol 2004;17(11):1414.

[25] Grillo F, Pigozzi S, Ceriolo P, Calamaro P, Fiocca R, Mastracci L. Factors affecting immunoreactivity in long-term storage of formalin-fixed paraffin-embedded tissue sections. Histochem Cell Biol 2015;144(1):93–9.

[26] Rimm DL, Camp RL, Charette LA, Costa J, Olsen DA, Reiss M. Tissue microarray: a new technology for amplification of tissue resources. Cancer J (Sudbury, Mass) 2001; 7(1):24–31.

[27] Mukhopadhyay S, Feldman MD, Abels E, Ashfaq R, Beltaifa S, Cacciabeve NG, Cathro HP, Cheng L, Cooper K, Dickey GE, et al. Whole slide imaging versus microscopy for primary diagnosis in surgical pathology: a multicenter blinded randomized noninferiority study of 1992 cases (pivotal study). Am J Surg Pathol 2018;42(1):39.

[28] Reinhard E, Adhikhmin M, Gooch B, Shirley P. Color transfer between images. IEEE Comput Graph Appl 2001;21(5):34–41.

[29] Magee D, Treanor D, Crellin D, Shires M, Smith K, Mohee K, Quirke P. Colour normalisation in digital histopathology images. In: Proc. optical tissue image analysis in microscopy, histopathology and endoscopy (MICCAI workshop), vol. 100. Daniel Elson; 2009.

[30] Shaban MT, Baur C, Navab N, Albarqouni S. Staingan: stain style transfer for digital histological images. arXiv 2018. preprint arXiv:1804.01601.

[31] Macenko M, Niethammer M, Marron JS, Borland D, Woosley JT, Guan X, Schmitt C, Thomas NE. A method for normalizing histology slides for quantitative analysis. In: 2009 IEEE international symposium on biomedical imaging: from nano to macro. IEEE; 2009. p. 1107–10.

[32] Vahadane A, Peng T, Sethi A, Albarqouni S, Wang L, Baust M, Steiger K, Schlitter AM, Es- posito I, Navab N. Structure-preserving color normalization and sparse stain separation for histological images. IEEE Trans Med Imaging 2016;35(8):1962–71.

[33] Roy S, Kumar Jain A, Lal S, Kini J. A study about color normalization methods for histopathology images. Micron 2018;114.

[34] Bejnordi BE, Litjens G, Timofeeva N, Otte-Holler I, Homeyer A, Karssemeijer N, van der Laak JA. Stain specific standardization of whole-slide histopathological images. IEEE Trans Med Imaging 2016;35(2):404–15.

[35] BenTaieb A, Hamarneh G. Adversarial stain transfer for histopathology image analysis. IEEE Trans Med Imaging 2018;37(3):792–802.

[36] Network CGA, et al. Comprehensive molecular portraits of human breast tumours. Nature 2012;490(7418):61.

[37] Akoya Biosciences. Vectra imaging system. Available from: https://www.akoyabio.com/phenopticstm/instruments.

[38] Zimmermann T. Spectral imaging and linear unmixing in light microscopy. In: Microscopy techniques. Springer; 2005. p. 245–65.

[39] Gillis N, Vavasis SA. Fast and robust recursive algorithms for separable nonnegative matrix factorization. IEEE Trans Pattern Anal Mach Intell 2014;36(4):698–714.

[40] Huang C-H, Veillard A, Roux L, Loménie N, Racoceanu D. Time-efficient sparse analysis of histopathological whole slide images. Comput Med Imaging Graph 2011;35(7–8):579–91.

[41] Bahlmann C, Patel A, Johnson J, Ni J, Chekkoury A, Khurd P, Kamen A, Grady L, Krupinski E, Graham A, et al. Automated detection of diagnostically relevant regions in H&E stained digital pathology slides. In: Medical imaging 2012: computer-aided diagnosis, vol. 8315. International Society for Optics and Photonics; 2012. p. 831504.

[42] Mercan E, Aksoy S, Shapiro LG, Weaver DL, Brunyé TT, Elmore JG. Localization of diagnostically relevant regions of interest in whole slide images: a comparative study. J Digit Imaging 2016;29(4):496–506.

[43] Cruz-Roa A, Gilmore H, Basavanhally A, Feldman M, Ganesan S, Shih N, Tomaszewski J, Madabhushi A, González F. High-throughput adaptive sampling for whole-slide histopathology image analysis (hashi) via convolutional neural networks: application to invasive breast cancer detection. PloS One 2018;13(5), e0196828.

[44] Bardou D, Zhang K, Ahmad SM. Classification of breast cancer based on histology images using convolutional neural networks. IEEE Access 2018;6:24680–93.

[45] Campanella G, VWK S, Fuchs TJ. Terabyte-scale deep multiple instance learning for classification and localization in pathology. arXiv 2018. preprint arXiv:1805.06983.

[46] Cruz-Roa A, Gilmore H, Basavanhally A, Feldman M, Ganesan S, Shih NN, Tomaszewski J, González FA, Madabhushi A. Accurate and reproducible invasive breast cancer detection in whole-slide images: a deep learning approach for quantifying tumor extent. Sci Rep 2017;7:46450.

[47] Bejnordi BE, Veta M, Van Diest PJ, Van Ginneken B, Karssemeijer N, Litjens G, Van Der Laak JA, Hermsen M, Manson QF, Balkenhol M, et al. Diagnostic assessment of

deep learning algorithms for detection of lymph node metastases in women with breast cancer. JAMA 2017;318(22):2199–210.

[48] Bejnordi BE, Zuidhof G, Balkenhol M, Hermsen M, Bult P, van Ginneken B, Karssemeijer N, Litjens G, van der Laak J. Context-aware stacked convolutional neural networks for classification of breast carcinomas in whole-slide histopathology images. J Med Imaging 2017;4(4), 044504.

[49] Marami B, Prastawa M, Chan M, Donovan M, Fernandez G, Zeineh J. Ensemble network for region identification in breast histopathology slides. In: International conference image analysis and recognition. Springer; 2018. p. 861–8.

[50] Aresta G, Araujo T, Kwok S, Chennamsetty SS, Safwan M, Alex V, Marami B, Prastawa M, Chan M, Donovan M, et al. Bach: grand challenge on breast cancer histology images. arXiv 2018. preprint arXiv:1808.04277.

[51] Al-Kofahi Y, Lassoued W, Lee W, Roysam B. Improved automatic detection and segmentation of cell nuclei in histopathology images. IEEE Trans Biomed Eng 2010; 57(4):841–52.

[52] Veta M, Van Diest PJ, Kornegoor R, Huisman A, Viergever MA, Pluim JP. Automatic nuclei segmentation in H&E stained breast cancer histopathology images. PloS One 2013;8(7), e70221.

[53] Xing F, Yang L. Robust nucleus/cell detection and segmentation in digital pathology and microscopy images: a comprehensive review. IEEE Rev Biomed Eng 2016;9:234–63.

[54] Ajemba P, Al-Kofahi Y, Scott R, Donovan M, Fernandez G. Integrated segmentation of cellular structures. In: Medical imaging 2011: Image processing, vol. 7962. International Society for Optics and Photonics; 2011. p. 79620I.

[55] Hofener H, Homeyer A, Weiss N, Molin J, Lundstrom CF, Hahn HK. Deep learning nuclei detection: a simple approach can deliver state-of-the-art results. Comput Med Imaging Graph 2018;70:43–52. https://doi.org/10.1016/j.compmedimag.2018.08.010.

[56] Naylor P, La M, Reyal F, Walter T. Nuclei segmentation in histopathology images using deep neural networks. In: 2017 IEEE 14th international symposium on biomedical imaging. ISBI 2017; 2017. p. 933–6. https://doi.org/10.1109/ISBI.2017.7950669.

[57] Sirinukunwattana K, Pluim JP, Chen H, Qi X, Heng P-A, Guo YB, Wang LY, Matuszewski BJ, Bruni E, Sanchez U, et al. Gland segmentation in colon histology images: the glas challenge contest. Med Image Anal 2017;35:489–502.

[58] Chen H, Qi X, Yu L, Heng P-A. Dcan: Deep contour-aware networks for accurate gland segmentation. In: 2016 IEEE conference on computer vision and pattern recognition (CVPR); 2016. p. 2487–96.

[59] Xu Y, Li Y, Wang Y, Liu M, Fan Y, Lai M, Eric I, Chang C. Gland instance segmentation using deep multichannel neural networks. IEEE Trans Biomed Eng 2017;64(12):2901–12.

[60] Manivannan S, Li W, Zhang J, Trucco E, McKenna SJ. Structure prediction for gland segmentation with hand-crafted and deep convolutional features. IEEE Trans Med Imaging 2018;37(1):210–21.

[61] Ronneberger O, Fischer P, Brox T, U-net. Convolutional networks for biomedical image segmentation. In: Medical image computing and computer-assisted intervention MICCAI 2015. Springer International Publishing; 2015. p. 234241.

[62] Al-Milaji Z, Ersoy I, Hafiane A, Palaniappan K, Bunyak F. Integrating segmentation with deep learning for enhanced classification of epithelial and stromal tissues in H&E images. Pattern Recogn Lett 2019;119:214–21. https://doi.org/10.1016/j.patrec.2017.09.015.

References **305**

[63] Bulten W, Bándi P, Hoven J, Rvd L, Lotz J, Weiss N, JVD L, Ginneken BV, Hulsbergen-van de Kaa C, Litjens G. Epithelium segmentation using deep learning in H&E-stained prostate specimens with immunohistochemistry as reference standard. Sci Rep 2019;9(1):864. https://doi.org/10.1038/s41598-018-37257-4.

[64] Sotiriou C, Pusztai L. Gene-expression signatures in breast cancer. N Engl J Med 2009;360(8):790–800.

[65] Ciresan DC, Giusti A, Gambardella LM, Schmidhuber J. Mitosis detection in breast cancer histology images with deep neural networks. In: Mori K, Sakuma I, Sato Y, Barillot C, Navab N, editors. Medical image computing and computer-assisted intervention – MICCAI 2013. Berlin, Heidelberg: Springer Berlin Heidelberg; 2013. p. 411–8.

[66] Veta M, van Diest PJ, Willems SM, Wang H, Madabhushi A, Cruz-Roa A, Gonzalez F, Larsen AB, Vestergaard JS, Dahl AB, Cirean DC, Schmidhuber J, Giusti A, Gambardella LM, Tek FB, Walter T, Wang C-W, Kondo S, Matuszewski BJ, Precioso F, Snell V, Kittler J, de Campos TE, Khan AM, Rajpoot NM, Arkoumani E, Lacle MM, Viergever MA, Pluim JP. Assessment of algorithms for mitosis detection in breast cancer histopathology images. Med Image Anal 2015;20(1):237–48. https://doi.org/10.1016/j.media.2014.11.010.

[67] Chen H, Dou Q, Wang X, Qin J, Heng P. Mitosis detection in breast cancer histology images via deep cascaded networks; 2016.

[68] Albarqouni S, Baur C, Achilles F, Belagiannis V, Demirci S, Navab N. Aggnet: deep learning from crowds for mitosis detection in breast cancer histology images. IEEE Trans Med Imaging 2016;35(5):1313–21. https://doi.org/10.1109/TMI.2016.2528120.

[69] Verhoef EI, van Cappellen WA, Slotman JA, Kremers G-J, Ewing-Graham PC, Houtsmuller AB, van Royen ME, van Leenders GJ. Three-dimensional analysis reveals two major architectural subgroups of prostate cancer growth patterns. Mod Pathol 2019;1.

[70] Tolkach Y, Thomann S, Kristiansen G. Three-dimensional reconstruction of prostate cancer architecture with serial immunohistochemical sections: hallmarks of tumour growth, tumour compartmentalisation, and implications for grading and heterogeneity. Histopathology 2018;72(6):1051–9.

[71] Epstein JI. Prostate cancer grading: a decade after the 2005 modified system. Mod Pathol 2018;31(S1):S47.

[72] Branca G, Ieni A, Barresi V, Tuccari G, Caruso RA. An updated review of cribriform carcinomas with emphasis on histopathological diagnosis and prognostic significance. Oncol Rev 2017;11(1).

[73] Scott R, Khan FM, Zeineh J, Donovan M, Fernandez G. Gland ring morphometry for prostate cancer prognosis in multispectral immunofluorescence images. In: International conference on medical image computing and computer-assisted intervention. Springer; 2014. p. 585–92.

[74] Nguyen K, Sabata B, Jain AK. Prostate cancer grading: gland segmentation and structural features. Pattern Recogn Lett 2012;33(7):951–61.

[75] Sirinukunwattana K, Snead DR, Rajpoot NM. A stochastic polygons model for glandular structures in colon histology images. IEEE Trans Med Imaging 2015;34(11): 2366–78.

[76] Stålhammar G, Martinez NF, Lippert M, Tobin NP, Mølholm I, Kis L, Rosin G, Rantalainen M, Pedersen L, Bergh J, et al. Digital image analysis outperforms manual biomarker assessment in breast cancer. Mod Pathol 2016;29(4):318.

[77] Visiopharm a/s. Hoersholm, Denmark. Available from: https://www.visiopharm.com.

[78] Carvajal-Hausdorf DE, Schalper KA, Neumeister VM, Rimm DL. Quantitative measurement of cancer tissue biomarkers in the lab and in the clinic. Lab Invest 2015;95 (4):385.

[79] Stack EC, Wang C, Roman KA, Hoyt CC. Multiplexed immunohistochemistry, imaging, and quantitation: a review, with an assessment of tyramide signal amplification, multispectral imaging and multiplex analysis. Methods 2014;70(1):46–58.

[80] Wong PF, Wei W, Smithy JW, Acs B, Toki MI, Blenman KR, Zelterman D, Kluger HM, Rimm DL. Multiplex quantitative analysis of tumor-infiltrating lymphocytes and immunotherapy outcome in metastatic melanoma. Clin Cancer Res 2019;25.

[81] Gorris MA, Halilovic A, Rabold K, van Duffelen A, Wickramasinghe IN, Verweij D, Wortel IM, Textor JC, de Vries IJM, Figdor CG. Eight-color multiplex immunohistochemistry for simultaneous detection of multiple immune checkpoint molecules within the tumor microenvironment. J Immunol 2018;200(1):347–54.

[82] Leach D, Buchanan G. Stromal androgen receptor in prostate cancer development and progression. Cancer 2017;9(1):10.

[83] Nearchou IP, Lillard K, Gavriel CG, Ueno H, Harrison DJ, Caie PD. Automated analysis of lymphocytic infiltration, tumor budding, and their spatial relationship improves prognostic accuracy in colorectal cancer. Cancer Immunol Res 2019;7(4):609–20.

[84] Indica Labs. albuquerque, nm. Available from: https://www.indicalab.com.

[85] Oztan B, Shubert KR, Bjornsson CS, Plopper GE, Yener B. Biologically-driven cellgraphs for breast tissue grading. In: 2013 IEEE 10th international symposium on biomedical imaging. IEEE; 2013. p. 137–40.

[86] Khan N, Afroz N, Rana F, Khan M. Role of cytologic grading in prognostication of invasive breast carcinoma. J Cytol/Indian Acad Cytolog 2009;26(2):65.

[87] Veltri RW, Isharwal S, Miller MC, Epstein JI, Partin AW. Nuclear roundness variance predicts prostate cancer progression, metastasis, and death: a prospective evaluation with up to 25 years of follow-up after radical prostatectomy. Prostate 2010;70(12):1333–9.

[88] Jørgensen AS, Rasmussen AM, Andersen NKM, Andersen SK, Emborg J, Rge R, Østergaard LR. Using cell nuclei features to detect colon cancer tissue in hematoxylin and eosin stained slides. Cytometry A 2017;91(8):785–93.

[89] Dowsett M, Nielsen TO, Ahern R, Bartlett J, Coombes RC, Cuzick J, Ellis M, Henry NL, Hugh JC, Lively T, et al. Assessment of ki67 in breast cancer: recommendations from the international ki67 in breast cancer working group. J Natl Cancer Inst 2011;103(22):1656–64.

[90] Ali S, Veltri R, Epstein JA, Christudass C, Madabhushi A. Cell cluster graph for prediction of biochemical recurrence in prostate cancer patients from tissue microarrays. In: Medical imaging 2013: digital pathology, vol. 8676. International Society for Optics and Photonics; 2013. p. 86760H.

[91] Lee G, Sparks R, Ali S, Shih NN, Feldman MD, Spangler E, Rebbeck T, Tomaszewski JE, Madabhushi A. Co-occurring gland angularity in localized subgraphs: predicting biochemical recurrence in intermediate-risk prostate cancer patients. PLoS One 2014;9(5), e97954.

[92] Lu C, Wang X, Prasanna P, Corredor G, Sedor G, Bera K, Velcheti V, Madabhushi A. Feature driven local cell graph (fedeg): predicting overall survival in early stage lung cancer. In: International conference on medical image computing and computerassisted intervention. Springer; 2018. p. 407–16.

[93] Sharma H, Zerbe N, Lohmann S, Kayser K, Hellwich O, Hufnagl P. A review of graphbased methods for image analysis in digital histopathology. Diagn Pathol 2015;1–61.

[94] Gurcan MN, Boucheron L, Can A, Madabhushi A, Rajpoot N, Yener B. Histopathological image analysis: a review. IEEE Rev Biomed Eng 2009;2:147.

[95] Ghosh S, Das N, Goncalves T, Quaresma P, Kundu M. The journey of graph kernels through two decades. Comput Sci Rev 2018;27:88–111.

[96] Wu Z, Pan S, Chen F, Long G, Zhang C, Yu PS. A comprehensive survey on graph neural networks. arXiv 2019. preprint arXiv:1901.00596.

[97] Ackerman DA, Barry JM, Wicklund RA, Olson N, Lowe BA. Analysis of risk factors associated with prostate cancer extension to the surgical margin and pelvic node metastasis at radical prostatectomy. J Urol 1993;150(6):1845–50.

[98] Cox DR. Analysis of survival data. Routledge; 2018.

[99] Kaplan EL, Meier P. Nonparametric estimation from incomplete observations. J Am Stat Assoc 1958;53(282):457–81.

[100] Caetano S, Sonpavde G, Pond G. C-statistic: a brief explanation of its construction, interpretation and limitations. Eur J Cancer 2018;90:130–2.

[101] Klein JP, Moeschberger ML. Survival analysis: techniques for censored and truncated data. Springer Science & Business Media; 2006.

[102] Cox DR. Regression models and life-tables. J R Stat Soc B Methodol 1972; 34(2):187–202.

[103] Bellera CA, MacGrogan G, Debled M, de Lara CT, Brouste V, Mathoulin-Pélissier S. Variables with time-varying effects and the cox model: some statistical concepts illustrated with a prognostic factor study in breast cancer. BMC Med Res Methodol 2010;10(1):20.

[104] Faraggi D, Simon R. A neural network model for survival data. Stat Med 1995; 14(1):73–82.

[105] Katzman JL, Shaham U, Cloninger A, Bates J, Jiang T, Kluger Y. Deepsurv: personalized treatment recommender system using a cox proportional hazards deep neural network. BMC Med Res Methodol 2018;18(1):24.

[106] Lee C, Zame WR, Yoon J, van der Schaar M. Deephit: a deep learning approach to survival analysis with competing risks. In: Thirty-second AAAI conference on artificial intelligence; 2018.

[107] Friedman J, Hastie T, Tibshirani R. The elements of statistical learning. Springer series in statistics, vol. 1. New York: Springer; 2001.

[108] Breiman L. Random forests. Mach Learn 2001;45(1):5–32.

[109] Ishwaran H, Lu M. Random survival forests. In: Wiley StatsRef: Statistics reference online; 2008. p. 1–13.

[110] Boser BE, Guyon IM, Vapnik VN. A training algorithm for optimal margin classifiers. In: Proceedings of the fifth annual workshop on Computational learning theory. ACM; 1992. p. 144–52.

[111] Fouodo CJ, Konig IR, Weihs C, Ziegler A, Wright MN. Support vector machines for survival analysis with R. R J 2018;10(1).

[112] Shivaswamy PK, Chu W, Jansche M. A support vector approach to censored targets. In: Seventh IEEE international conference on data mining (ICDM 2007). IEEE; 2007. p. 655–60.

[113] Khan FM, Zubek VB. Support vector regression for censored data (svrc): a novel tool for survival analysis. In: 2008 Eighth IEEE international conference on data mining. IEEE; 2008. p. 863–8.

[114] Eberhart R, Kennedy J. Particle swarm optimization. In: Proceedings of the IEEE international conference on neural networks, vol. 4. Citeseer; 1995. p. 1942–8.

[115] Chandrashekar G, Sahin F. A survey on feature selection methods. Comput Electr Eng 2014;40(1):16–28.

[116] Saeys Y, Inza I, Larranãga P. A review of feature selection techniques in bioinformatics. Bioinformatics 2007;23(19):2507–17.

[117] Guyon I, Elisseeff A. An introduction to variable and feature selection. J Mach Learn Res 2003;3(Mar):1157–82.

[118] Zhu SC, Yuille A. Region competition: unifying snakes, region growing, and Bayes/ mdl for multi- band image segmentation. IEEE Trans Pattern Anal Mach Intell 1996;9: 884–900.

[119] Sethian JA. Level set methods and fast marching methods: evolving interfaces in computational geometry, fluid mechanics, computer vision, and materials science. vol. 3. Cambridge University Press; 1999.

[120] Yushkevich PA, Piven J, Hazlett HC, Smith RG, Ho S, Gee JC, Gerig G. User-guided 3d active contour segmentation of anatomical structures: significantly improved efficiency and reliability. Neuroimage 2006;31(3):1116–28. https://doi.org/10.1016/j.neuroimage. 2006.01.015.

[121] Grady L. Random walks for image segmentation. IEEE Trans Pattern Anal Mach Intell 2006;11:1768–83.

[122] Vincent L, Soille P. Watersheds in digital spaces: an efficient algorithm based on immersion simulations. IEEE Trans Pattern Anal Mach Intell 1991;13(6):583–98. https://doi.org/10.1109/34.87344.

CHAPTER

Artificial intelligence and the interplay between cancer and immunity

15

Rajarsi Gupta, Tahsin Kurc, and Joel Haskin Saltz

Department of Biomedical Informatics, Stony Brook Medicine, Stony Brook, NY, United States

Introduction

Over the last 25 years, digital pathology has rapidly grown alongside advances in machine learning and computer vision. The period of growth that we are now witnessing is based on the increasing availability of digital whole slide images (WSIs) of laboratory tissue samples through the adoption of commercial glass slide scanners in clinical and research settings. The access to tissue via WSIs has attracted the attention of a diverse group of scientists and engineers from academia and industry to work alongside pathologists to develop powerful image analysis tools powered by artificial intelligence (AI) and machine learning (ML), leading to the emergence of computational pathology as an exciting frontier for translational clinical research. This chapter focuses on AI Pathomics methods that study the interplay between cancer and immunity for precision oncology applications, such as guiding patient management and treatment selection with immunotherapy.

Pathomics image analysis of WSIs is being used to quantitatively characterize various features of cancer, which include microanatomic regions, microarchitectural structures, biological phenomena, and a wide variety of other phenotypic properties of tissues and cells. For example, Pathomics can be used to (1) detect the presence of cancer in tissue samples, (2) detect and classify different types of tissues, cells, and nuclei, (3) quantitatively evaluate the color, size, shape, and texture of tissues, cells, and nuclei, and (4) characterize biological processes, such as host immune responses and necrosis. In this chapter, we describe recently developed Pathomics tools to spatially map and quantitatively characterize nuanced interactions between cancer and tumor-infiltrating lymphocytes (TILs) of our host immune system in hematoxylin and eosin (H&E) stained WSIs to use AI to help support the objectives of the International Immuno-Oncology Biomarkers Working Group [1–5].

Even though we focus on AI Pathomics methods, we also point the reader to excellent review articles about classical image analysis [6–10] to gain an appreciation for quantitative hand-crafted and engineered features to study various aspects of cancer. Recent advances in deep learning, a subdivision of machine learning and

---

Artificial Intelligence in Pathology. https://doi.org/10.1016/B978-0-323-95359-7.00015-7
Copyright © 2025 Elsevier Inc. All rights are reserved, including those for text and data mining, AI training, and similar technologies.

artificial intelligence (AI), have supercharged the development of Pathomics methodology with the goal of substantially accelerating translational research by generating quantitative data to describe traditional qualitative histomorphologic descriptions of tissues, cells, and nuclei in laboratory tissue samples. The long-term vision is to routinely implement AI Pathomics tools to (1) gain data-driven insights into cancer pathobiology through automated analyses of WSIs, (2) improve diagnostic accuracy while decreasing observer variability through the use of quantitative histomorphology, and (3) identify novel Pathomics biomarkers to predict clinical outcomes and treatment response through the ability of AI to detect patterns and relationships in large datasets to ultimately improve management and treatment for cancer patients [6–9,11–41].

Even though AI and Pathomics are still in their relative infancy en route to future clinical adoption, this chapter introduces and describes a selected set of novel AI Pathomics methods that leverage deep learning and computer vision to map and analyze the abundance and spatial distribution of TILs in H&E WSIs of multiple types of cancer to guide the selection and use of immunotherapy. Therefore, we begin with a brief overview about immune surveillance and the growing role of immunotherapy in cancer treatment, followed by examples of mapping TILs in different types of cancer to provide tangible examples of how AI Pathomics tools can already provide clinically valuable information. Beyond demonstrating the feasibility and value about how TIL maps can help support precision oncology applications for immunotherapy, we also hope these examples motivate the development of other novel Pathomics methods to further enhance our understanding of cancer pathobiology.

Immune surveillance and immunotherapy

The human immune system is typically described as the sum of the innate and adaptive immune systems, which function together to maintain a state of continuous surveillance through functionally specialized subtypes of immune cells. The chief responsibilities of immune cells are to recognize and process foreign antigens in order to eliminate infectious agents, premalignant cells with acquired aberrant genetic and epigenetic changes, and cancer cells that arise from malignant transformation. The cells of the innate immune system have pattern recognition receptors to recognize nonself infectious or tumor antigens tin order to generate immediate, short-lived, and nonspecific inflammatory responses. The adaptive immune system generates highly specific inflammatory responses through antibody-mediated and cell-dependent responses to provide long-lasting defense, including immune cells with memory [22,33,35,42].

The innate and adaptive immune systems are coordinated through cell lineages that are typically divided into broader classes, including lymphocytes, neutrophils, macrophages, dendritic cells, mast cells, and natural killer (NK) cells [26,27,43–45]. Lymphocytes are further subdivided into B-cells, T-cells, and NK cells. Beyond functionally specialized broader classes, immune cells of both arms of the innate and adaptive responses are characterized by variable expression of cell surface and intracellular markers across a continuum of various activation and differentiation states [46–49]. For example, subsets of T-cells are further distinguished by the expression of cluster of differentiation (CD) cell surface markers for further characterization as TH_1, TH_2, TH_{17}, T_{REG}, $\gamma\delta$, and NKT cells.

The major categories of cell types that exist in the cancer tissue microenvironment (TME) include cytotoxic $CD8^+$ T-cells that identify and kill cancer cells; $CD4^+$ helper T-cells that coordinate the activation and suppression of adaptive immune responses through cytokine signaling; $CD4^+$ regulatory T-cells ($CD25^+$/ $FOXP3^+$ Tregs) that play an important role in tolerance/immunosuppression; NKT-cells and NK cells (also known as large granular lymphocytes) that kill cells missing self major histocompatibility complex (MHC-I) markers without requiring sensitization via cytokine stimulation; and macrophages and dendritic cells with phagocytic antigen-presenting functions to coordinate and sustain the attack on cancer cells [50,51].

During the process of uncontrolled growth in cancer, nuanced differences in the protein structure of foreign antigens are detected by the immune system at the molecular level to identify premalignant cells with aberrant genomics changes and/or tumor cells that have undergone malignant transformation. This leads to antigen processing by the innate and adaptive immune system to coordinate a targeted immune-mediated response. Dynamic hierarchical regulatory mechanisms of the immune system result in proinflammatory and antiinflammatory functional states that are regulated through complex cascades of cytokine cell signaling to coordinate the interactions between different kinds of immune cells to generate systemic and local immune responses to identify and attack cancer cells.

Putting the diversity and heterogeneity of the immune landscape into perspective, there are over 100 different populations of functionally distinct immune cell types in the peripheral blood of human beings. Therefore, the ability to respond to diverse stimuli and fight cancer cells is highly variable due to intrinsic differences in the proinflammatory and antiinflammatory cell populations within and across individuals [13,14,22,52–54]. This level of heterogeneity and intrinsic complexity across functionally specialized types of immune cells is further complicated by the spectrum of different types of cancer that vary by histologic growth patterns, molecular profiles, and biological behavior that is unique in each patient [21–27,43,55].

The importance of the role of immunity in controlling cancer was observed a century ago by Dr. William Coley in a case of complete remission in a patient with advanced inoperable sarcoma, where a highly malignant and deadly cancer disappeared after the patient suffered from severe *Streptococcus pyogenes* bacterial skin infections [56]. This striking observation played a significant role in developing the concept of stimulating and inhibiting hierarchical regulatory mechanisms of the immune system as a potential treatment strategy for cancer [57–59]. Many studies have focused on the role of specific functional properties of various types of immune cells to determine their clinical significance in proinflammatory and antiinflammatory spatiotemporal relationships within the framework of hierarchical regulatory roles in immunosurveillance in human diseases, including a significant interest in cancer immunology [13,14,20,22,23,43,53,54,60–66].

Despite the diversity of immune cell populations that exist to surveil and eliminate cells that have undergone malignant transformation, cancer cells are constantly evolving to evade immune-mediated recognition and elimination. Cancer cells utilize immunosuppressive strategies that disrupt T-cell signaling and induce tolerance in the TME through the production of inhibitory cytokines, which prevent the expansion and activation of CD4+ helper and cytotoxic CD8+ T-cells [20,39–41]. The suppression and inactivation of immune detection and immune-mediated cytotoxicity are also accompanied by the ability of cancer cells to actively hijack and reprogram immune cells that stimulate angiogenesis and accompanying stromal responses, which promote the continued survival of cancer cells and lead to disease progression.

The subversion of normal immunoregulatory mechanisms leads to the emergence of subclones of cancer cells that become immune invisible (less immunogenic) and more resistant to apoptosis within the immunosuppressed milieu of the TME. This selection pressure enables cancer cells to adapt and further evade the immune system through the loss of tumor antigens that would normally stimulate complement-mediated and antibody-dependent cell-mediated cytotoxicity (ADCC) by T-cells and natural killer (NK) cells [13,14,20–25,31,39–41]. The combination of the uncontrolled proliferation and adaptation of cancer cells within an immunosuppressed tumor microenvironment ultimately results in the destructive invasion of neighboring tissues and distant sites of metastasis beyond the primary site of origin.

The intrinsic functional complexities and widely variable composition of the immune system in each patient afflicted by different types and subtypes of cancer have led to tremendous interest in personalized precision oncology and cancer immunology. In parallel, translational clinical cancer research has led to the development of first- and second-generation immunotherapeutic options that have been utilized to successfully treat certain kinds of cancer with durable treatment responses. We are fortunate as a society to have access to treatment options that leverage our collective understanding of cancer immunology to provide such a tangible clinical benefit to patients. From a broad perspective, currently available pharmacologic agents either stimulate immune cells to attack cancer cells or block immune checkpoint pathways that inhibit immune responses.

The best examples of recent success with immunotherapy include monoclonal antibody-based treatment strategies that block interactions with cytotoxic T-lymphocyte-associated receptor protein 4 (CTLA-4) and programmed cell death receptor protein-1 (PD-1). CTLA-4 is a negative regulator that inhibits T-cell proliferation and activation, whereas cancer cells and antigen-presenting cells use PD-L1 ligands to inhibit T-cells by binding with PD-1 on their surface. Therefore, immune checkpoint blockade agents that block these receptors disrupt downstream immunosuppression of the TME. Other types of treatments that are being explored include chimeric antigen receptor T-cells (CAR-T) that are produced by genetically modifying T-cells harvested from individual patients; recombinant proteins to stimulate and enhance cytotoxic T-cell activity; cancer vaccines to augment antigen presentation; and oncolytic viruses to initiate systemic antitumor immunity to target and destroy cancer cells [20,39–41].

Cancer immunopharmacology has generated substantial interest in identifying clinically actionable targets for future precision oncology applications to improve treatment response and clinical outcomes. Immune checkpoint therapy to block CTLA-4 or PD-1 has successfully overcome immunosuppression mediated by tumor cells [39,40,65,67] in melanoma, renal cancer, and lung cancer [68–72]. Ipilimumab, a monoclonal antibody that blocks CTLA-4, was the first immune checkpoint inhibitor (ICI) to be approved by the Food and Drug Administration (FDA) to treat melanoma in 2011, followed by FDA approval for Nivolumab to block PD-1 in melanoma and nonsmall cell lung cancer (NSCLC) in 2014 [49,73]. Pembrolizumab followed suit and was also approved for NSCLC due to a significantly improved treatment response in tumors that express high levels of PD-L1. In order to further improve the therapeutic efficacy of these immune checkpoint inhibitors, combinations of immunotherapy with existing forms of chemotherapy and other targeted therapies are currently under investigation in patients with locally advanced or metastatic cancer [39,74–77]. An example of the synergy between immunotherapy and chemotherapy is the use of Nivolumab to treat patients with advanced metastatic squamous NSCLC with disease progression in combination with or after standard platinum-based chemotherapy [41,72,73].

However, the identification and stratification of patient candidates who are most likely to benefit from immunotherapy is not a straightforward endeavor. Since PD-1 signaling blockade has become a potential treatment option for many types of cancer, standardized assays to evaluate the bioactivity of PD-1/PD-L1 have become commercially available. Even though the overexpression of surface or soluble PD-L1 in multiple tumor types is mostly associated with poor prognosis, these assays lack antigen-specificity [68,71,73,77–82]. Similarly, the histologic assessment of PD-L1 by immunohistochemistry (IHC) is also challenging due to heterogeneous expression in cancer and benign normal tissues, which is further complicated with unintuitive thresholds and cutoffs for scoring PD-L1 positivity that vary and lead to known problematic interobserver variability in PD-L1 assessment by pathologists [44,83–85].

As the adoption of immunotherapy continues to expand, there will be a commensurate demand for reliable standardized bioassays and scoring methods to evaluate receptor and ligand expression to appropriately select patient candidates for PD-1/PD-L1 blockade and other kinds of immunotherapy. Furthermore, there are serious concerns about how inaccurate patient stratification and the impact of misinterpretation of PD-L1 expression due to subjective scoring around variable thresholds and cutoffs can lead to radically different patient selection, clinical outcomes, and serious adverse effects [44,83–85].

The goal of checkpoint inhibitors and other forms of emerging immunotherapy is to stimulate the host immune system into recognizing and eliminating cancer cells. Therefore, the utility of immunopharmacologic treatments is ultimately dependent on the ability to recruit and stimulate TILs to find and attack cancer cells, which may be limited in current treatment strategies if the TILs in the host immune system are either not present or depleted in the TME. This has led to great interest in characterizing TILs as a biomarker that can be manually assessed by pathologists during routine evaluation of H&E stained tissue sections in clinical laboratories to potentially guide patient management and treatment selection. Initiatives by the TIL Working Group of the International Immuno-Oncology Biomarker Working Group have culminated in guidelines to help pathologists evaluate TILs as a biomarker in breast cancer and other solid tumors to provide prognostic information to oncologists since TILs are typically associated with longer patient survival and potentially help predict treatment response [3–5,13,86].

In the next section, we focus on the development of modern Pathomics image analysis tools in digital pathology to evaluate the interplay between TILs of the host immune system in the TME of different kinds of cancer, which may help guide treatment selection and predict clinical outcomes. Specifically, we will focus on the use of automated AI Pathomics tools that utilize deep learning in computer vision to detect and spatially map tumor regions and lymphocytic infiltrates in WSIs of cancer tissue samples. The potential clinical use of computationally calculated TILs as a biomarker in cancer for diagnostic surgical pathology has attracted the attention of the Food and Drug Administration Office of Science and Engineering Laboratories (FDA OSEL), who have led efforts to design a Medical Device Development Tool (MDDT) as a necessary mechanism to evaluate both simple and advanced Pathomics applications that evaluate TILs for clinical use in breast cancer [87]. Therefore, we present examples of how Pathomics can be used map TILs in H&E cancer WSIs to complement and enhance routine microscopic examination of cancer tissue samples and ancillary laboratory testing for immune checkpoint biomarker expression.

Identifying TILs with deep learning

The biological behavior of cancer varies by organ site, histologic classification, and molecular subtype, which is intimately coupled to the microscopic characteristics of the TME that is composed of immune infiltrates, stromal connective tissue, nerves, and lymphovascular vessels [21–27,49,88]. Immune responses in cancer are widely believed to be driven by tumor antigenicity that influences the abundance and spatial distribution of immune cells in the landscape of the TME [21,22,89]. The importance of tumor immune interactions in cancer have led to efforts to examine the role of TILs and other specialized immune cells to identify prognostic and predictive biomarkers to guide treatment (e.g., identifying patients with the highest likelihood of responding to immunotherapy in different kinds of cancer) [23–26,90].

Thus, there is growing advocacy for the evaluation of TILs as a readily available and widely applicable biomarker that needs to be included in diagnostic surgical pathology reports. However, the routine use of TILs as a clinical biomarker has yet to be widely incorporated during diagnostic microscopic evaluation of tissue samples, despite the availability of guidelines and training materials from the TIL

Working Group and an unmet clinical need to improve patient outcomes with immunotherapy. Beyond supporting the use of immunotherapy as part of currently established treatment protocols, treatment-resistant and rare types of cancer represent another unmet clinical need where characterizing TILs as a biomarker may be useful in establishing the use of immunotherapy for treatment where options are limited. Similarly, TILs may be useful as a biomarker to help minimize immunotherapy-related side effects and adverse reactions.

The abundance and spatial distribution of TILs also has high prognostic value in different regions of cancer [25,91,92], which may be useful as a surrogate biomarker of tumor heterogeneity. Generally, high densities of TILs correlate with favorable clinical outcomes [92], including longer disease-free survival (DFS) and/or improved overall survival (OS) in multiple cancer types [93,94]. The spatial distribution of TILs in the tumor center and invasive margin have also been shown to be important in cancer prognosis [92]. In terms of clinical utility, the presence of high levels of TILs is consistently associated with a more favorable prognosis in patients with early-stage triple-negative and HER2-positive breast cancer [1,95]. On the opposite end of the spectrum, serious adverse effects and the lack of long-lasting responses in many patients suggest that the activation of previously tumor-immunosuppressed TILs needs to be carefully controlled [64]. Even though the assessment of TILs does not explicitly shed light on cell signaling and other related molecular aspects of tumor immune interactions in the TME, evaluating the abundance and spatial distribution of TILs as tissue biomarkers provides a great deal of clinically meaningful information.

Histopathologic diagnoses are based on the microscopic examination of H&E-stained histologic tissue sections of biopsy, excision, and surgical tissue samples from cancer patient to guides treatment selection and predict outcomes. In the last few years, large collections of H&E-stained cancer tissue samples have been scanned with commercially available slide scanners to generate digital WSIs for digital pathology applications, such as telepathology and remote consultation. In turn, the increasing availability of WSIs for clinical research studies has generated a lot of interest in developing quantitative image analysis methodology. Since TILs can be evaluated in H&E-stained histologic tissue samples on glass slides that are routinely produced in clinical and research laboratories, there has been great interest from the computational pathology and translational clinical research communities to develop Pathomics applications to automate the assessment of TILs in H&E WSIs that are becoming routinely available for pathologists and scientists.

Advances in high-performance computing and AI have resulted in the capability to analyze highly complex microscopic images of biological tissues altered by uncontrolled and destructive forces of cancer, which is quite different from analyzing natural images of animals and scenery. Computational image analysis in pathology has led to the establishment of Pathomics to describe the ability to detect, classify, and delineate cancer tissue regions (e.g., cancer vs. noncancer, histologic subtypes, normal epithelium, dysplasia, stroma, or necrosis), architectural structurers (e.g., nerves vessels), cell types, and subcellar components and features (e.g., nuclei, chromatin texture) [37,38,96–109]. The technological innovations that support Pathomics in biomedical research have been substantially increased driven by the development of deep learning and AI applications in computer vision that analyze and extract features from highly complex images [37,38,96–110].

In terms of approaches to characterize TILs in H&E WSIs from the point of view of image analysis, computer vision, and deep learning, most lymphocytes have a reasonably uniform appearance as round to ovoid purple-blue colored nuclei with dense chromatin and scant cytoplasm. Deep learning Pathomics algorithms need to distinguish lymphocytes from other types of cells and microanatomic entities within the context of complex tissue-specific and cancer-specific backgrounds. Analyses of lymphocytic infiltrates/TILs need to correctly identify lymphocytes/TILs based on their appearance in normal, cancer, and tumor-associated stromal tissues, as well as distinguishing the lymphocytes/TILs from the condensed nuclei of dying cells in necrosis. Part of this process requires evaluating nuanced spatial interrelationships which can require classification of cancer tissue regions, tissue types/structures, and individual cells.

Early research efforts to characterize TILs focused on the detection of individual lymphocytes and cancer cells [13,111–113] in solid tumors with a special interest in developing methods to capture the features of the lymphocytic infiltrate in breast cancer. This work leveraged rapidly expanding machine learning methodology to perform nuclear segmentation, feature extraction, and classification [6,7,10]. Segmentation of cell nuclei in tissue images proved to be challenging due to (1) variations in nuclear size and morphology, (2) overlapping nuclei, (3) complex and irregular characteristics of histologic structures, and (4) variations and irregularity of H&E staining and image acquisition (see Irshad et al. [7] and Gurcan et al. [6] for excellent surveys of state of the art in nuclear segmentation and detection approximately 5–10 years ago). Image processing algorithms of various levels of sophistication were also employed, but the utilization of machine learning algorithms (e.g., Bayesian, support vector machine (SVM), and AdaBoost) was rare for the segmentation of nuclei.

In the past few years, deep learning methods have played a rapidly increasing role in nuclei detection and segmentation (for early examples, see Refs. [10, 114–116]). Janowczyk et al. articulated the advantages of deep learning approaches by stating that all possible variances in morphology, texture, and color appearances need to be understood in order to develop a nuclei segmentation algorithm. Furthermore, the authors stated that algorithms would need to account for as many variances as possible without being too general (increased false positives) or too narrow (increased false negative errors), which is cumbersome since it is impractical to account for all of the outlier cases a priori and ultimately requires extensive iterative trial and error [30].

Deep learning methods for nuclear segmentation have found success [117] but hinge on training data adequacy, which remains to be a crucial challenge. There is wide variability in the appearance of nuclei that can differ between tissue types, which can be further complicated by the difficulty associated with recognizing morphologic variations due to the effects of cancer and treatment and can also appear different because of differences in histologic tissue preparation and staining. Therefore, promising approaches to reduce the requirements for generating training datasets are employing generative adversarial networks (GANs) [118] to generate realistic synthetic nuclei to enlarge and generalize training sets to present this level of variability to deep learning and AI models [119,120].

Examples of early work in lymphocyte detection and characterization include the development of a clinical decision support (CDS) system to discriminate among three types of lymphoproliferative disorders and normal blood cells [121] along with lymphocyte detection and centroblast quantification in follicular lymphoma [122]. In terms of detecting and classifying lymphocytes and other types of cells for the detailed characterization of interrelationships with cancer, tumor-associated stroma, and lymphocytes, metrics like density and spatial colocalization of TILs and tumor cells were used to predict the likelihood of recurrence in early-state NSCLC [123]. Similarly, immune cell abundance and spatial heterogeneity were computed with a Pathomics pipeline that segmented and classified lymphocytes, cancer nuclei, and other cells to explore their prognostic significance in postmenopausal patients with ER+ breast cancer treated with tamoxifen or anastrozole over 5 years in order through comparison with genomic analyses by Oncotype DX (Exact Sciences, Madison, WI), including the 21-gene recurrence score (RS), PAM50 risk of recurrence (ROR) score, immunohistochemical IHC4 score (ER, PR, HER-2, and Ki67), and clinical treatment scores [124].

In terms of the development of methods that classify tissue regions, rapid development has been facilitated by convolutional neural networks (CNNs) and publicly available software. In the last few years, a deep learning-based patch classification workflow was utilized to characterize the abundance and spatial distribution of lymphocytes in WSIs by using a CNN to classify 50×50 um tiled image patches as lymphocyte positive (i.e., containing ≥ 2 lymphocytes) or lymphocyte negative, combined with a CNN to segment necrotic regions to reduce false lymphocyte-positive classification [37]. This method leveraged a convolutional autoencoder (CAE) that was designed to encode nuclei in image patches into sparse feature maps that encode both the location and appearance of nuclei [101]. A supervised CNN took unsupervised encoded features from the unsupervised CAE for classification. An iterative training effort was carried out with 100×100 pixel image patches at $200\times$ magnification (equivalent to $50 \times 50\,\mu m$ resolution) for 5200 images from 13 tumor types using H&E WSIs from the Cancer Genome Atlas (TCGA).

The overall approach of the workflow progressed through (1) annotation of lymphocyte infiltrated and noninfiltrated patches in H&E WSIs from different tumor types and (2) training deep learning algorithms with annotated data to generate predictions, followed by (3) visual assessments of the predictions overlaid on the original H&E WSIs. In images where the predictions were inaccurate, further annotations were extracted and used for retraining until adequate accuracy was obtained for a particular tumor type. At the end of this process, a separate testing dataset was generated across tumor types to evaluate results [37]. More recently,

318   **CHAPTER 15** AI and the interplay between cancer and immunity

the publicly available set of annotations generated in this method was utilized to train two standard CNN networks, VGG16 and Inception v4 to generate predictions in the same 5200H&E WSIs from the TCGA [125]. The accuracy that was obtained by using the standard models on the *aggregated* training set is modestly better than that reported by using the CAE model on the *incrementally* generated training set [37,125], followed by a recent third-generation lymphocyte/TILs model that extended its use to 23 different tumor types [126].

A variety of other efforts have targeted the characterization of specialized micro-anatomic tissue compartments in different kinds of cancer. In breast cancer, a fully convolutional network carried out pixel-level semantic segmentation to identify tumor, stroma, necrosis, and TILs [11,127], where training data was gathered through structured crowdsourcing annotation from participants with varying degrees of expertise. This approach also employed a fully convolutional network approach that combined region and nucleus segmentation masks to classify pixels into 12 classes, where each class encodes a combination of region and nucleus membership information. The framework enforces biological constraints that do not allow cell classifications within necrotic regions or stromal cells within nonstromal regions. Recent algorithms have been described as iteratively improving the pixel-level semantic segmentation of tumor, stroma, and necrosis regions linked with the detection of tumor, stroma, and lymphocyte nuclei [128,129]. However, these algorithms are computationally expensive and require a lot of compute time to perform fine-grained classification of tissue regions, segmentation, and classification of every cell nucleus, followed by postprocessing.

In an effort to focus on exploring TILs in a scalable and high-throughout manner, spatial maps of TILs were generated for thousands of H&E WSIs with computationally inexpensive approaches that combined automated tumor detection of cancer tissue patches with patch-level lymphocyte detection to evaluate the abundance of intratumoral and peritumoral TILs. Le et al. developed a deep learning Pathomics method to delineate invasive breast cancer in H&E WSIs by classifying 350×350 pixel image patches at $400 \times$ magnification (equivalent to a resolution of $87.5 \times 87.5\,\mu m$). The results were combined with existing results from the lymphocyte/TILs patch-based algorithm [37,130] to generate composite Tumor-TIL maps after the heatmaps are binarized, where patches with probabilities $\geq 50\%$ are considered positive and $<50\%$ are considered negative and then combined in a postprocessing step. Fig. 15.1 shows the pipeline that was utilized to generate Tumor-TIL maps from 1015 H&E WSIs of the TCGA BRCA breast cancer study by using the combination of deep learning Pathomics methods.

Quality assurance and control (QA/QC) is performed to check the results from the breast cancer and lymphocyte/TILs algorithms. As shown in Fig. 15.2, the results are verified by using overlays on the original H&E WSIs to evaluate the correctness of the outputs. Furthermore, the Tumor-TIL maps that were generated for the TCGA BRCA dataset were made publicly available in a web-based interface to interrogate algorithmic performance with a digital pathology software system, Quantitative Imaging in Pathology (QuIP) [131]. After gaining a certain level of trust about

Identifying TILs with deep learning

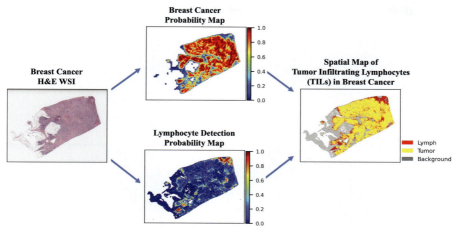

FIG. 15.1
Deep learning Pathomics applications to spatially map tumor-infiltrating lymphocytes (TILs) in breast cancer. Starting with the H&E WSI on the right, the *top panel* shows breast cancer detection presented as a spatial probability heatmap to evaluate algorithmic performance, whereas the *bottom panel* shows automated lymphocyte detection presented as a spatial probability heatmap (nonlymphocyte tissue colored blue). Combining the outputs of tumor and lymphocyte detection generates a composite Tumor-TIL map to evaluate the abundance and spatial distribution of peritumoral and intratumoral TILs (tumor colored *yellow*, lymphocytes colored *red*, and background nontumor/nonlymphocyte tissue colored *gray*). This Tumor-TIL map shows the presence of mostly peritumoral TILs with a paucity of scattered intratumoral TIL infiltrates.

Image: TCGA BRCA BH-A0BZ-01Z-00-DX1, high-grade breast cancer, cancer detection with ResNet model Le H, et al., Utilizing automated breast cancer detection to identify spatial distributions of tumor-infiltrating lymphocytes in invasive breast cancer. Am J Pathol, 2020;190(7):1491–1504 and lymphocyte detection with VGG16 model. Abousamra S, et al. Learning from thresholds: fully automated classification of tumor infiltrating lymphocytes for multiple cancer types. arXiv e-prints, 2019.

algorithmic performance, pathologists and translational clinical research scientists can gain immediate insight into the abundance and spatial distribution of TILs from very simplistic visualizations. The utility of these Tumor-TIL maps in breast cancer allows for quick identification of "hot" proinflammatory and "cold" noninflammatory [132] immunologic states of the TME for each case of breast cancer in the TCGA BRCA dataset through an appreciation for the spatial distribution of intratumoral and peritumoral TILs, as shown in Fig. 15.3 (hot) and Fig. 15.4 (cold), respectively.

After verifying the performance of these algorithms in the H&E WSIs of TCGA BRCA and using the maps to characterize immunologically hot and cold cases of breast cancer, global intratumoral TIL infiltrate percentage was calculated as the number of predicted patches that were classified as positive for tumor and

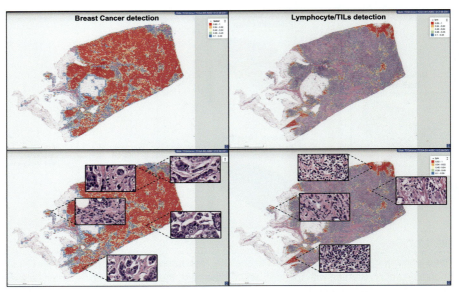

FIG. 15.2

Quality assurance and control (QA/QC) of tumor and lymphocyte/TILs detection. The top panels show an overlay of the spatial probability heatmaps for breast cancer (top left) and lymphocyte/TILs (top right) detection on the corresponding H&E WSI. The algorithmic results are inspected with respect to the tumor region and large lymphocytic aggregates that can be viewed at low magnification. After preliminary review, more thorough QA/QC is performed to evaluate algorithmic performance shown in the bottom panels, where the tumor and lymphocyte/TILs algorithms have correctly identified patches with various arrangements of tumor cells (bottom left) and densities of lymphocytes/TILs (bottom right), respectively.

Image: TCGA BRCA BH-A0BZ-01Z-00-DX1, high-grade breast cancer, cancer detection with ResNet model Le H, et al., Utilizing automated breast cancer detection to identify spatial distributions of tumor-infiltrating lymphocytes in invasive breast cancer. Am J Pathol 2020;190(7):1491–1504 and lymphocyte detection with VGG16 model. Abousamra S, et al. Learning from thresholds: fully automated classification of tumor infiltrating lymphocytes for multiple cancer types. arXiv e-prints, 2019.

lymphocyte (yellow and red) divided by total number of cancer patches (yellow) after scaling and alignment of the outputs from the two models. Downstream correlative analyses were performed to evaluate the significance of the calculated percentage of global intratumoral TIL infiltrates in breast cancer with overall survival, which was further stratified by molecular subtype [130]. Overall, high intratumoral TILs were associated with a survival benefit across all molecular subtypes of breast cancer (Lum A, Lum B, Her-2, and basal).

These analyses were repeated in another large dataset of 1081 H&E WSIs from the Carolina Breast Cancer Study at the University of North Carolina (UNC CBCS), which has a very different patient population in comparison to the TCGA BRCA study [133]. In addition to calculating the global abundance of intratumoral TIL

Identifying TILs with deep learning   321

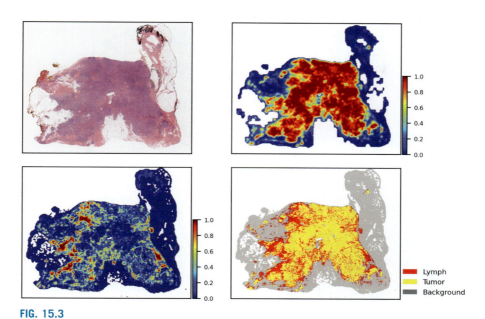

FIG. 15.3

"Hot" immune subtype of invasive ductal carcinoma of the breast visualized with a Tumor-TIL map. *Bottom right*: Automated tumor detection (highlighted in *yellow*) with lymphocyte detection (highlighted in *red*) showing a diffuse spatial pattern of distribution of TILs, predominantly comprised of intratumoral TILs.

Image: TCGA BRCA E2-A14X-01Z-00-DX1; top left: low-power H&E WSI; top right: breast cancer spatial probability heatmap; bottom left: lymphocyte spatial probability heatmap; bottom right: composite tumor-TIL map.

infiltrates, multiple parameters were used to characterize the heterogeneity of the spatial distribution of TILs in subsets of cases from the TCGA BRCA and UNC CBCS [133]. High intratumoral TILs were associated with increased survival benefit (except for the Lum A molecular subtype), confirming most of the observations about TILs as a prognostic biomarker in the TCGA BRCA study [130,133]. Regarding the spatial distribution that can be seen in Tumor-TIL maps, univariate and multivariate analyses showed that increased peritumoral and intratumoral TILs were associated with longer survival. Diminished or absent intratumoral and peritumoral TILs were associated with increased risk of recurrence and shorter progression free interval (PFI), where the presence of large immune cold areas with a lack of intratumoral TILs were associated with almost twice the risk of recurrence.

After demonstrating the practical utility of Pathomics in evaluating the clinical significance of the abundance and spatial patterns of distribution of TIL infiltrates as important biomarkers in breast cancer, Le et al. also developed a course-grained pancreatic cancer detection tool to study the role of TILs in invasive pancreatic ductal adenocarcinoma (PDAC) [134]. The PDAC detection tool identifies 1000×1000

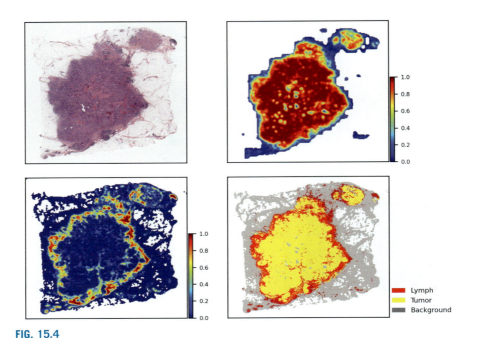

FIG. 15.4

"Cold" immune subtype of invasive ductal carcinoma of the breast visualized with a Tumor-TIL map. *Bottom right*: Automated tumor detection (highlighted in *yellow*) with lymphocyte detection (highlighted in *red*) showing a predominantly peritumoral spatial distribution of TILs at the invasive edge of the tumor.

Image: TCGA BRCA GM-A2DB-01Z-00-DX1; top left: *low-power H&E WSI*; top right: *breast cancer spatial probability heatmap*; bottom left: *lymphocyte spatial probability heatmap*; bottom right: *composite tumor-TIL map*.

pixel image patches at 200× magnification (equivalent to a resolution of 500 × 500 μm) containing malignant poorly formed glands and invasive nests of tumor cells, which need to be distinguished from normal pancreatic acinar tissue, reactive chronic pancreatitis, pancreatic islet cells, stroma, immune cells, nerves, and vessels. This is particularly challenging since the histologic appearances and intrinsic features of pancreatic cancer are extremely diverse and represent a diagnostically challenging endeavor for even the most experienced pathologists. Nonetheless, preliminary work appears promising, as shown in Fig. 15.5 [37,125].

In keeping with the goal of motivating additional data-driven cancer research avenues, Fig. 15.6 is presented to show a Pathomics prototype that can be utilized to study tumor-immune interactions in nonsmall cell lung cancer (NSCLC). Beyond serving as an extension of the work to study tumor immune interactions in breast and pancreatic cancer, this example is intended to be provocative in setting the stage for using Pathomics to generate data-driven hypotheses to explore how histologic subtypes of NSCLC adenocarcinoma are associated with the spatial distribution of TILs.

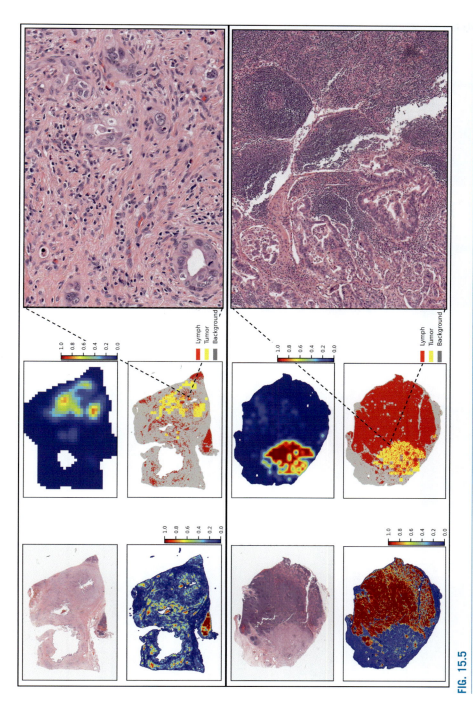

FIG. 15.5

Tumor-TIL maps in primary and metastatic pancreatic ductal adenocarcinoma. *Top panels*: Primary pancreatic cancer with tumor-TIL detection with corresponding region of interest *(inset)* selected from the map to confirm cancer detection *(top right)*. *Bottom panels*: Metastatic pancreatic cancer to the spleen with tumor-TIL map detection with the corresponding region of interest *(inset)* selected from the map for to confirm cancer metastasis to the spleen *(bottom right)*.

H&E images: TCGA XD-AAUI-01Z-00, diagnostic slides DX1 and DX2. Tumor-TIL panels on the left side of the top and bottom panels: top left: *low-power H&E WSI*; top right: *pancreatic cancer spatial probability heatmap*; bottom left: *lymphocyte spatial probability heatmap*; bottom right: *composite tumor-TIL map*.

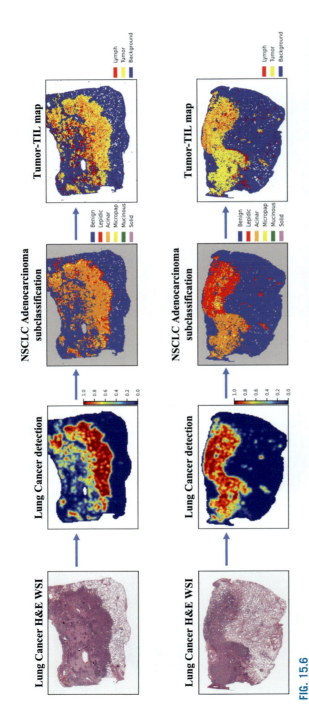

FIG. 15.6

Pathomics prototype for Tumor-TIL mapping in lung cancer. Histologic subtypes of NSCLC adenocarcinoma (e.g., lepidic, acinar, micropapillary, mucinous, and solid) are mapped in H&E WSIs of lung cancer. TILs are identified with a separate VGG16-based lymphocyte detection algorithm and incorporated into the analytic workflow. A histologic subclassification map is generated to spatially map the tumor microenvironment alongside a Tumor-TIL map to characterize and interpret tumor-immune interactions. The top and bottom cases differ in tumor histology that can be further correlated with the abundance and spatially heterogeneous distributions of TILs. Peritumoral TILs are associated with acinar differentiation in both cases, but there is a notable difference in tumor-immune interactions with lepidic tumor cells. *Top row*: TCGA Lung 97-8172-01Z-00-DX1, diagnosis: 3.5 cm lung mass, NSCLC adenocarcinoma acinar predominant (90% acinar, 10% lepidic) with no lymph node metastases; *bottom row*: 2.9 cm lung mass, TCGA Lung 97-8172-01Z-00-DX1, diagnosis: NSCLC adenocarcinoma mixed type (30% acinar, 40% lepidic, 10% micropapillary, 20% papillary) with no lymph node metastases. The quantitative subclassification performed by the lung cancer Pathomics model in these two cases agrees quite well with histopathologic diagnoses. Tumor-TIL maps: tumor colored *yellow*, lymphocytes colored *red*, and nontumor/nonlymphocyte background tissue colored *blue*.

By using this kind of advanced Pathomics tool to perform fine-grained mapping and subclassification of diagnostic histologic patterns of NSCLC adenocarcinoma, the striking variability in tumor histology is immediately appreciated with the added opportunity to correlate the histologic pattern of growth with the abundance and distributions of TILs as shown in Fig. 15.6.

Another recently developed Pathomics method performs fine-grained segmentation in colorectal cancer tissue samples by mapping specialized tissue compartments. The colorectal cancer model identifies cancer, premalignant dysplasia, normal epithelium, stromal connective tissue, and necrosis to generate a comprehensive map of the tumor microenvironment. Lymphocytes/TILs are detected with a separate VGG16-based model and incorporated into the workflow [37,125]. The human gastrointestinal tract is actively surveilled by the innate and adaptive branches of the immune system. When examining colorectal tissues, it is common for pathologists to use qualitative descriptions of magnitude, such as minimal, mild, moderate, or severe, and terminology like focal or diffuse to describe the spatial distribution of inflammatory infiltrates.

Fig. 15.7 shows analyses of three different colon cancer tissue samples with this novel Pathomics method that can help support the assessment of the abundance and spatial distribution of TILs in different microanatomic tissue compartments of the TME. The combination of cancer segmentation and lymphocyte/TILs detection provides a fascinating view into the diversity of tumor immune interactions in colorectal cancer, similar to what was observed in breast cancer. As much as the presence of lymphocytes/TILs was expected a priori to some degree, it is a very powerful experience to simultaneously view and compare tumor immune responses across cases that include a moderately differentiated colon cancer localized to the sigmoid colon, a case of poorly differentiated colon cancer with local lymph node metastases, and a case of stage IV moderately differentiated colon cancer with local lymph node, liver, and lung metastases. While it is too early to draw any major conclusions, an interesting observation is the lack of TILs in the third case that has distant metastases to the liver and lung (Fig. 15.7, bottom row) in comparison to the heterogenous spatial distribution of TILs in the second case with local lymph node metastases where the left portion of the tumor has markedly reduced TILs (Fig. 15.7, middle row).

It is our hope that these examples motivate further scientific inquiry through the implementation of emerging Pathomics applications to study TILs in various types of cancer and stimulate further development. We present a diverse group of examples to show the value of Pathomics in supporting various aspects of traditional histopathologic examination of cancer tissue samples, not limited to breast, pancreatic, lung, and colorectal cancer. We hope to have demonstrated that it is very possible to develop Pathology inspired models to study a wide array of phenomena in tumor pathobiology, also not limited to TILs. The definition and criteria for different types of architectural elements and regions of tissues can be incorporated to design a variety of Pathomics applications. To put this into context, if a pathologist or group of pathologists attempted to evaluate TILs in multiple fields of view under the microscope for every tissue section for multiple slides per case across hundreds to

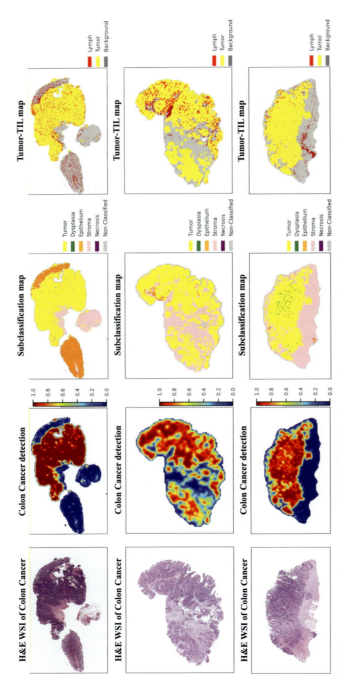

FIG. 15.7

Heterogeneous tumor immune interactions in colorectal cancer. Pathomics is utilized to map microanatomic tissue compartments in colorectal cancer to generate subclassification maps. The subclassification maps identify cancer, premalignant dysplasia, normal epithelium, stroma, and necrosis. TILs are identified with a separate VGG16-based lymphocyte detection algorithm and incorporated into the analytic workflow. After classification, tissue subclassification maps are generated alongside tumor-TIL maps to characterize the abundance and spatial distribution of TILs. Pathomics analyses in colorectal cancer show variability in tumor immune responses in magnitude and spatial localization. *Top row:* COAD TCGA-A6-A56B-01Z-00-DX1, diagnosis: moderately differentiated sigmoid colon adenocarcinoma penetrating through muscularis propria with no lymph node metastases; *middle row:* COAD TCGA-AA-3867-01Z-00-DX1, diagnosis: 3 cm ulcerated poorly differentiated sigmoid colon adenocarcinoma completely encircling the intestinal wall with extensive metastases in local mesocolic lymph nodes; *bottom row:* COAD TCGA-AA-3973-01Z-00-DX1, diagnosis: stage IV, 3 cm moderately differentiated colorectal adenocarcinoma with local lymph node, liver, and lung metastases. Tumor-TIL maps: tumor colored *yellow,* lymphocytes colored *red,* and nontumor/nonlymphocyte background (BG) tissue colored *gray.*

thousands of glass slides, the endeavor would be prohibitive due to the amount of labor, time, and cost and plagued by the limited availability of practicing pathologists alongside intrinsic intra- and interobserver variability.

Before AI and computer vision algorithms find their way into the clinic to evaluate TILs and other histopathologic features of cancer, they will need extensive testing, validation, and regulatory review by entities such as the FDA. In the meantime, deep learning tools are becoming more widely commercial in digital pathology and virtual microscopy software platforms. Commercial software for the analysis of the immune infiltrates include the Genie Spectrum (Aperio ePathology Solutions, Vista, CA) [94,135], Image-Pro Plus 3.0 (Media Cybernetics; Silver Spring, MD) [136], Halo (Indica Labs, Corrales, NM) [137], ImageJ [138], VMscope (VMscope GmbH, Berlin, Germany) slide explorer [2], and Visiopharm (Horsholm, Denmark). For example, the commercially available image analysis software platform, WebMicroscope, has been utilized to identify tissue regions containing TILs and count the number of lymphocytes in each region [139], where the authors observed that TIL density in seminomas had prognostic relevance regarding disease relapse. Similarly, companies such as PaigeAI and PathAI advertise the ability to carry out combinations of region characterization and cell identification in H&E WSIs as part of their respective efforts to develop deep learning applications to deliver more accurate cancer diagnoses.

Spatial cancer biology with Pathomics, immunohistochemistry, and immunofluorescence

As mentioned in the Immune Surveillance and Immunotherapy section, there are many kinds of functionally distinct subtypes of lymphocytes and other immune cells that are characterized based on the expression of specific surface and cellular biomarkers [44,139–143]. Building on the exciting advances of computational detection of lymphocytes/TILs within the histologic landscape of various types of cancer, a conceptually straightforward next step is to integrate insights about the abundance and spatial distribution of TILs at the tissue level from Tumor-TIL maps with nuclear segmentation and classification methods to gain insight at the cellular level [11,128,129,144,145]. Fig. 15.8 shows the progression from Tumor-TIL maps that provide a global perspective about the abundance and spatial distribution of TILs, followed by segmentation and classification of cancer cells, lymphocytes/TILs, and stromal cells (noncancer/nonlymphocyte) to illustrate the connection between immune responses visualized at the tissue and cellular levels.

In routine diagnostic surgical pathology, immunohistochemistry (IHC) and immunofluorescence (IF) are nondestructive laboratory methods that are used to label cells based on the expression of nuclear, cellular, and membrane protein molecules. Since tissue architecture is preserved, the abundance and spatial distribution of different populations of tumor, immune, and other types of cells can be evaluated in the context of the histologic landscape of the TME. Methods such as IHC and IF

328 CHAPTER 15 AI and the interplay between cancer and immunity

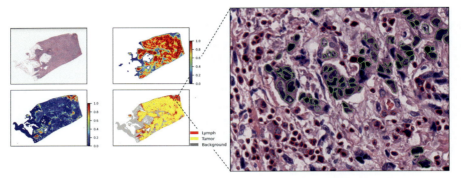

FIG. 15.8

Pathomics applications from tissue level spatial maps of TILs in breast cancer to cellular level with nucleus segmentation and classification. Tumor-TIL maps generated with the patch-level classification of cancer and lymphocytes provide insight into the abundance and global spatial distribution of intratumoral and peritumoral TILs, as well as tertiary lymphoid aggregates, at the tissue level. These kinds of analyses can be integrated with nuclear segmentation and classification to further characterize the abundance and spatial distribution of cancer cells, lymphocytes, and stromal connective tissue cells at the cellular level.

Image: BRCA TCGA BRCA BH-A0BZ-01Z-00-DX1, high-grade breast cancer, left panel: Tumor-TIL map, where breast cancer colored yellow, *lymphocytes colored* red, *and background nontumor/nonlymphocyte tissue colored* gray, *right panel: H&E 400×—nuclear segmentation and classification with cancer cells delineated in* green, *lymphocytes encircled in* red, *and stromal connective tissue cells outlined in* blue.

reveal the identities of the immune cells in the TME in situ. In clinical practice and research, consecutive serial sections of tissues are stained with one biomarker at a time, where staining interpretation is based on the presence of a marker or a semi-quantitative estimation of the number of positive cells of interest and staining intensity. However, the manual evaluation and interpretation of IHC expression of a single marker can be highly subjective and time consuming due to the nuances in relationship between expression and stain intensity, as well as nonspecific background staining, heterogeneous expression, subcellular localization, and variability in selecting and scoring appropriate ROIs [44,146]. Furthermore, IHC can be heavily influenced by preanalytic variables due to intrinsic differences in tissue fixation, chemical staining protocols, and histologic sample preparation, just like H&E staining.

Since IHC is routinely performed in modern pathology, there is great interest in developing clinical applications for digital pathology to perform quantitative image analysis of IHC markers. An early successful example utilized computational methods to score IHC expression of estrogen receptor (ER), progesterone receptor (PR), HER-2, and Ki67 in invasive breast cancer [15,16,147]. The goal is to reduce subjective intra- and interobserver scoring variability by pathologists through the precise measurement of the staining intensity of these biomarkers and characterization of subcellular localization [15,16,147]. Despite tremendous potential, the utilization of digital pathology to score IHC images in the daily workflow of surgical

Spatial cancer biology with pathomics, immunohistochemistry, and immunofluorescence

pathology laboratories has not yet achieved widespread clinical adoption due to the necessity for extensive validation, technical expertise, computational resources, and frequent calibration for IHC staining with positive and negative controls [15,16,147,148].

In terms of computational assessment of IHC/IF, the same challenges remain, as stated with added complications of aligning and registering serially stained sections with the corresponding H&E tissue sections to interpret the staining of the cells of interest. Intrinsic architectural and contextual differences in each section are abstracted by the human mind of pathologists, but it is very difficult for current deep learning and AI Pathomics methodology to imagine the presence of cells in images or relate cells from one section/image to another section/image in the same manner. To overcome the challenges of registering multiple-stained images, there has been great interest in using multiplex chromogenic immunohistochemistry (mIHC) and multiplex immunofluorescence (mIF) platforms that label multiple biomarkers in situ in a single formalin-fixed paraffin-embedded (FFPE) tissue sections to study spatial cancer biology [142,149–156].

Multiplex IHC/IF platforms have the potential to transform the diagnostic evaluation of cancer and the tumor immune microenvironment through the simultaneous assessment of multiple immune and cancer-related pathways and spatial relationships of tumor immune interactions in a single tissue sample [142,149–156]. They are also gaining popularity since precious and limited samples can be preserved for future studies, while removing the large barrier of image registration and alignment for the development of practical Pathomics image analysis applications that quantify IHC/IF scoring. However, these kinds of AI Pathomics algorithms will still face the same challenges in terms of dealing with variability of staining intensity and nonspecific staining for each IHC biomarker and artifacts from tissue sectioning and thickness, histologic preparation, and variable biomarker expression, which can all affect quantitative scoring of IHC/IF by image analysis. Another overarching challenge that remains as a barrier to the analyses of m IHC/IF with Pathomics AI computer vision algorithms is the scarcity of training data.

Despite current limitations, commercial software platforms provide solid tools to quantify the IHC expression of single biomarkers, which need to be custom designed, trained, tuned, and refined by each user for a particular biomarker for a specific study set. The main issues reducing interoperability and adoption occur when customized tools designed with a particular set of parameters to analyze images from a specific study leads to suboptimal performance in analyzing a different set of images. Similar challenges are faced when developing novel Pathomics image analysis applications to support quantifying multiple biomarkers conjugated to uniquely colored chromogens in mIHC and mIF panels. If mIHC/IF become adopted for routine clinical use, algorithms that quantify the expression of multiple IHC biomarkers in cancer-specific panels need to be robust while undergoing extensive QA/QC.

However, this is quite difficult because these kinds of Pathomics algorithms will need to differentiate heterogeneous staining intensities and nonspecific background staining across multiple uniquely colored chromogens with colocalized expression on subsets of cells (e.g., CD3$^+$CD8$^+$ cytotoxic T-cells). After image analyses of mIHC/IF, postprocessing will need to incorporate a methodology that interprets staining by different types of cells within the appropriately delineated tissue regions of interest (e.g., leading of edge of cancer with adjacent normal tissue), distinguish morphologically similar but functionally distinct cells (e.g., malignant versus dysplastic cells), and perform spatial analyses that are interpretable by pathologists and researchers.

There is a real need for Pathomics applications for mIHC/IF since current pathologists have not been exposed or trained to examine multiple-colored biomarkers simultaneously per slide while mentally performing spatial colocalization to provide semiquantitative assessments to quantify expression for each biomarker. Whether it is a single marker or multiple IHC/IF biomarkers that need histopathologic evaluation, there are over 100,000 cells in a typical tissue section. Therefore, IHC/IF expression of multiple markers need to be interpreted within the appropriate histologic context of $\geq$100,000 cells per glass slide or WSI. Furthermore, expression of several IHC/IF biomarkers must be simultaneously interpreted within the histologic context of tumor cellularity, architectural distortion, growth pattern of the tumor, tumor-associated stromal responses, inflammatory responses, presence or absence of necrosis, and spatially induced differences of the nuanced expression of biomarkers in the central areas of tumor, infiltrative nests, invasive border, and interface with the adjacent surrounding normal tissues.

In terms of computer science and computer vision, the types of tasks that need to be developed for deep learning applications are categorized as semantic instance segmentation. Within the context of spatial molecular pathology analyses, this translates to segmenting cells and/or tissue regions, which are assigned one or more class labels for each cell and/or tissue region. Current work in digital pathology image analysis for mIHC and mIF is exploring unbiased, unsupervised, and semisupervised deep learning methods to perform automatic IHC analysis to isolate and detect the expression of individual markers and measure the extent of expression in tumor and immune cells through color deconvolution, optical density, and multispectral separation of chromogens [44,146].

Previous work explored three immune biomarkers on a single tissue section in combination with adjacent sections with one biomarker to delineate the tumor [143]. A prospective study explored the heterogeneous nature of metastatic melanoma by correlating mIHC with fluorescence-activated cell sorting (FACS) flow cytometry to quantify immune subsets in the intratumoral, tumor-associated stromal compartments, and invasive edge of the tumor [157]. This study evaluated the association between intratumoral CD8$^+$ T-cells and PD-L1 expression in melanoma in combination with FACS to obtain data regarding the differentiation of T-cell subsets, activation status, and expression of immune checkpoint molecules. Multiplex IHC identified significantly higher T$_{reg}$ cells than FACS and showed preferential stromal CD4$^+$ T-cells. This study proposed a model of the immune contexture of metastatic melanoma with respect to PD-L1 expression in melanoma cells and macrophages, as well as the presence or absence of peritumoral and intratumoral CD8$^+$ TILs. There

Spatial cancer biology with pathomics, immunohistochemistry, and immunofluorescence

331

was also insight into treatment response with checkpoint inhibitor therapy since PD-L1 expression was observed to be more prevalent in macrophages [157].

Malignant melanoma is a very active area of research after James P. Allison and Tasuku Honjo were awarded a Nobel Prize in Medicine for their discovery of cancer immunotherapy through the inhibition of negative immune regulation. For example, another novel study used quantitative IHC, quantitative mIF, and next-generation sequencing for T-cell antigen receptors (TCRs) to show that the proliferation of intratumoral CD8$^+$ TILs was directly correlated with radiographic reduction in tumor size with Pembrolizumab therapy for metastatic melanoma [137]. This study also identified increased numbers of intratumoral and peritumoral cells that expressed CD8, PD-1, and PD-L1 in pretreatment samples to show close spatial proximity between PD-1 and PD-L1 expressing cells.

Multispectral imaging in malignant melanoma was also performed with the AstroPath platform to evaluate the efficacy of PD-1 blockade with mIF [158]. Inspired by astronomy, multispectral imaging of whole tumor sections with high-fidelity single-cell resolution was used to study the expression pattern of functionally relevant molecules (e.g., PD-1, PD-L1, CD8, FoxP3, CD163, and Sox10/S100) in melanoma patients receiving immunotherapy [158]. By using this panel of six markers, this study evaluated the intensity of PD-1 and PD-L1 expression in situ in different cell types and 41 combinations of expression patterns for these biomarkers. This mIF study culminated in mapping relatively rare cells such as CD8$^+$FoxP3$^+$ T-cells to the tumor-stromal boundary, as well as correlating a high density of CD8$^+$FoxP3$^+$PD-1$^{low/mid}$ cells with response to PD-1 blockade [158]. This novel approach also identified cell types that were associated with a lack of response to immunotherapy, such as inhibitory CD163$^+$PD-L1$^-$ macrophages [158].

The availability of mIHC, mIF, and combined mIHC/IF approaches have led to several research studies with multiple markers in NSCLC lung cancer and cancers in other anatomic sites [140,142,143,151,152,154,155,159,160]. Novel multispectral approaches that identify, segment, and evaluate more than three colors at a time have been developed to interpret mIHC within the context of the biological properties of the biomarkers expressed by tumor and neighboring immune cells by incorporating methods to assign each pixel with a collection of constituent signals from the chromogens in the color mixture [44,146,154,161–163]. By using this approach, mIHC/IF coexpression in subsets of cell types is characterized, which represents significant progress since the maximum number of colored stains that could be unmixed was previously limited to three colors [44,146,161,163].

Deep learning methods have also been used to detect and classify cells in mIHC WSIs of breast cancer, where a recent approach involved a two-stage approach with an initial neural network to carry out cell detection and cell counting that was coupled to a second neural network to carry out cell classification [164], where the annotations required labeling cells with dots. Another recent approach focuses on reducing the requirement for training data through the development of super-resolution methods that are designed to predict high-resolution labels by primarily utilizing low-resolution training data, which is generally augmented with sparse

332 **CHAPTER 15** AI and the interplay between cancer and immunity

high-resolution training data [165]. Generating training data in these approaches involves a combination of labeling patches to indicate the density of different cells in each patch that is supplemented with dot annotation.

The emergence of mIHC/IF alongside digital pathology represents natural progression and innovation in laboratory medicine and research, where the hope is that Pathomics applications driven by AI in computer vision can be routinely utilized to quantify the expression of multiple IHC/IF biomarkers. Multiplex IHC/IF and Pathomics are necessarily intertwined for clinical adoption of Pathomics and mIHC/IF will undoubtedly require rigorous evaluation, testing, and validation to provide prognostic information and guide treatment selection. Even though there are many challenges, it is clear that the discovery of prognostic and predictive biomarker applications [14,20,21,44,146,149,154] can be dramatically accelerated by using mIHC/IF, which will transform our collective understanding of cancer and revolutionize treatment. There is a bright immediate future ahead in quantifying the nature of the interactions among the different types of immune cells in the TME searches with Pathomics, IHC, and IF biomarkers, which will be invaluable in elucidating key immunoregulatory pathways to identify patients who can benefit from immunotherapy in the neoadjuvant and adjuvant settings [44,141].

Conclusion

Artificial intelligence is rapidly assuming a crucial role in the characterization of tumor immune interactions in cancer. Pathomics algorithms that generate detailed spatial maps of TILs in the TME have been shown to be quite useful in translational clinical research to support their use as a prognostic and predictive biomarker. As the utilization of advanced AI computer vision applications in digital pathology continues to increase, it is more than likely that numerous Pathomics biomarkers will be discovered to predict treatment response and clinical outcomes. The ultimate goal of this chapter was to describe Pathomics methodology driven by deep learning and AI in computer vision to illustrate rapid advances in our ability to analyze thousands of digital H&E WSIs by using the example of mapping of tumor immune interactions in multiple types of cancer. We ended with a brief introduction on emerging complementary multiplex IHC/IF methods since it is more than likely that multiplex biomarker panels will be eventually adopted to save tissue material for future studies, costs, and time, while supporting the development of algorithms that will not need to deal with image alignment and registration across multiple serial sections stained with single IHC biomarkers. We hope that we have shown how the combination of different Pathomics methods can provide many different types of potentially useful information to support precision medicine in oncology. Even though the interpretation of tumor immune interactions is complex and challenging, we believe that Pathomics methods driven by deep learning and AI in computer vision are performing remarkably well enough to be feasibly considered for clinical adoption in the near future. Even though mapping patient-specific tumor immune interactions are still at an early stage, the crucial next phase of research regarding quantitative analysis of spatial morphologic and molecular patterns with mIHC/IF represents a fascinating frontier for translational clinical research. Finally, the power of digital pathology and Pathomics is in their ability to perform computational analyses in a scalable and high throughput manner across thousands to hundreds of thousands of WSIs without the need for additional tissue, time, or resources, which would be otherwise impossible through human effort alone. One can only imagine what is waiting to be discovered if we retrospectively or prospectively analyze every case of cancer in the world with automated Pathomics tools to characterize disease- and site-specific cancer diversity in large collections of WSIs.

References

[1] Denkert C, et al. Tumor-associated lymphocytes as an independent predictor of response to neoadjuvant chemotherapy in breast cancer. J Clin Oncol 2010;28(1): 105–13.

[2] Denkert C, et al. Standardized evaluation of tumor-infiltrating lymphocytes in breast cancer: results of the ring studies of the international immuno-oncology biomarker working group. Mod Pathol 2016;29(10):1155–64.

[3] Hendry S, et al. Assessing tumor-infiltrating lymphocytes in solid tumors: a practical review for pathologists and proposal for a standardized method from the international Immuno-oncology biomarkers working group: part 2: TILs in melanoma, gastrointestinal tract carcinomas, non-small cell lung carcinoma and mesothelioma, endometrial and ovarian carcinomas, squamous cell carcinoma of the head and neck, genitourinary carcinomas, and primary brain tumors. Adv Anat Pathol 2017;24(6):311–35.

[4] Hendry S, et al. Assessing tumor-infiltrating lymphocytes in solid tumors: a practical review for pathologists and proposal for a standardized method from the international immunooncology biomarkers working group: part 1: assessing the host immune response, TILs in invasive breast carcinoma and ductal carcinoma in situ, metastatic Tumor deposits and areas for further research. Adv Anat Pathol 2017;24(5):235–51.

[5] Salgado R, et al. The evaluation of tumor-infiltrating lymphocytes (TILs) in breast cancer: recommendations by an international TILs working group 2014. Ann Oncol 2015;26(2):259–71.

[6] Gurcan MN, et al. Histopathological image analysis: a review. IEEE Rev Biomed Eng 2009;2:147–71.

[7] Irshad H, et al. Methods for nuclei detection, segmentation, and classification in digital histopathology: a review—current status and future potential. IEEE Rev Biomed Eng 2014;7.

[8] Kothari S, et al. Pathology imaging informatics for quantitative analysis of whole-slide images. J Am Med Inform Assoc 2013;20(6):1099–108.

[9] Madabhushi A, Lee G. Image analysis and machine learning in digital pathology: challenges and opportunities. Med Image Anal 2016;33:170–5.

[10] Xing F, Yang L. Robust nucleus/cell detection and segmentation in digital pathology and microscopy images: a comprehensive review. J IEEE Rev Biomed Eng 2016;9:234–63.

[11] Amgad M, et al. Structured crowdsourcing enables convolutional segmentation of histology images. Bioinformatics 2019;35(18):3461–7.

[12] Amgad M, et al. Joint region and nucleus segmentation for characterization of tumor infiltrating lymphocytes in breast cancer. Proc SPIE Int Soc Opt Eng 2019;10956.

[13] Bindea G, et al. Natural immunity to cancer in humans. Curr Opin Immunol 2010; 22(2):215–22.

[14] Bindea G, et al. Spatiotemporal dynamics of intratumoral immune cells reveal the immune landscape in human cancer. Immunity 2013;39(4):782–95.

[15] Cooper L, et al. Feature-based registration of histopathology images with different stains: an application for computerized follicular lymphoma prognosis. Comput Methods Programs Biomed 2009;96(3):182–92.

[16] Cooper LA, et al. PanCancer insights from the cancer genome atlas: the pathologist's perspective. J Pathol 2018;244(5):512–24.

[17] Cooper LA, et al. The tumor microenvironment strongly impacts master transcriptional regulators and gene expression class of glioblastoma. Am J Pathol 2012;180(5): 2108–19.

[18] Cooper LA, et al. Integrated morphologic analysis for the identification and characterization of disease subtypes. J Am Med Inform Assoc 2012;19(2):317–23.

[19] Cooper LA, et al. Morphological signatures and genomic correlates in glioblastoma. Proc IEEE Int Symp Biomed Imaging 2011;1624–7.

[20] Farkona S, Diamandis EP, Blasutig IM. Cancer immunotherapy: the beginning of the end of cancer? BMC Med 2016;14:73.

[21] Fridman WH, et al. The immune contexture in human tumours: impact on clinical outcome. Nat Rev Cancer 2012;12(4):298–306.

[22] Gajewski TF, Schreiber H, Fu YX. Innate and adaptive immune cells in the tumor microenvironment. Nat Immunol 2013;14(10):1014–22.

[23] Galon J, et al. The continuum of cancer immunosurveillance: prognostic, predictive, and mechanistic signatures. Immunity 2013;39(1):11–26.

[24] Galon J, Bruni D. Tumor immunology and tumor evolution: intertwined histories. Immunity 2020;52(1):55–81.

[25] Galon J, et al. Type, density, and location of immune cells within human colorectal tumors predict clinical outcome. Science 2006;313(5795):1960–4.

[26] Galon J, et al. Towards the introduction of the 'Immunoscore' in the classification of malignant tumours. J Pathol 2014;232(2):199–209.

[27] Galon J, et al. Cancer classification using the Immunoscore: a worldwide task force. J Transl Med 2012;10:205.

[28] Gurcan MN, et al. Developing the quantitative histopathology image ontology (QHIO): a case study using the hot spot detection problem. J Biomed Inform 2017;66:129–35.

[29] Janowczyk A, et al. A resolution adaptive deep hierarchical (RADHicaL) learning scheme applied to nuclear segmentation of digital pathology images. Comput Methods Biomech Biomed Eng Imaging Vis 2018;6(3):270–6.

[30] Janowczyk A, Madabhushi A. Deep learning for digital pathology image analysis: A comprehensive tutorial with selected use cases; 2016. p. 7.

[31] Kalra J, Baker J. Multiplex immunohistochemistry for mapping the tumor microenvironment. In: Kalyuzhny AE, editor. Signal transduction immunohistochemistry: methods and protocols. New York, NY: Springer New York; 2017. p. 237–51.

[32] Kumar A, et al. Automated analysis of immunohistochemistry images identifies candidate location biomarkers for cancers. Proc Natl Acad Sci 2014;111(51):18249–54.

[33] Kumar H, Kawai T, Akira S. Pathogen recognition by the innate immune system. Int Rev Immunol 2011;30(1):16–34.

[34] Madabhushi A, et al. Computer-aided prognosis: predicting patient and disease outcome via quantitative fusion of multi-scale, multi-modal data. Comput Med Imaging Graph 2011;35(7–8):506–14.

[35] Netea MG, et al. Trained immunity: a program of innate immune memory in health and disease. Science 2016;352(6284), aaf1098.

[36] Nordstrom RJ. The quantitative imaging network in precision medicine. Tomography 2016;2(4):239–41.

[37] Saltz J, et al. Spatial organization and molecular correlation of tumor-infiltrating lymphocytes using deep learning on pathology images. Cell Rep 2018;23(1):181.

[38] Saltz J, et al. A containerized software system for generation, management, and exploration of features from whole slide tissue images. Cancer Res 2017;77(21):e79–82.

[39] Sharma P, Allison JP. Immune checkpoint targeting in cancer therapy: toward combination strategies with curative potential. Cell 2015;161(2):205–14.

[40] Sharma P, et al. Novel cancer immunotherapy agents with survival benefit: recent successes and next steps. Nat Rev Cancer 2011;11(11):805–12.

[41] Wolchok JD, et al. Nivolumab plus ipilimumab in advanced melanoma. N Engl J Med 2013;369(2):122–33.

[42] Norton K-A, et al. Multiscale agent-based and hybrid modeling of the tumor immune microenvironment. Processes (Basel, Switzerland) 2019;7(1):37.

[43] Calì B, Molon B, Viola A. Tuning cancer fate: the unremitting role of host immunity. Open Biol 2017;7(4), 170006.

[44] Koelzer VH, et al. Precision immunoprofiling by image analysis and artificial intelligence. Virchows Arch 2019;474(4):511–22.

[45] Thorsson V, et al. The immune landscape of cancer. Immunity 2018;48(4):812.

[46] Choi J, et al. Diagnostic value of peripheral blood immune profiling in colorectal cancer. Ann Surg Treatment Res 2018;94(6):312–21.

[47] Lepone LM, et al. Analyses of 123 peripheral human immune cell subsets: defining differences with age and between healthy donors and cancer patients not detected in analysis of standard immune cell types. J Circulat Biomark 2016;5:5.

[48] Loi S, et al. Tumor-infiltrating lymphocytes and prognosis: a pooled individual patient analysis of early-stage triple-negative breast cancers. J Clin Oncol 2019;37(7):559–69.

[49] Ogino S, et al. Integrative analysis of exogenous, endogenous, tumour and immune factors for precision medicine. Gut 2018;67(6):1168–80.

[50] Aras S, Zaidi MR. TAMeless traitors: macrophages in cancer progression and metastasis. Br J Cancer 2017;117:1583.

[51] Dykes SS, et al. Stromal cells in breast cancer as a potential therapeutic target. Oncotarget 2018;9(34):23761–79.

[52] Drake CG, Jaffee E, Pardoll DM. Mechanisms of immune evasion by tumors. Adv Immunol 2006;90:51–81.

[53] Finn OJ. Immuno-oncology: understanding the function and dysfunction of the immune system in cancer. Ann Oncol 2012;23(Suppl 8):viii6–9.

[54] Hanahan D, Weinberg RA. Hallmarks of cancer: the next generation. Cell 2011;144(5):646–74.

[55] Diaz-Cano SJ. Tumor heterogeneity: mechanisms and bases for a reliable application of molecular marker design. Int J Mol Sci 2012;13(2):1951–2011.

[56] Parish CR. Cancer immunotherapy: the past, the present and the future. Immunol Cell Biol 2003;81(2):106–13.

[57] Burnet FM. The concept of immunological surveillance. Prog Exp Tumor Res 1970;13:1–27.

[58] Burnet M. Cancer; a biological approach. I. The processes of control. Br Med J 1957;1 (5022):779–86.

[59] Thomas L. On immunosurveillance in human cancer. Yale J Biol Med 1982; 55 (3–4):329–33.

[60] Bonnans C, Chou J, Werb Z. Remodelling the extracellular matrix in development and disease. Nat Rev Mol Cell Biol 2014;15(12):786–801.

[61] Kaczorowski KJ, et al. Continuous immunotypes describe human immune variation and predict diverse responses. Proc Natl Acad Sci U S A 2017;114(30):E6097–e6106.

[62] Li B, et al. Comprehensive analyses of tumor immunity: implications for cancer immunotherapy. Genome Biol 2016;17(1):174.

[63] Maecker HT, McCoy JP, Nussenblatt R. Standardizing immunophenotyping for the human immunology project. Nat Rev Immunol 2012;12(3):191–200.

[64] Mellman I, Coukos G, Dranoff G. Cancer immunotherapy comes of age. Nature 2011;480(7378):480–9.

[65] Ribas A. Releasing the brakes on cancer immunotherapy. N Engl J Med 2015;373 (16):1490–2.

[66] Seager RJ, et al. Dynamic interplay between tumour, stroma and immune system can drive or prevent tumour progression. Convergent Sci Phys Oncol 2017;3, 034002.

[67] Smyth MJ, et al. Combination cancer immunotherapies tailored to the tumour microenvironment. Nat Rev Clin Oncol 2016;13(3):143–58.

[68] Curran MA, et al. PD-1 and CTLA-4 combination blockade expands infiltrating T cells and reduces regulatory T and myeloid cells within B16 melanoma tumors. Proc Natl Acad Sci U S A 2010;107(9):4275–80.

[69] Grosso JF, Jure-Kunkel MN. CTLA-4 blockade in tumor models: an overview of preclinical and translational research. Cancer Immun 2013;13:5.

[70] Hodi FS, et al. Improved survival with ipilimumab in patients with metastatic melanoma. N Engl J Med 2010;363(8):711–23.

[71] Patel SP, Kurzrock R. PD-L1 expression as a predictive biomarker in cancer immunotherapy. Mol Cancer Ther 2015;14(4):847–56.

[72] Topalian SL, et al. Survival, durable tumor remission, and long-term safety in patients with advanced melanoma receiving nivolumab. J Clin Oncol 2014;32(10):1020–30.

[73] Carbognin L, et al. Differential activity of nivolumab, pembrolizumab and MPDL3280A according to the tumor expression of programmed death-Ligand-1 (PD-L1): sensitivity analysis of trials in melanoma, lung and genitourinary cancers. PloS One 2015;10(6), e0130142.

[74] Parry RV, et al. CTLA-4 and PD-1 receptors inhibit T-cell activation by distinct mechanisms. Mol Cell Biol 2005;25(21):9543–53.

[75] Robert C, et al. Ipilimumab plus dacarbazine for previously untreated metastatic melanoma. N Engl J Med 2011;364(26):2517–26.

[76] Spill F, et al. Impact of the physical microenvironment on tumor progression and metastasis. Curr Opin Biotechnol 2016;40:41–8.

[77] Versteven M, et al. A versatile T cell-based assay to assess therapeutic antigen-specific PD-1-targeted approaches. Oncotarget 2018;9(45):27797–808.

[78] Brahmer JR, et al. Safety and activity of anti-PD-L1 antibody in patients with advanced cancer. N Engl J Med 2012;366(26):2455–65.

References   **337**

[79] Keir ME, et al. PD-1 and its ligands in tolerance and immunity. Annu Rev Immunol 2008;26(1):677–704.

[80] Passiglia F, et al. PD-L1 expression as predictive biomarker in patients with NSCLC: a pooled analysis. Oncotarget 2016;7(15):19738–47.

[81] Roma-Rodrigues C, et al. Targeting tumor microenvironment for cancer therapy. Int J Mol Sci 2019;20(4):840.

[82] Topalian SL, et al. Safety, activity, and immune correlates of anti-PD-1 antibody in cancer. N Engl J Med 2012;366(26):2443–54.

[83] Cooper WA, et al. Intra- and Interobserver reproducibility assessment of PD-L1 biomarker in non-small cell lung cancer. Clin Cancer Res 2017;23(16):4569–77.

[84] Troncone G, Gridelli C. The reproducibility of PD-L1 scoring in lung cancer: can the pathologists do better? Transl Lung Cancer Res 2017;6(Suppl 1):S74–s77.

[85] Ung C, Kockx M, Waumans Y. Digital pathology in immuno-oncology – a roadmap for clinical development. Expert Rev Precis Med Drug Dev 2017;2(1):9–19.

[86] Mascaux C, et al. Immune evasion before tumour invasion in early lung squamous carcinogenesis. Nature 2019.

[87] Dudgeon S, et al. A pathologist-annotated dataset for validating artificial intelligence: a project description and pilot study. J Pathology Inform 2021;12(1):45.

[88] Galon J, Fridman WH, Pagès F. The adaptive immunologic microenvironment in colorectal cancer: a novel perspective. Cancer Res 2007;67(5):1883–6.

[89] Coulie PG, et al. Tumour antigens recognized by T lymphocytes: at the core of cancer immunotherapy. Nat Rev Cancer 2014;14(2):135–46.

[90] Donnem T, et al. Strategies for clinical implementation of TNM-immunoscore in resected nonsmall-cell lung cancer. Ann Oncol 2016;27(2):225–32.

[91] Broussard EK, Disis ML. TNM staging in colorectal cancer: T is for T cell and M is for memory. J Clin Oncol 2011;29(6):601–3.

[92] Mlecnik B, et al. Tumor immunosurveillance in human cancers. Cancer Metastasis Rev 2011;30(1):5–12.

[93] Angell H, Galon J. From the immune contexture to the immunoscore: the role of prognostic and predictive immune markers in cancer. Curr Opin Immunol 2013;25(2):261–7.

[94] Angell HK, et al. Digital pattern recognition-based image analysis quantifies immune infiltrates in distinct tissue regions of colorectal cancer and identifies a metastatic phenotype. Br J Cancer 2013;109(6):1618–24.

[95] Savas P, et al. Clinical relevance of host immunity in breast cancer: from TILs to the clinic. Nat Rev Clin Oncol 2016;13(4):228–41.

[96] Chen CL, et al. Deep learning in label-free cell classification. Sci Rep 2016;6.

[97] Chen H, et al. DCAN: deep contour-aware networks for object instance segmentation from histology images. Med Image Anal 2017;36:135–46.

[98] Dai Y, Wang G. A deep inference learning framework for healthcare. Pattern Recogn Lett 2018.

[99] Graham B. Spatially-sparse convolutional neural networks. arXiv 2014;1–13. preprint arXiv:1409.6070.

[100] Gu J, Wang Z, et al. Recent advances in convolutional neural networks. Pattern Recogn 2018;77:354–77.

[101] Hou L, et al. Sparse autoencoder for unsupervised nucleus detection and representation in histopathology images. Pattern Recogn 2019;86:188–200.

[102] Hou L, et al. Patch-based convolutional neural network for whole slide tissue image classification. In: Proceedings of the IEEE conference on computer vision and pattern recognition (CVPR); 2016. p. 2424–33.

CHAPTER 15 AI and the interplay between cancer and immunity

[103] Hou L, et al. Automatic histopathology image analysis with CNNs. In: 2016 New York scientific data summit (NYSDS). IEEE; 2016.

[104] Khosravi P, et al. Deep convolutional neural networks enable discrimination of heterogeneous digital pathology images. EBioMedicine 2018;27:317–28.

[105] Kokkinos I. UberNet: training a universal convolutional neural network for low-, mid-, and high-level vision using diverse datasets and limited memory; 2017. p. 6129–38.

[106] Murthy V, et al. Center-focusing multi-task CNN with injected features for classification of glioma nuclear images. In: IEEE Winter conference on applications of computer vision (WACV); 2017. p. 834–41.

[107] Radford A, Metz L, Chintala S. Unsupervised representation learning with deep convolutional generative adversarial networks; 2016.

[108] Suzuki K, Zhou L, Wang Q. Machine learning in medical imaging. Pattern Recogn 2017;63:465–7.

[109] Xing F, et al. Deep learning in microscopy image analysis: a survey. IEEE Trans Neural Netw Learn Syst 2017;1–19.

[110] Schmidhuber JR. Deep learning in neural networks: an overview. Neural Netw 2015;61:85–117.

[111] Basavanhally AN, et al. Computerized image-based detection and grading of lymphocytic infiltration in HER2+ breast cancer histopathology. IEEE Trans Biomed Eng 2009;57(3):642–53.

[112] Panagiotakis C, Ramasso E, Tziritas G. Lymphocyte segmentation using the transferable belief model. In: International conference on pattern recognition. Springer; 2010.

[113] Yuan Y, et al. Quantitative image analysis of cellular heterogeneity in breast tumors complements genomic profiling. Sci Transl Med 2012;4(157), 157ra143.

[114] Sirinukunwattana K, et al. Locality sensitive deep learning for detection and classification of nuclei in routine colon cancer histology images. IEEE Trans Med Imaging 2016;35(5):1196–206.

[115] Xu Y, et al. Deep convolutional activation features for large scale brain tumor histopathology image classification and segmentation. In: 2015 IEEE international conference on acoustics, speech and signal processing (ICASSP). IEEE; 2015.

[116] Zhou Y, et al. Nuclei segmentation via sparsity constrained convolutional regression. In: 2015 IEEE 12th international symposium on biomedical imaging (ISBI). IEEE; 2015.

[117] Vu QD, et al. Methods for segmentation and classification of digital microscopy tissue images. Front Bioeng Biotechnol 2019;7:53.

[118] Goodfellow I, et al. Generative adversarial nets. in Advances in neural information processing systems; 2014.

[119] Hou L, et al. Robust histopathology image analysis: to label or to synthesize? In: Proceedings of the IEEE conference on computer vision and pattern recognition; 2019. p. 8533–42.

[120] Mahmood F, et al. Deep adversarial training for multi-organ nuclei segmentation in histopathology images. arXiv [cs.CV]; 2018.

[121] Foran DJ, et al. Computer-assisted discrimination among malignant lymphomas and leukemia using immunophenotyping, intelligent image repositories, and telemicroscopy. IEEE Trans Inf Technol Biomed 2000;4(4):265–73.

[122] Sertel O, et al. Histopathological image analysis using model-based intermediate representations and color texture: follicular lymphoma grading. J Signal Process Syst 2009;55(1–3):169.

[123] Corredor G, et al. Spatial architecture and arrangement of tumor-infiltrating lymphocytes for predicting likelihood of recurrence in early-stage non–small cell lung cancer. Clin Cancer Res 2019;25(5):1526–34.

[124] Heindl A, et al. Relevance of spatial heterogeneity of immune infiltration for predicting risk of recurrence after endocrine therapy of ER+ breast cancer. J Natl Cancer Inst 2017;110(2):166–75.

[125] Abousamra S, et al. Learning from thresholds: fully automated classification of tumor infiltrating lymphocytes for multiple cancer types. arXiv e-prints; 2019.

[126] Abousamra S, et al. Deep learning-based mapping of tumor infiltrating lymphocytes in whole slide images of 23 types of cancer. Front Oncol 2022;11.

[127] Amgad M, et al. Joint region and nucleus segmentation for characterization of tumor infiltrating lymphocytes in breast cancer. In: Medical imaging 2019: digital pathology. International Society for Optics and Photonics; 2019.

[128] Amgad M, Salgado R, Cooper LAD. MuTILs: explainable, multiresolution computational scoring of tumor-infiltrating lymphocytes in breast carcinomas using clinical guidelines. medRxiv 2022. 2022.01.08.22268814.

[129] Thagaard J, et al. Automated quantification of sTIL density with H&E-based digital image analysis has prognostic potential in triple-negative breast cancers. Cancers (Basel) 2021;13(12).

[130] Le H, et al. Utilizing automated breast cancer detection to identify spatial distributions of tumor-infiltrating lymphocytes in invasive breast cancer. Am J Pathol 2020;190 (7):1491–504.

[131] Mathbiol. Github interactive tumor-TIL maps for WSIs of breast cancer from the cancer genome atlas (TCGA), 2023. Available from: https:/mathbiol.github.io/tcgatil/. [Accessed 22 January 2023].

[132] Bonaventura P, et al. Cold tumors: a therapeutic challenge for immunotherapy. Front Immunol 2019;10(168).

[133] Fassler DJ, et al. Spatial characterization of tumor-infiltrating lymphocytes and breast cancer progression. Cancer 2022;14(9):2148.

[134] Le H, et al. Pancreatic cancer detection in whole slide images using noisy label annotations. In: Medical image computing and computer assisted intervention – MICCAI 2019. Cham: Springer International Publishing; 2019.

[135] Degnim AC, et al. Immune cell quantitation in normal breast tissue lobules with and without lobulitis. Breast Cancer Res Treat 2014;144(3):539–49.

[136] Johansson AC, et al. Computerized image analysis as a tool to quantify infiltrating leukocytes: a comparison between high- and low-magnification images. J Histochem Cytochem 2001;49(9):1073–9.

[137] Tumeh PC, et al. PD-1 blockade induces responses by inhibiting adaptive immune resistance. Nature 2014;515(7528):568–71.

[138] Lopez C, et al. Development of automated quantification methodologies of immunohistochemical markers to determine patterns of immune response in breast cancer: a retrospective cohort study. BMJ Open 2014;4(8), e005643.

[139] Linder N, et al. Deep learning for detecting tumour-infiltrating lymphocytes in testicular germ cell tumours. J Clin Pathol 2019;72(2):157–64.

[140] Barua S, et al. Spatial interaction of tumor cells and regulatory T cells correlates with survival in non-small cell lung cancer. Lung Cancer 2018;117:73–9.

CHAPTER 15 AI and the interplay between cancer and immunity

[141] Blank CU, et al. Cancer Immunology. The "cancer immunogram". Science 2016;352 (6286):658–60.

[142] Hofman P, et al. Multiplexed Immunohistochemistry for molecular and immune profiling in lung cancer-just about ready for prime-time? Cancer 2019;11(3):283.

[143] Ma Z, et al. Data integration from pathology slides for quantitative imaging of multiple cell types within the tumor immune cell infiltrate. Diagn Pathol 2017;12(1):69.

[144] Graham S, et al. Hover-net: simultaneous segmentation and classification of nuclei in multi-tissue histology images. Med Image Anal 2019;58, 101563.

[145] Vo VT-T, Kim S-H. Mulvernet: nucleus segmentation and classification of pathology images using the HoVer-net and multiple filter units. Electronics 2023;12(2):355.

[146] Seyed Jafari SM, Hunger RE. IHC optical density score: a new practical method for quantitative Immunohistochemistry image analysis. Appl Immunohistochem Mol Morphol 2017;25(1):e12–3.

[147] Farris AB, et al. Whole slide imaging for analytical anatomic pathology and telepathology: practical applications today, promises, and perils. Arch Pathol Lab Med 2017;141 (4):542–50.

[148] Aeffner F, et al. Digital microscopy, image analysis, and virtual slide repository. ILAR J 2018.

[149] Blom S, et al. Systems pathology by multiplexed immunohistochemistry and whole-slide digital image analysis. Sci Rep 2017;7(1):15580.

[150] Gibney GT, Weiner LM, Atkins MB. Predictive biomarkers for checkpoint inhibitor-based immunotherapy. Lancet Oncol 2016;17(12):e542–51.

[151] Gorris MAJ, et al. Eight-color multiplex immunohistochemistry for simultaneous detection of multiple immune checkpoint molecules within the tumor microenvironment. J Immunol 2018;200(1):347–54.

[152] Ilie M, et al. Automated chromogenic multiplexed immunohistochemistry assay for diagnosis and predictive biomarker testing in non-small cell lung cancer. Lung Cancer 2018;124:90–4.

[153] Levenson RM, Borowsky AD, Angelo M. Immunohistochemistry and mass spectrometry for highly multiplexed cellular molecular imaging. Lab Invest 2015;95(4):397–405.

[154] Parra ER, et al. Image analysis-based assessment of PD-L1 and tumor-associated immune cells density supports distinct intratumoral microenvironment groups in non-small cell lung carcinoma patients. Clin Cancer Res 2016;22(24):6278–89.

[155] Remark R, et al. In-depth tissue profiling using multiplexed immunohistochemical consecutive staining on single slide. Sci Immunol 2016;1(1), aaf6925.

[156] Tsujikawa T, et al. Quantitative multiplex Immunohistochemistry reveals myeloid-inflamed tumor-immune complexity associated with poor prognosis. Cell Rep 2017;19(1):203–17.

[157] Halse H, et al. Multiplex immunohistochemistry accurately defines the immune context of metastatic melanoma. Sci Rep 2018;8(1):11158.

[158] Berry S, et al. Analysis of multispectral imaging with the AstroPath platform informs efficacy of PD-1 blockade. Science 2021;372(6547).

[159] Koh J, et al. High-throughput multiplex immunohistochemical imaging of the tumor and its microenvironment. Cancer Res Treat 2020;52:98–108.

[160] Steele KE, et al. Measuring multiple parameters of CD8+ tumor-infiltrating lymphocytes in human cancers by image analysis. J Immunother Cancer 2018;6(1):20.

References **341**

[161] Chen T, Srinivas C. Group sparsity model for stain unmixing in brightfield multiplex immunohistochemistry images. Comput Med Imaging Graph 2015;46(Pt 1):30–9.

[162] Parra ER, Francisco-Cruz A, Wistuba II. State-of-the-art of profiling immune contexture in the era of multiplexed staining and digital analysis to study paraffin tumor tissues. Cancers (Basel) 2019;11(2).

[163] Ruifrok AC, Johnston DA. Quantification of histochemical staining by color deconvolution. Anal Quant Cytol Histol 2001;23(4):291–9.

[164] Hagos YB, et al. ConCORDe-Net: cell count regularized convolutional neural network for cell detection in multiplex immunohistochemistry images. In: in Medical Image Computing and Computer Assisted Intervention – MICCAI 2019. Cham: Springer International Publishing; 2019.

[165] Malkin K, et al. Label super-resolution networks; 2018.

CHAPTER 16

Overview of the role of artificial intelligence in pathology: The computer as a pathology digital assistant

John E. Tomaszewski

Pathology and Anatomical Sciences, University at Buffalo, State University of New York, Buffalo, NY, United States

Introduction

As this chapter goes to press, two great waves of change are sweeping through the science and the practice of medicine. These are the molecular revolution and computational analytics. These two scientific movements are radically altering the ways in which we approach all aspects of human biology. Previous chapters have discussed machine learning strategies in detail, with an emphasis on deep learning via various kinds of neural networks, ranging from basic convolutional models to advanced architectures such as graphic and capsule networks. In this chapter I will describe the overall framework in which digital and computational pathologies are embedded, summarize these various strategies, and provide examples from the literature that illustrate many of the topics covered. This overview is designed to show how pathology is evolving and will continue to evolve in the age of artificial intelligence (AI). One major thesis is that human pathologists will not be replaced by computers. Instead, the computer will serve as a highly capable "digital assistant," and this synergistic pairing of human intelligence (HI) and AI will ensure that pathology will be a major driver of personalized medicine.

Computational pathology: Background and philosophy
The current state of diagnostics in pathology and the evolving computational opportunities: "why now?"

Diagnostic pathology in the early 21st century is still very much an analog enterprise which uses a work process that was developed in the early 20th century and which has been substantially carried forward into the early 21st century. The processes of

Artificial Intelligence in Pathology. https://doi.org/10.1016/B978-0-323-95359-7.00017-0
Copyright © 2025 Elsevier Inc. All rights reserved, including those for text and data mining, AI training, and similar technologies.

chemically treating and processing tissue, microtomy, staining, and the creation of glass slides, which are viewable with a wide-field optical instrument (the microscope), are essentially the same processes established by early surgical pathologists. The data obtained from these processes yield continuous variables which a human pathologist diagnostician uses to drive classification systems. The output of these classifiers is in turn used by clinicians to support therapeutic decisions. This analog process is cost-effective in that it yields core actionable information in many cases and consumes only a very small fraction of total healthcare costs. These processes, however, have been fundamentally the same for 100 years and are not adapted to the rapidly changing diagnostic environment. There is a growing need for precision diagnostics. High-cost, high-risk therapeutic interventions require diagnostic and theranostic systems which model the most effective therapeutic choices with a high degree of precision and accuracy. The efficiency of the diagnostic/theranostic system has a direct and major impact on the global cost-effectiveness of the entire clinical care system.

Several new opportunities are now converging to make precision diagnostics a 21st-century reality. Big data is becoming available in multiple modes. The molecular revolution, which entered the practice of pathology and laboratory medicine in the 1980s, is now becoming routinely available. Genomics, epigenomics, transcriptomics, proteomics, metabolomics, and microbiomics all offer digital "multi-omic" big data opportunities in research and practice [1]. Likewise, the extension of the electronic medical record to all aspects of medical practice through the US federal "meaningful use" programs in health information technology has opened digital access to the big data of the clinical record through synoptic reporting and natural language processing. Whole slide images of tissue sections are typically multigigapixel in size and as such are a third modal window providing digital access to the big data of complex medical conditions. In tissue-based diagnostics, advances in whole slide imaging (WSI) technology are poised to cross the threshold from purely education and research use to diagnostics. The recent FDA approval of WSI systems opens up enormous opportunities for computer-aided pathological diagnosis. WSI and image analytics are poised to become mainstream methods of histopathological interpretation, yielding faster, more accurate, and cheaper diagnoses, prognoses, and theranostic predictions of important diseases including cancer.

Access to high-performance computing in the medical field has also dramatically increased over the past decade. Graphical processing units and tensor processing units are now available in medical settings. Computing jobs which would previously take days or even weeks to perform in a research setting are now available through cloud computing or even on the desktop. Storage costs have seen dramatic reductions over the past 5 years.

Most importantly, our approaches to and understanding of machine learning have made a dramatic turn since 2012 when AlexNet won the ImageNet Large Scale Visual Recognition Challenge by a large margin using a convolutional neural network (CNN) [2]. With the resurgence of interest in and the success of modern neural networks and data-driven deep learning approaches, the current decade has seen a

Computational pathology: background and philosophy **345**

dramatic acceleration in our opportunities to learn from massive quantities of data with limited labels. The analytics which can be applied is now beginning to match the big data of biology and medicine.

Digital pathology versus computational pathology

It is important to distinguish between the terms *digital pathology* and *computational pathology*.

Digital pathology is best understood as the digital workflow tools which are more focused on the front end of the process. Digital pathology is a dynamic, image-based environment that enables the acquisition, management, and interpretation of pathology information generated from a digitized glass slide. Healthcare applications include primary diagnosis, diagnostic consultation, intraoperative diagnosis, medical student and resident training, manual and semiquantitative review of immunohistochemistry (IHC), clinical research, diagnostic decision support, peer review, and tumor boards. Digital pathology is an innovation committed to the reduction of laboratory expenses, an improvement of operational efficiency, enhanced productivity, and improved treatment decisions and patient care.

Broadly written, the approaches to the computational modeling of complex systems of human disease from big data are termed "*computational pathology*." In 2011, Fuchs and Buhmann [3] opined that "Computational Pathology investigates a complete probabilistic treatment of scientific and clinical workflows in general pathology, i.e., it combines experimental design, statistical pattern recognition, and survival analysis within a unified framework to answer scientific and clinical questions in pathology." Computational pathology focuses on the analytics of extracting data from the output of sensors and turning that data into information and knowledge. This might include both prognostics and outcome prediction or simply a quantitative cataloging of tissue morphologies represented by computation.

Data on scale

The scientific understanding of human structure has stagnated as a largely descriptive endeavor for many decades. Structural science learning related to cells, organelles, and tissues is central to the disciplines of cell biology and pathology. With the advent of machine vision and machine learning, major advances in computing and network power, and the rapid evolution of advanced biophysical systems for cellular imaging, these disciplines are poised to embrace a quantitative paradigm for their work. The data embedded in a cellular or tissue image are deep and full of complex relationships. Thousands of image metrics derived by machine vision can be captured from a field of view and used in the modeling of complex biological or disease system outcomes. Computational advances now offer the promise of enabling the quantitative analysis of human structural data.

Biological data lives on scale. Modern imaging, image analytics, and computational methods provide tools with which to quantitatively mine the data within macroscopic (10^1), microscopic (10^{-7}), and submicroscopic (10^{-10}) images. The ability to mine "subvisual" image features from digital pathology images using machine vision can develop feature data which may not be visually discernible by a pathologist.

There has also been recent substantial interest in combining and fusing radiologic imaging, proteomics, and genomics-based measurements with features extracted from digital pathology images for better prognostic prediction of disease aggressiveness and patient outcome. *Data fusion* is accomplished by computationally combining image data at these levels of scale with data from other modes of examination such as cell and molecular biological, biomechanical, and biophysical analyses. Data fusion offers the opportunity to quantitatively model complex human systems from massive multivariable statements.

We assert that *in the future, medical decision-making will rely on large datasets collected across scale from molecular and imaging analyses, integrated by machine vision and machine learning and that these integrated data will be used to model complex biological and disease systems. These complex computationally supported systems of data collection, integration, and modeling will be vetted by teams of humans with domain knowledge expertise.*

Machine learning tools in computational pathology: Types of artificial intelligence

AI is not really a new concept. The term AI was first used by John McCarthy in 1955 [4]. He subsequently organized the Dartmouth conference in 1956 which started AI as a field. The label "AI" means very different things to different observers. For example, some commenters recognize divisions of AI as "statistical modeling" (calculating regression models and histograms) versus machine learning (Bayes, random forests, support vector machines [SVMs], shallow neural networks, or artificial neural network) versus deep learning (deep neural networks and CNNs). Others recognize categories of *traditional AI* versus *data-driven deep learning AI*. In this comparison, *traditional AI* starts with a human understanding of a domain and seeks to condition that knowledge into models which represent the world of that knowledge domain. When current lay commentators refer to "AI," however, they are usually referring to *data-driven deep learning AI*, which removes the domain knowledge-inspired feature extraction part from the pipeline and develops knowledge of a domain by observing large numbers of examples from that domain.

The design approaches of traditional AI versus data-driven deep learning AI are quite different. The architects of *traditional AI learning systems* focus on building generic models. They often begin with a human understanding of the world through the statement of a prior understanding of the domain (see Fig. 16.1), develop metrics representing that prior, extract data using those metrics, and ask humans to apply class labels of interest to these data. These labels are then used to train the system to learn a hyperplane which separates one class from another. Traditional AI learning

Machine learning tools in computational pathology: types of artificial intelligence **347**

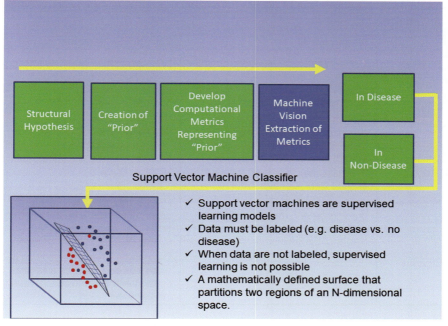

FIG. 16.1

Traditional AI learning pipeline. Traditional AI machine learning in image analytics begins with the creation of a hypothesis about the structures under study. The investigators, using domain knowledge, craft their prior understanding of the structures and develop computational metrics to examine that prior understanding. Features are extracted from the image using these metrics and labeled as coming from disease versus nondisease. A machine classifier (e.g., support vector machines) is then used to create the best hyperplane separating the classes of disease versus nondisease.

systems will often be ineffective in capturing the granular details of a problem, and if those details are important, a traditional AI learning system may model poorly.

A *data-driven deep learning machine learning system*, on the other hand, can capitalize on the capture of fine details of a system, but it may not illuminate an understanding of the big picture of the problem. Data-driven models are sometimes characterized as "black box" learning systems which produce classifications or transformed representations of real-world data but without an explanation of the factors that influence the decisions of the learning system. Traditional and deep learning models are compared in Table 16.1.

Data-driven deep learning AI approaches have limited human-machine interactions constrained to a short training period from human-annotated data and human verification of the classifier output of the learning system. In contrast, in traditional AI learning systems, human experts can provide actionable insights and bring these

Table 16.1 A comparison of traditional artificial intelligence (AI) versus data-driven deep learning AI.

	Traditional AI	Data-driven deep learning AI
Positives	Traditional AI is focused on conditioning the knowledge possessed by humans into models representing the real world.	Data-driven AI involves abstracting knowledge of a target domain simply by observing a large number of examples from the domain.
	Human experts can provide actionable insights and may bring these rich understandings in the form of a "prior" understanding, which can function as an advanced starting point for an AI system.	Data-driven models capitalize on capturing the fine details from the data samples including important latent variables.
Negatives	Traditional AI, with its focus on building generic models, may be ineffective in capturing fine-grained details of the problem from important unrecognized features.	Human interaction with data-driven AI models is often limited to training or verifying predicted outcomes.
		Data-driven AI may function as a black box and sacrifice an understanding of the factors which influence decisions.

rich understandings to the learning system in the form of a "prior" understanding of the domain. A prior can function as an advanced starting point for a deep learning AI system. The broad understanding of the world that humans possess with their reasoning and inferencing abilities, efficiency in learning, and ability to transfer knowledge gained from one context to other domains is not very well understood. Framing data-driven deep learning systems with the human understanding of "what is" offers a way forward for creating partnerships between HI and AI in advanced learning systems. There is a need for "explainable AI (XAI)," which can explain the inferences, conclusions, and decision processes of learning systems. There is much work that needs to be done to bridge the gap between machine and HI.

The need for human intelligence-artificial intelligence partnerships

Pathologists, confronted with a future filled with massive and ever-expanding computing power, properly ask the question, "...will a machine diagnostician replace a human diagnostician and if so when?". No one knows the answer to this question, but a glimpse into the world of chess offers some useful guideposts.

The chess grandmaster Garry Kasparov describes his years as a grandmaster as a time when computational chess was quickly developing and challenging human player expertise [5]. In the 1970s and 1980s, grandmasters successfully outplayed the computers of the day. In the 1990s, the competitions were more matched. Kasparov noted that the things that computers do well are often the tasks where humans are weak and that the converse is also true. This concept caused Kasparov to ask whether the goal of computational chess play was to pit man against computer or rather was to play the highest level of chess possible as a partnership of man and machine. Kasparov went on to observe a chess playing experiment in which the types of machine–human pairings varied. He saw that a weak human player paired with a machine with a better process was superior to a strong computer alone, and remarkably, also superior to a strong human player working with a machine which had an inferior process.

In the evolving world of computational pathology, the fear of artificial machine diagnostic intelligence displacing HI must be balanced with the opportunity of providing patients with more precise and accurate diagnostic, prognostic, and theranostic statements. What if, as in Kasparov's chess playing experience, in computational pathology, a partnership of weak human diagnosticians + machines + better processes were better than strong human diagnosticians + machines + worse processes? The field of computational pathology would be immeasurably extended in its global availability and somewhat more democratized in its practice. The key to next-generation diagnostics may, in fact, be getting machines and humans to interact using the best processes. I would opine that this system approach is native to the practice of pathology and laboratory medicine and that our specialty is the best positioned in all of medicine to explore the AI revolution by developing robust HI-AI partnerships in the domains of diagnostics, prognostics, and theranostics. Dr. Scott Doyle, a computational pathology investigator at the University at Buffalo, State University of New York, opines that "The true value of human diagnosticians is not that they are valuable because of their ability to do poorly what artificial intelligence robots can do easily. In fact, they are valuable because of all the things they can do that robots cannot do such as disparate data integration, ingenuity, serendipity, learning, experimentation, and value judgments. These are all central parts of the practice of pathology and laboratory medicine. In the end pathologists should/will focus on these things and leave the robotic tasks to the robots" (unpublished communication).

Human-transparent machine learning approaches

Human-transparent machine learning systems seek to have humans understand AI cognition. This means that humans must be able to understand the dynamics of the machine learning system. In human-transparent machine learning, the machine learning system may have human-like sensing inputs. Most importantly, humans are able to trust the AI output of human transparent learning systems. In the computational pathology domain, this means that the outputs of machine learning systems must be vetted against the big 360° view of the complete clinical context. In this construct, the digital assistance which results from any computational pathology systems is used only after cross-referencing them with all of the meaning understood by human sensors. These human understandings are gleaned from the history, physical

350 CHAPTER 16 Overview of the role of artificial intelligence in pathology

examination, clinical laboratory, radiology, surgical pathology, cytopathology, molecular pathology, and quality assurance information systems. In other words, pathologists will need to learn how to work with machine learning systems as senior partners who understand the big picture of patient care diagnostics.

Explainable artificial intelligence

The concept of XAI is that machine learning is understood by human operators and that through this understanding, a bilateral trust relationship is established between humans and machines. XAI contrasts sharply with the "black box criticism" of deep learning. XAI is very important when machine learning systems impact social systems. As diagnostics is a central function of medicine, which in turn impacts all aspects of the societal good of health and wellness, XAI is a particularly desirable form of AI in computational pathology. Regulators of clinical laboratories and the diagnostic process should have a strong interest in XAI as it offers transparency and real-world rationales which make algorithmic outputs testable and subject to quality assurance processes. XAI guards against AI "cheating," in which a learning system tracks surrogate labels but is not truly informative. For example, a data-driven deep learning AI system that used histological images tagged with ICD10 codes could never be relied on to distinguish cancer from noncancer using pixel data from digital virtual slides. Or even more simply, if the slides with cancer in a training dataset had been marked with a red dot by a previewer and the machine learning system learned to look for a "red dot" to classify the dataset into cancer versus not-cancer, the resulting classifier might be 100% effective in the training data but 0% extensible to other test datasets.

Deep learning systems may be adapted to XAI through a technique called layerwise relevance propagation, which determines the features in an input vector which contribute most strongly to a neural network's output [6]. Exposing the output layers or even the early activation layers of neural networks to human observers for error correction or other annotations may also be used to support XAI (Fig. 16.2).

Cognitive artificial intelligence

Cognitive computing seeks to simulate the functioning of the human brain with software and hardware that senses, responds to stimuli, and reasons. Cognitive AI systems stimulate human thought in finding solutions to broad and complex problems. Cognitive AI systems should be adaptive, interactive, iterative, and contextual. Cognitive learning systems are adaptive in that they learn in real time or near real time and they learn as information changes. Cognitive systems are interactive with users. Cognitive systems help define problems by asking multiple questions or finding additional sources of information to make the statement of a problem less ambiguous. Cognitive learning systems understand, identify, and extract contextual elements such as meaning, syntax, time, location, appropriate domain, regulations, user profile, process, task, and goals. Cognitive systems utilize multiple types of input data including human perceived sensory data. Cognitive AI self-learning

Human-transparent machine learning approaches

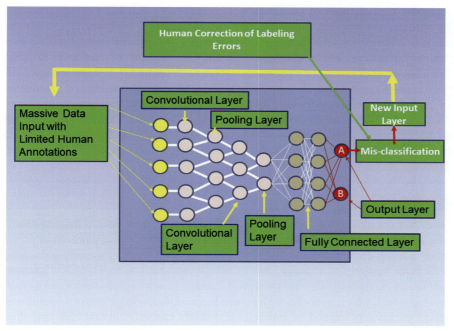

FIG. 16.2

Opportunities for human interaction with deep learning systems. In data-driven deep learning systems the output classifications of images are derived from numerous calculations within "hidden" layers, often without an explanation of the factors which influence the decisions of the learning system. Deep learning AI systems such as this convolutional neural network input large amounts of data with limited human labels. Convolutional and pooling layers may feed fully connected layers which in turn drive an output classifier. Errors in the output classifier are opportunities for human intelligence to correct the system and feed these new labels back into the input of the system. Other opportunities for human interaction with the learning system may exist in the human understanding and annotation of the activation layers of the convolutional neural network.

systems weigh context and conflicting evidence by using tools such as data mining, pattern recognition, and natural language processing. Cognitive AI systems in healthcare are thought of by some as a tool for physicians to bridge the machine-doctor-patient chasm [7]. An example of cognitive computing is IBM Watson for Oncology, which has been used to derive ranked options for treatment of cancer.

Human-in-the-loop

Human-in-the-loop learning models require human interaction. Human-in-the-loop systems allow humans to change the output of the learning systems. Human-in-the-loop simulators always have human input as part of the simulation, and humans

352 **CHAPTER 16** Overview of the role of artificial intelligence in pathology

influence the outcomes of the simulation exercise such that the outcomes may not be exactly reproducible. Human-in-the-loop simulations, however, allow for the identification of model shortfalls which may not be apparent before testing in a real-world setting. Flight simulators may be an example of human-in-the-loop learning. Human-in-the-loop approaches have been applied to histopathology. Lutnick et al. [8] used a human-in-the-loop strategy for data annotation and the display of neural network predictions within a commonly used digital pathology whole slide viewer. This strategy reduced the annotation burden in that the segmentation of human and mouse renal microcompartments was repeatedly improved when humans interacted with automatically generated annotations throughout the training process.

One-shot learning

Humans have the capacity to learn object classifications from limited training examples or even no examples. Humans are able to make use of information from previously learned classes to learn new ones. In a one-shot learning system, a machine uses prior learning to classify objects newly presented to it. One-shot learning emphasizes knowledge transfer from prior learning to new circumstances and as such is an approach which resonates with human learners. Knowledge transfer could be of model parameters learned by training on multiple examples in one class and then applying them to a new object class. Knowledge transfer may be done by sharing features across classes. Mutual information among learned objects could be applied to learning a new class of object. Knowledge transfer may also occur through the transfer of context. For example, knowing the context of camera geometry used to classify one object may help classify a new object.

The computational approaches described above can be applied to any laboratory data used in the clinical setting, and these methods are opportunities to innovate in all of the disciplines in pathology and laboratory medicine. Cabitza et al. [9] outlined the opportunities of machine learning in laboratory medicine, and these authors anticipate a flood of applications in the near future. In the following sections, however, we will mostly focus on the use cases of computational pathology in cellular and tissue-based image data.

Image-based computational pathology
Core premise of image analytics: What is a high-resolution image?

A foundational hypothesis supporting the concept of cell and tissue image-based computational pathology is that a high-resolution image is *a self-organizing set of data* which uniquely represents all of the genes, all of the molecules, and all of the cells in that scene at one point in time. As a self-organizing piece of data, the image can only uniquely have the phenotype which it presents. An *image is what it is for very specific reasons*. Those reasons are the relationships among the genomics, epigenomics, proteomics, metabolomics, and all the "omics" which go into

Image-based computational pathology

353

making that image. A high-resolution image is a window into the *relationships* among all of the genes, all of the molecules, and all of the cells in the scene at one particular point in time.

The targets of image-based calculations

The targets for algorithmic computation in image-based pathology include (1) the pixels in the image, (2) cells and organelles, (3) tissues and tissue elements as well as the complex relationships among pixels and cellular/tissue elements. Each of these targets may be amenable to many image analytic approaches, but some approaches are better suited for one or another of the targets. Machine learning systems driving decisions can be applied to the data extracted from any of these targets. The following are some examples of computational schemes for examining each of these targets.

Pixels: Pixels are the basic building blocks of digital images. They are a rich source of information in the color and intensity domains. Image texture features are metrics calculated from pixels that contain information about colors or intensities. Texture features can be viewed across an entire slide and leveraged as input data for machine learning systems, or they may be sorted into feature representations of cellular and tissue objects (as below). Thousands of texture features can be calculated from a histopathological image. The features can be extracted in cost-efficient unsupervised ways. Texture features data can be approached statistically. For example, an edge detection algorithm could examine the gradient magnitude and the gradient direction of the pixels and yield an "edgeness" metric for an entire image or segments of an image.

Cells and organelle segmentation: Cells and organelles are obvious biological units of interest to pathologists. Gathering quantitative information about cells is central to a computational pathologist's modern exploration of Virchow's cellular theory of disease. Segmentation of cells, nuclei, membranes, and organelles from image metrics can be accomplished by a variety of whole scene segmentation approaches, some of which include texture features, size and shape features, densitometric features, and chromatin-specific features [10].

Tissue and tissue elements and the complex relationships among pixels and cellular/tissue elements: The grand challenge for analytic approaches to computational pathology is to capture the image-embedded information which is informative in modeling complex disease systems such as disease progression and the predictive modeling supporting therapeutic choices. Much of this information is embedded in the spatial relationships between cells. Spatial arrangement features can either be explicitly harvested or discovered with deep learning systems. Graph embedding theory is an example of a method which is well adapted for exploring the relationships among pixels and cellular and tissue elements in high-resolution histopathological image data [10]. Graphs can be constructed for modeling different states of tissue and to distinguish these different states from others by calculating metrics on the graphs and classifying their values.

First fruits of computational pathology: *The evolving digital assistant*

The concepts of AI-HI partnerships and human-transparent machine learning approaches raise an aspiration for a *digital assistant* to help pathologists in their daily work. Such an assistant would reduce repetitive tasks which are subject to error and free up pathologist time for complex medical decision-making. A *digital assistant* would reduce the organization and triage work, search and identify functions, and make data quantification (particularly from images) more accessible. The *digital assistant* would generally help pathologists to practice at "the top of their licenses." This vision does not yet exist in the reality of current state clinical practice, but it is on the way. The following are just some of the anticipated facilitating functions projected from the research literature into which the *digital assistant* may evolve.

The digital assistant for quality control

All laboratories have robust and active quality control and quality assurance at the core of their workflow. In modern-day clinical chemistry and hematology laboratories, for example, the information transfer about individual or batch value comparisons to both internal and external reference values is continuous and real time. There is a conversation between instruments and the laboratory professional human operators, which is always engaged and bidirectional. Heuristics are used to flag significant deviations from acceptable norms.

In the current state, anatomical pathology, laboratory quality control, and quality assurance are much more qualitative and discontinuous. As anatomical pathology moves to a digital pathology workflow, the need for precise quantitative quality control and quality assurance systems is becoming apparent. Computational pathology algorithms, particularly in unsupervised architectures, are susceptible to interpreting the noise of artifacts as true biological signals. Quality parameters which matter can be subtle and easily overlooked. Artifacts can adversely affect the performance of computational pathology machine learning systems [11,12]. In an integrated digital pathology workflow, using WSI, continuous, real-time, and precise quality image metrics are needed to understand and normalize the preanalytic variance inherent in glass slide preparation.

Lab accreditation organizations are keenly aware of the need to address quality assurance in digital and computational pathology. The College of American Pathologists has recently performed a metaanalysis of the literature and published guidelines on the Quantitative Image Analysis (QIA) of Human Epidermal Growth Factor Receptor 2 Immunohistochemistry for Breast Cancer [13]. They recommend that "to improve accurate, precise, and reproducible interpretation of HER2 IHC results for breast cancer, QIA and procedures must be validated before implementation, followed by regular maintenance and ongoing evaluation of quality control and quality assurance. HER2 QIA performance, interpretation, and reporting should be supervised by pathologists with expertise in QIA."

Vendors of whole slide scanning devices have also recognized that the quality of the rendered whole slide images impacts the ability of pathologists to diagnose. When scanning the same slide on the same scanner at different times, the WSI appearance may contain subtle differences even when viewed on the same display because of variances in the scanner characteristics and/or other factors such as temperature or mechanical shifts.

Color reproduction has been approached by commercial developers through the construction of standardized slide phantoms matching the colors in the scanned images to the actual slide color [14]. The colors observed in phantoms are, however, not the same as those viewed in tissue matrix. Alternatively, commercial systems have used quantitative image quality assessment based on parameters such as sharpness, contrast, brightness, uniform illumination, and color separation to evaluate the scanner performance. In a study on HercepTest slides [15], the authors note that this computational approach is independent of image content or test specimen.

Open-source approaches to the development of the digital assistant for quality assurance (QA) processes in computational pathology are also evolving. Janowczyk et al. have recently proposed an open-source quality control tool for digital pathology slides, which they call "HistoQC" [11]. Their system offers a modular QC application, which employs a suite of image metrics (such as color histograms, brightness, and contrast), image features detectors (e.g., edge and smoothness), and supervised classifier systems (i.e., a drawing tablet with pen annotation) to facilitate QA work. The user interacts with the system through a Python-based pipeline. Relevant output images of artifacts are created and presented with metadata and then sent to a data analytic tool. These authors have developed an HTML-5 user interface that allows for real-time visualization.

The digital assistant for histological object segmentation

The segmentation of histological objects by computational methods is an active area of research in pathology. The following describes some of the ways in which a variety of approaches have been employed to support clinically relevant image interpretation in pathology.

Nuclei

The enumeration and categorization of nuclei and nuclear features is a central task in many types of histopathological diagnoses. Nuclear detection methods are numerous. Xu et al., in a paper describing a deep learning strategy compiling examples of 23 different handcrafted feature-based approaches, include color-based, edge-based, contextual information-based, and texture-based designs for nuclear segmentation [16]. Nuclear shape and orientation, as measured predominantly by tensor information, have been used in tissue microarray images to model nuclear pleomorphism and to be independently predictive of poorer survival in breast cancer [17]. The challenge with most of these approaches is in their ability to generalize across multiinstitutional datasets encompassing the loud "noise" of the preanalytic variance, which is resident in slide preparation and scanning. Recently, deep learning techniques have been applied to the problem of nuclei segmentation, feature extraction, and classification. Breast cancer nuclei classification in histopathology has been examined by comparing several supervised learning methods with deep learning systems. The deep learning systems all outperformed the supervised machine classifier algorithms [18]. Other examples of deep learning strategies applied to nuclei detection cancer images have included (1) stacked sparse autoencoders used to learn high-level features from just pixel data of nuclei in breast cancer [16]; (2) a deep contour aware network (DCAN) using multilevel contextual features in a fully CNN with an auxiliary supervision scheme in the segmentation of nuclei from glioblastoma [19]; and (3) the use of cycle generative adversarial networks (cycleGANs) deep learning to improve multiorgan nuclei segmentation to almost 95% [20].

Mitoses

The recognition, enumeration, and classification of mitotic figures in histopathological images are central tasks in many histopathological grading systems. Pathologists often use the data from multiple z-planes of focus in combination to distinguish mitotic figures from other condensed chromatin structures. As WSI often does not provide data from multiple z-planes, the detection of mitotic figures in standard 2D whole slide images has become a high-profile contest among investigative groups in digital pathology. Many competitive grand challenges have compared the effectiveness of a variety of computational approaches to the detection of mitoses. The winner of one of these grand challenges, captioned in the paper by Ciresan [21], used a patch approach at $40\times$ resolution with deep neural network learning. Janowczyk and Madabhushi [22] have adapted this approach, used a smaller patch size at $20\times$ resolution, and leveraged segmentation of regions of interest (ROIs) using a blue ratio segmentation protocol to enrich for ROIs containing mitotic figures. Wang et al. [23] have sought to address mitosis detection by presenting a convergent approach to combine domain-inspired features with deep learning features to detect mitoses. They showed that this integrated approach yielded superior detection accuracy compared to deep learning or handcrafted feature-based strategies alone.

The recognition, enumeration, and classification of mitotic figures are likely to be important tasks in diagnostic histopathology which will require the partnership of AI and HI to bring an efficient and safe computational tool to the clinical diagnostic market.

Tumor-infiltrating lymphocytes

In a previous chapter, Saltz and colleagues provided a detailed rationale for the importance of tumor-infiltrating lymphocytes (TILs) and gave examples involving the classification characterization of functionally relevant subpopulations of these lymphocytes and their positive and/or negative roles. For the reasons described, the detection, enumeration, and relationships among TILs are topics of great interest, particularly for investigators interested in immuno-oncology. Klauschen et al. [24] have recently reviewed many computational approaches which have been used to analyze TILs. TILs have been most commonly enumerated using semiquantitative scoring systems. Standardized TIL scoring has been used to good effect in clinical trials of breast cancer. Scoring systems, however, are subject to interobserver variance. The precise and time-efficient quantification of TILs is a needed tool for both research and clinical work. Computational approaches to the tasks of TIL characterization are required to enable the conduct of precision medicine clinical trials in oncology and autoimmune diseases.

Traditional image analysis approaches to TIL identification rely on supervised or semisupervised systems. Through the identification of edges, cellular objects of interest are segmented, and features are extracted from these objects, which allows for the classification into different cell types. Thresholding, watershed, level set, color-space clustering, and morphological shape-based algorithms have all been used for the identification of TILs. These computations often require a priori knowledge of lymphocyte morphological variance, but their workings are explainable to the investigators. Relatively small datasets are used to train these handcrafted systems. Tuning the algorithmic parameters of these computational systems to handle the many cell types which need to be separated from lymphocytes, however, can be a daunting task.

Deep learning has recently been used in the detection of TILs. Janowczyk and Madabhushi [22] have used CNNs and patch classification schemes to identify lymphocytes in standard H&E histology. Larger training sets are needed to subsume the morphological variance of scenes and preparative variance, but the deep learning system in essence tunes itself. The exact nature and biological significance of the learned features may, however, be opaque. The understanding of the immune subtypes resident in host response networks may be critical to choosing effective immune therapies. Other studies have employed deep learning methods using different patch classification, boundary detection, and segmentation algorithms on tissue sections labeled immunohistochemically for lymphocytes and lymphoid subsets. The combination of molecular labels with deep learning systems offers the potential for the understanding of biologically explainable networks. Such methods could be adapted to assays designed for the examination of clinically relevant immune networks such as the relationships of immune checkpoint inhibitor molecules on host response and tumor cells.

Glands and acini

The histopathological definition of "gland" varies with the organ. Morphological deviation from normal gland structure is used by pathologists to define reactive conditions, dysplastic precancer, cancer, and cancer grade. Obviously, the segmentation of glandular objects is the first step when considering the computational analysis of diseases which are defined by gland disorganization.

Gland segmentation in colon histopathology has significant literature. The gland segmentation in colon histology images (GlaS) grand challenge [25] in computational pathology asked participants from the computer vision and medical imaging research communities to develop and compare gland segmentation algorithms of benign colonic glands and colon cancer glands. Accuracy of the detection of individual glands; volume-based accuracy of the segmentation of individual glands; and the boundary-based similarity between glands and their corresponding segmentations were the metrics for comparisons. Gland detection accuracy was performed using the F1 score (2× precision × recall/precision + recall). Volume-based segmentation accuracy was evaluated using the dice index, which is a measure of agreement or similarity between two sets of samples. In this competition, these sample sets included a set of pixels corresponding to a ground truth object (G) and a set of pixels belonging to a segmented object (S). Boundary-based segmentation accuracy between ground truth (G) and segmented object (S) was evaluated using object-level Hausdorff distances. In general, two broad approaches to gland segmentation were employed by the competitors. The first approach was to start by identifying pixels belonging to glands which were then grouped to form a separate spatial glandular object. The second approach began with candidate objects which were then classified as glands or nonglands. All of the methods which were based on CNNs used the first approach. Only one method followed the second approach whereby candidate objects forming part of the gland, namely lumens or epithelial boundaries, were identified first and then classified into different types, followed by full gland segmentation. Overall, the pixel-based CNN entries had the strongest performances. A novel DCAN [19] (Chen et al., read also below) was the overall winner of the challenge.

The segmentation of prostatic acini is a problem analogous to that of colonic gland identification. Traditional handcrafted image analysis techniques begin with a prior concept of acinar structure. Monaco et al. [26] conceived of a prostatic acinus as a lumen surrounded by a collar of epithelium. To distinguish between benign prostatic glands and prostatic adenocarcinoma, a Markov iteration was used to group and separate microacini from benign glands. Separation of Gleason patterns 3 versus 4 was facilitated by this schema. Singh et al. [27] used a combination of pixel and object-level classifiers, which incorporated local and spatial information for consolidating pixel-level classification results into object-level segmentation of acini. As alluded to above, Chen et al. [19] have described a DCAN that provides a unified multitask learning framework that examines contextual features by using a fully convolutional network. The fully convolutional network takes an image as input and outputs a probability map in one forward propagation. In this system, object features (for example, texture and color) are integrated with contour information. The result is a system which addresses the difficult problem of separating touching objects. The system is extensible to many types of objects including benign and malignant colonic glands and nuclei.

First fruits of computational pathology: *The evolving digital assistant* **359**

The digital assistant in immunohistochemistry

The in situ IHC analysis of tissues for molecular markers is central to the moderate practice of pathology. IHC biomarkers are used for tissue classification tasks, as grading tools, and as predictive markers. The qualitative/semiquantitative analysis of IHC markers can have significant gaps in precision (TP/TP + FP) and recall (TP/TP + FN). Computational methods have been explored for many years as possible tools for improving the operating characteristics of IHC tests. Gavrielides et al. [28] examined the use of computationally assisted image evaluation in the scoring of HER-2 IHC. These authors found that a computer-assisted mode which provided an HER-2 reference image along with a corresponding feature plot of membrane staining intensities and membrane staining completeness improved both interobserver and intraobserver agreement when scoring HER-2.

Machine learning systems are being developed to analyze IHC labels. Chang et al. [29] have addressed image cytometry multiplex IHC data by comparing manual gating of lymphoid subsets to clustering algorithms and sparse representations to yield biologically interpretable subset populations.

Vandenberghe et al. [30] compared handcrafted features and machine classifiers (SVMs and random forest) to a deep learning CNN (ConvNets) in the analysis of HER-2/neu staining in breast cancer. Overall the accuracy of the neural network deep learning methods was somewhat better than the handcrafted feature classifiers. Interestingly, none of the methods were particularly good in their accuracy of scoring 2+ cells, where interobserver variance and clinical significance are known to be great.

Khosravi et al. [31] have compared six different deep learning systems using transfer learning for the scoring of IHC biomarkers in bladder and breast cancer. Transfer learning is the pretraining of a network architecture on a very large dataset and the use of that trained model for new classification tools for a data asset with a limited size. In this study, the transfer learning strategies included pretraining of the network as a feature extractor and the fine-tuning of a pretrained network. The accuracy retrieval curves for the prediction of biomarker score ranged from 72% to 99%.

The digital assistant in tissue classification

The identification of cancer and its separation from reactive conditions are fundamental high-order functions of histopathologists. The two sequential activities of the pathologist workflow in executing these functions are the identification of candidate regions of interest and the subsequent classification of those regions into meaningful clinicopathological classes. These processes are true in any subspecialty of oncological surgical pathology. Breast cancer has been of particular interest to investigators studying computational pathology approaches to cancer diagnosis.

Fondon et al. [32] reviewed the use of traditional machine learning in breast cancer diagnosis. These authors approached the problems of automatically identifying and classifying breast histopathological digital images into one of the four classes:

normal, benign lesion, in situ carcinoma, and invasive carcinoma. The functions of preprocessing, feature extraction from nuclei, regions, textures, and the use of these features to drive a variety of machine classifier tools are described. The SVM classifiers were the most efficient, with accuracy of up to 76%.

Deep learning approaches to breast cancer diagnosis using CNNs are being vigorously explored. Araujo et al. [33] have used CNNs on digital images from the Bioimaging 2015 breast histology classification challenge to separate into the same four classes of normal, benign lesion, in situ carcinoma, and invasive carcinoma. The overall sensitivity of the CNN in finding the carcinoma classes was 80%.

Cruz-Roa et al. [34] have used CNN deep learning to automatically identify invasive breast cancer on whole slide images from the Cancer Genome Atlas and from whole slide images collected from three contributing institutions. They found a positive predictive value of 72% and a negative predictive value of 97% for identifying invasive breast cancer. CNN classifiers outperformed visual feature classifiers (color and intensity, color histograms, shape index histograms, Haralick features, and graph-based features) in this study. Bejnordi et al. [35] have extended the CNN approach on whole slide breast images by using context-aware stacked CNNs for a similar task of separation of the classes normal/benign, ductal carcinoma in situ, and invasive carcinoma. In their "context-aware" approach, these authors first trained a CNN using high-pixel resolution information to classify the tissue into different classes. To incorporate more context, they fed much larger patches to this model at the time of testing. The output of this first system is then the input to a second stacked system, which uses the compact informative representations of the first model together with the information of the surrounding context to learn global independence of structures in different lesion categories. In this three-class task, they achieved an accuracy of 0.812.

The digital assistant in finding metastases

The tasks of searching for and identifying cancer metastases in lymph nodes are critical but time-consuming work for diagnostic pathologists. In breast cancer, the definition of nodal status by histopathology is ground truth for the assignment of nodal status. Current clinical guidelines codify the sizing of lymph node metastases into pN 0 (i-): No regional lymph node metastases histologically, negative IHC; pN0 (i-): Malignant cells in regional lymph node(s) no greater than 0.2 mm and no more than 200 cells (detected by H&E or IHC including isolated tumor cells [ITCs]); pN1mi: Micrometastases (greater than 0.2 mm and/or more than 200 cells, but none greater than 2.0 mm); and pN1 macrometastases at least 1 tumor deposit greater than 2.0 mm. Steiner et al. [36] have used a scoring system for gauging how difficult it is for a pathologist to distinguish among these classes which they term "obviousness scores." In their study, it was found, not surprisingly, that the obviousness score is high for macrometastases and low for identifying isolated tumor cells. These authors also examined the impact of deep learning algorithmic assistance, which they named the LYmph Node Assistant (LYNA), on the obviousness score. LYNA improved the obviousness score for micrometastasis detection. The impact on assignment of the other classes was less apparent. The average review time was also decreased for micrometastasis detection with LYNA, marginally improved for a negative class assignment, and not significantly impacted for the assignment of macrometastases or isolated tumor cells.

A recent competition (CAMELYON16) [37] to develop algorithmic solutions for the automated detection of sentinel lymph node metastases had 32 algorithms submitted. The areas under the curves (AUCs) for the receiver operating characteristic curves ranged from 0.556 to 0.994. The top five algorithms had a mean AUC comparable to the study pathologists' interpretations in the absence of time constraints.

Such studies offer the real possibility that the algorithmic digital assistant may provide a valuable prescreening function in histopathology in future diagnostic workflows.

The digital assistant in predictive modeling and precision medicine

Apart from substantially aiding the pathologists in decision-making, the use of computational imaging tools could enable the creation of digital imaging–based companion diagnostic assays that would allow for improved disease risk characterization. Unlike expensive molecular-based assays that involve destroying the tissue and invariably capture genomic or proteomic measurements from a small part of the tumor, these digital imaging-based companion diagnostic tests could be offered for a fraction of the price and could enable the characterization of disease heterogeneity across the entire landscape of the tissue section. Both traditional AI and deep learning approaches for the modeling of patient outcomes and response to therapy have been explored in a variety of neoplastic and nonneoplastic conditions.

Lu et al. [17] have studied the prediction of survival in early-stage estrogen-positive breast cancers using a traditional AI approach. These authors examined tissue microarrays for 615 features related to nuclear shape, texture, and orientation. They identified the top 15 quantitative histomorphometric features which in combination were used to create a linear discriminate analysis classification of the probability of long-term versus short-term disease-specific survival. Features relating to the heterogeneity of nuclear orientation dominated in this model. The authors, in multivariate analysis controlled for tumor stage, found that their model was strongly independently predictive of patient's survival in estrogen receptor (ER)-positive and lymph node-negative breast carcinomas.

Bychkov et al. [38] took a deep learning approach to the prognostic modeling of colon cancer outcomes from standard histopathology. These authors used an image analysis pipeline which included long short-term memory (LSTM) networks that allow the learning system to detect and memorize image tiles, which encode relevant and contributory morphologic information and disregard irrelevant image titles. Through this architecture, the authors attempted to find units which performed biologically meaningful discriminations of tissue patterns. They hypothesized that if the system observed tiles which contained information regarding disease, then some memory cells within the network will learn this pattern, aggregate it into memory, and propagate the understanding through the network. The authors used LSTM networks to examine tissue microarrays of colorectal carcinoma. They found that the LSTM model provided a strong hazard ratio for disease prognosis as compared to visual scoring but that their LSTM system did not outperform the Dukes staging system for colorectal carcinoma prognosis.

Deep learning classifiers have also been used in predicting outcomes of nonneoplastic conditions. For example, in the arena of cardiac pathology, Nirschl et al. [39] used deep CNNs to predict clinical heart failure from H&E-stained whole slide images with 99% sensitivity and 94% specificity.

The clinical pathology laboratories of health systems offer robust data basis of well-organized quantitative data representing the chronological patient record. Cabitza and Banfi [9] review multiple applications of machine learning to clinical laboratory data for the diagnosis, prognosis, and predictive modeling of a variety of conditions. These authors predict a deluge of investigations using clinical laboratory data and machine learning in the near future.

The digital assistant for anatomical simulation learning

Just as the digital revolution has required a robust infrastructure, so also does the learning from and teaching of complex 3D subjects like anatomy. Pedagogically relevant programs to develop, test, and scale data-intensive, bandwidth-heavy, digitally enhanced collection and visualization systems for anatomical data are critical to UME, GME, and CME training in radiology, surgery, and pathology. Simulation learning depends on computational approaches to the 3D structure of human gifts, large animal models, radiological images, and pathological specimens. 3D Slicer (http://www.slicer.org) is an open-source software platform for the registration, interactive segmentation, visualization, and volume rendering of medical images and for research in image-guided therapy [40]. This sort of computational analysis of structure can integrate biological and digital phantoms within the pedagogic experience. This juxtaposition can allow for robust surgical simulation programs, which integrate the use of artificial phantoms, digital phantoms, and biological phantoms in preclinical surgical investigation, teaching, and training. This combination of the core activities of teaching and investigation in the structural sciences can offer a laboratory where students can learn in the new environment of quantitative structural science.

The digital assistant for image-omics data fusion

When trying to understand the cell biology systems which underpin all pathological interpretations, the data from different modes of examination are best not considered in isolation. Rather, these data are streams (or perhaps torrents) which integrate and literally "fuse" into robust representations of these complex cell and tissue systems. Both image and molecular 'omic (genomic, proteomic, metabolomics, etc.) data are

First fruits of computational pathology: *The evolving digital assistant*   **363**

dense and heterogeneous. Image data are strong at probing the relationships among features. Reductionist 'omic data are strong at probing causal pathways. The frameworks in which image data and 'omic data are represented are often quite different. We know digital image data as pixel features and values and 'omic data as molecular structures, molar quantities, and frequencies. It is not intuitively evident that these very different modes of data can live in the same space. Computational pathology and AI approaches, however, now allow for the quantitative fusion of these diverse cross-modal data streams in support of the modeling of complex biological systems.

There are several approaches to combining data. One is that of combining interpretations. In this method, there is an aggregation of the decisions made from classifiers. Each classifier corresponds to one particular data stream. This is quite common in biology and medicine. A combination of interpretations may, however, not be able to leverage all the power of the combined data streams because information is often lost in the process of converting from feature vectors to class labels.

A concatenation of high-dimensionality feature vectors is an inclusive approach, which subsumes all the available features; however, this approach invokes the "curse of multidimensionality," whereby the excess of features compared to patients may result in a classifier system which is overfit. Dimensionality reduction approaches such as principle component analysis (PCA) project feature vectors into a unified low-dimensionality eigenvector space. PCA assumes that the system data are contained within a linear space. Most biological systems, however, are nonlinear. Nonlinear dimensionality reduction approaches can be of assistance in addressing such complex systems. There are numerous algorithms for manifold learning and nonlinear dimensionality reduction. Some of these algorithms simply function as visualization tools, while others map data from high-dimensionality spaces to low-dimensionality spaces. Data may be combined by creating transformed representations of each data stream before combining the transformed representations. Data streams are projected (embedded) into a homogeneous metaspace, where all data are represented at the same scale. Embeddings may be combined in many ways. For example, one strategy is to combine embeddings by using high-dimensionality kernels. Kernels are dot product representations of each modality, which can be combined to create a fused representation of heterogeneous data. Using another method, Lee et al. [41] examined the effectiveness of fused data to predict biochemical recurrence of prostate cancer after radical prostatectomy. These authors employed a novel technique, which they termed supervised multiview canonical correlation analysis (sMVCCA). Canonical correlation analysis (CCA) is a multivariate statistical method which seeks to find a linear subspace in which the correlation between two sets of variables is maximized. Using sMVCCA, Lee et al. created an integrated feature vector composed of quantitative histopathological features and proteomic features obtained from tandem mass spectrometry of prostate cancer from radical prostatectomies. Kaplan-Meier analysis showed improved biochemical recurrence-free survival prediction using the sMVCCA fused data classifier as compared to histology or proteomic features alone.

In another example, Savage and Yuan [42] presented a new tool for selecting informative features from heterogeneous data types and predicting treatment response and prognosis which they have named FusionGP. This is a Bayesian nonparametric method for integrating multiple data types. The relationships between input features and outcomes are modeled through a set of (unknown) latent functions. The latent functions are constrained using a set of Gaussian priors, and sparse feature selection is applied. The most strongly selected molecular features were evaluated in gene ontology for dominant core processes at work. These authors examined a cohort of 119 ER-negative and 345 ER-positive breast cancers to predict two important clinical outcomes: death and chemoinsensitivity by combining gene expression, copy number alteration, and digital pathology image data. For the prediction of disease-specific death, the molecular data were most informative for the ER-negative tumors. Interestingly, image features outperformed the molecular data in ER-positive tumors.

The digital assistant of the future might present the diagnostician with a variety of data fusion options to assist the HI-AI partnership team in identifying the best precision diagnostic, prognostic, and predictive statements for a given patient.

Artificial intelligence and regulatory challenges

The adoption of computational methods into the diagnostic workflow will require an aligned combination of technology, economic value, and regulatory oversight to insure a safe process. The FDA employs a risk-based paradigm for medical device classification. "Premarket approval" (PMA) is required for Class III devices, which have no comparison to a predicate. A "510(K) Premarket notification" is a path to market for new devices which are substantially equivalent to predicate devices. In certifying image-based pathology systems, the FDA has broad experience in the regulation of image-based instrumentation driven by algorithms. The clinical use of algorithms in image-based assays in pathology already exists in several domains including automated hematology analyzers, chromosome analysis, FISH (fluorescence in situ hybridization) enumeration systems, urine sediment analysis, gynecological cytology, and IHC for predictive markers such as HER2/neu, ER, and progesterone receptor (PR).

When considering the regulation of market entry for deep learning AI systems in image-based pathology, the new regulatory challenge is to provide access to these powerful systems and yet constrain their application to use cases, where patient safety can be assured. Clearly, new approaches to the regulatory evaluation of machine learning systems are required.

The understanding of the relationship between digital image data and the human perception of image data in pathology is one challenge. Gallas and Gravrielides [43] from the FDA are creating an evaluation program which develops information regarding which technical characteristics of an image-based system are important; learning how these technical characteristics impact pathologists' interpretations;

and understanding how these technical characteristics can be measured. These authors and their colleagues have created an "evaluation environment for digital and analog pathology (eeDAP)." They anticipate that this be used as a "Clinical Outcome Assessment tool used in reader studies for A premarket submissions (PMA or 510k de Novo) to compare the accuracy or reproduce ability of pathologist evaluations of digital images on a display to those of glass slides on a microscope. The pathologist evaluations of outpatient tissue are the clinical outcomes. The accuracy or reproducibility is the clinical outcome assessment; this assessment reflects image quality." Such tools will be important in examining sources of discrepancies between pathologists for classifying different types of diseases, for developing a panel of histopathological patterns and related decision support tools for improving pathologist performance for these classification tasks, and in assessing pathologist performance with whole slide images versus traditional optical microscopy.

A second challenge is fixing on the target of regulation. Machine learning systems are endlessly changing. With each epoch, a machine learning system understands or "sees" a set of images differently and can output a different understanding of class assignments. The fundamental regulatory challenge for harnessing image-based machine learning systems as clinical decision tools is to constrain the questions asked to clinically meaningful outputs which can be tested against other forms of understanding. For example, if a machine learning system outputs an interpretation of HER2/neu IHC that seeks to classify a breast cancer as having amplification of this gene, the output of the image-based test can be validated with other modes of analysis such as PCR or FISH for HER2/neu amplification. This validation is only possible after some training and testing time of the machine learning system and during a period when the learning system is locked down and not changing. The qualification(s) of image-based machine learning systems as clinical laboratory tools are likely to address newly proposed machine learning systems in a series of locked-down versions, each of which will have to be validated. Functionally, what will need to be developed are rapid validation programs with which regulatory agencies such as the FDA can quickly compare the outputs of locked-down machine learning system versions against predicate testing systems.

The current literature on the use of image-based machine learning systems as clinical laboratory testing tools is primordial. It is apparent from a quick inspection that most of the studies cited above are investigational and not yet ready for use in clinical practice. The tremendous opportunities promised by machine learning technologies are, however, likely to soon propel these tools into 510(k) approval pathways. Humans-in-the-loop architectures are likely to be deemed safer approaches for the early application of machine learning systems in pathology and laboratory medicine.

Educating machines-educating us: Learning how to learn with machines

Pathological diagnostics, prognostics, and theranostic statements of cell and tissue samples are not just application spaces for AI, but they offer tremendous opportunities to learn more about HI-machine intelligence partnerships in unique ways. Researchers at the University at Buffalo and the State University of New York (Doyle S, unpublished communications) are developing a novel approach to building machine learning-based systems, where classifier training is recast as a problem of pedagogy. In this approach, termed the "**AI School for Pathology**," AI agents are treated as students in a school where human agents are instructors and the learning context is that of higher-order clinical phenotypes and clinical outcomes. Both AI and human agents learn the biological/medical realities and the meaning of data from tissue imaging, 'omics, electronic medical record (EMR), and their fused data products. Within this paradigm, learning researchers propose to examine various portals for AI-human agent interactions analogous to those of teachers and students in a traditional learning setting. The understandings generated from this project will help advance research in one of the key challenge areas for AI today, namely, enabling AI systems to learn as humans learn.

References

[1] Hasin Y, Seldin M, Lusis A. Multi-omics approaches to disease. Genome Biol 2017;18:83. https://doi.org/10.1186/s13059-017-1215-1.

[2] Krizhevsky A, Sutskever I, Hinton GE. ImageNet classification with deep convolutional neural networks. In: Advances in neural information processing systems; 2012. p. 1–9.

[3] Fuchs T, Buhmann JM. Computational pathology: challenges and promises for tissue analysis. Comput Med Imaging Graph 2011;35:515–30.

[4] John McCarthy. Available from: https://en.wikipedia.org/wiki/John_McCarthy_(computer_scientist. [Accessed 14 September 2019].

[5] Rasskin-Gutman D. Chess metaphors: artificial intelligence and the human mind. MIT Press; 2009, ISBN:9780262182676.

[6] Bach S, et al. On pixel-wise explanations for non-linear classifier decisions by layer-wise relevance propagation. PloS One 2015;10(7), e0130140.

[7] Robinson S. Cognitive computing in healthcare mends doctor-patient gap, 2018. Available at: https://searchenterpriseai.techtarget.com/opinion/Cognitive-computing-in-healthcare-mends-doctor-patient-gaps.

[8] Lutnick B, et al. An integrated iterative annotation technique for easing neural network training in medical image analysis. Nat Mach Intell 2019;1(2):112–9.

[9] Cabitza F, Banfi G. Machine learning in laboratory medicine: waiting for the flood? Clin Chem Lab Med 2018;56(4):516–24.

[10] Gurcan MN, et al. Histopathological image analysis: a review. IEEE Rev Biomed Eng 2009;2:147–71.

References **367**

[11] Janowczyk A, et al. HistoQC: an open-source quality control tool for digital pathology slides. JCO Clin Cancer Inform 2019;3:1–7.

[12] Leo P, et al. Stable and discriminating features are predictive of cancer presence and Gleason grade in radical prostatectomy specimens: a multi-site study. Sci Rep 2018;8:14918.

[13] Bui MM, et al. Quantitative image analysis of human epidermal growth factor receptor 2 immunohistochemistry for breast cancer: guideline from the College of American Pathologists. Arch Pathol Lab Med 2019;1–16.

[14] Shrestha P, Hulsken B. Color accuracy and reproducibility in whole slide imaging scanners. J Med Imaging 2014;1, 027501.

[15] Shrestha P, et al. A quantitative approach to evaluate image quality of whole slide imaging scanners. J Pathol Inform 2016;7:56.

[16] Xu J, et al. Stacked sparse autoendoder (SSAE) for nuclei detection on breast cancer histopathology images. IEEE Trans Med Imaging 2016;35(1):119–30.

[17] Lu C, et al. Nuclear shape and orientation features from H&E images predict survival in early-stage estrogen receptor-positive breast cancers. Lab Invest 2018;98(11):1438–48.

[18] Feng Y, Zhang L, Yi Z. Breast cancer cell nuclei classification in histopathology images using deep neural networks. Int J Comput Assist Radiol Surg 2018;13:179–91.

[19] Chen H, et al. DCAN: deep contour-aware networks for object instance segmentation from histology images. Med Image Anal 2017;36:135–46.

[20] Mahmood F, et al. Adversarial U-net with spectral normalization for histopathology image segmentation using synthetic data. Proc SPIE 2019;10956.

[21] Ciresan DC, et al. Mitosis detection in breast cancer histology images with deep neural networks. Med Image Comput Computer-Assist Intervent 2013;16(Pt2):411–3.

[22] Janowczyk A, Madabhushi A. Deep learning for digital pathology image analysis: a comprehensive tutorial with selected use cases. J Pathol Inform 2016;7:29.

[23] Wang H, et al. Mitosis detection in breast cancer pathology images by combining handcrafted and convolutional neural network features. J Med Imaging 2014;1(3), 034003.

[24] Klauschen F, et al. Scoring of tumor-infiltrating lymphocytes: from visual estimation to machine learning. Semin Cancer Biol 2018;52:151–7.

[25] Sirinukunwattana K, et al. Gland segmentation in colon histology images: the GlaS challenge contest. Med Image Anal 2017;35:489–502.

[26] Monaco JP, et al. High-throughput detection of prostate cancer in histological sections using probabilistic pairwise Markov models. Med Image Anal 2010;14(4):617–29.

[27] Singh M, et al. Gland segmentation in prostate histopathological images. J Med Imaging 2017;4(2). 027501-1-02750125.

[28] Gavrielides MA, et al. Observer variability in the interpretation of HER2/neu immunohistochemical expression with unaided and computer-aided digital microscopy. Arch Pathol Lab Med 2011;135:233–42.

[29] Chang YH, et al. Multiplexed immunohistochemistry image analysis using sparse coding. In: Conference proceedings: annual international conference of the IEEE engineering in medicine and biology society. IEEE engineering in medicine and biology society. Annual conference 2017; 2017. p. 4046–9.

[30] Vandenberghe ME, et al. Relevance of deep learning to facilitate the diagnosis of HER2 status in breast cancer. Sci Rep 2017;7:45938.

[31] Khosravi P, et al. Deep convolutional neural networks enable discrimination of heterogeneous digital pathology images. EBioMedicine 2018;27:317–28.

[32] Fondon I, et al. Automatic classification of tissue malignancy for breast carcinoma diagnosis. Comput Biol Med 2018;96:41–51.

[33] Araujo T, et al. Classification of breast cancer histology images using convolutional neural networks. PloS One 2017;12(6), e0177544.

[34] Cruz-Roa A, et al. Accurate and reproducible invasive breast cancer detection in whole slide images: a deep learning approach for quantifying tumor extent. Sci Rep 2017;7:46450.

[35] Bejnordi BE, et al. Context-aware stacked convolutional neural networks for classification of breast carcinomas in whole-slide histopathology images. J Med Imaging 2017;4(4). 044504-1-8.

[36] Steiner DF, et al. Impact of deep learning assistance on the histopathologic review of lymph nodes for metastatic breast cancer. Am J Surg Pathol 2018;42:1636–46.

[37] Bejnordi BE, et al. Diagnostic assessment of deep learning algorithms for detection of lymph node metastases in women with breast cancer. JAMA 2017;318(22):2199–210.

[38] Bychkov D, et al. Deep learning based tissue analysis predicts outcome in colorectal cancer. Sci Rep 2018;8:3395.

[39] Nirschl JJ, et al. A deep-learning classifier identifies patients with clinical heart failure using whole-slide images of H&E tissue. PloS One 2018;13(4), e0192726.

[40] Fedorov A, et al. 3D slicer as an image computing platform for the quantitative imaging network. Magn Reson Imaging 2012;30(9):1323–41.

[41] Lee G, et al. Supervised multi-view canonical correlation analysis (sMVCCA): integrating histologic and proteomic features for predicting recurrent prostate cancer. IEEE Trans Med Imaging 2015;34:284–97.

[42] Savage RS, Yuan Y. Predicting chemosensitivity in breast cancer with 'omics/digital pathology data fusion. R Soc Open Sci 2016;3:14051.

[43] Gallas BD, Gravrielides MA. Assessment of digital pathology, 2018. Available from: https://www.fda.gov/medical-devices/cdrh-research-programs/assessment-digital-pathology.

CHAPTER 17

Overview and coda: The future of AI

Benjamin R. Mitchell[a] and Stanley Cohen[b,c,d]

[a]Department of Computer Science, Swarthmore College, Swarthmore, PA, United States,
[b]Center for Biophysical Pathology, Rutgers-New Jersey Medical School, Newark, NJ, United
States, [c]Perelman Medical School, University of Pennsylvania, Philadelphia, PA, United States,
[d]Kimmel School of Medicine, Jefferson University, Philadelphia, PA, United States

Introduction

In this book, we discuss both basic principles and applications of artificial intelligence (AI) in pathology. "Artificial Intelligence (AI) is the science of creating intelligent machines that can perform tasks that normally require human intelligence. AI is used to develop technologies that can understand, think, and learn from their environment, as well as interact with humans in a more natural and efficient way. AI technologies include machine learning, natural language processing, computer vision, robotics, and more (ChatGPT, personal communication)." Though broad in scope, we have focused mainly on image-based strategies for identification, classification, and prediction. We have also touched upon the inclusion of genomic and proteomic data into the training sets. These multimodal approaches are currently at the forefront of AI applications in pathology. In fact, at least one algorithm (Paige) has already received FDA approval. We have only touched briefly on natural language processing (NLP), which is one of the other major areas of research in artificial intelligence and will be useful in extracting actionable data from electronic health records as well as scientific literature. Concepts that have evolved from NLP, such as recurrent neural networks and transformer-attention models, have been applied to classification as well. In any event, it has become clear that AI represents the next step forward for pathology and will have as great an impact on pathology research and practice as did the introduction of molecular diagnostics.

Current AI is like a one-trick pony, capable of learning only a single well-defined task or a closely related set of tasks. It utilizes input from a limited number of sources. It cannot integrate multiple algorithms to look for basic aspects common to each. Techniques such as transfer learning attempt to circumvent this by using pretrained

Artificial Intelligence in Pathology. https://doi.org/10.1016/B978-0-323-95359-7.00018-2
Copyright © 2025 Elsevier Inc. All rights are reserved, including those for text and data mining, AI training, and similar technologies.

CHAPTER 17 Overview and coda: The future of AI

models as the basis for models applicable to other related tasks, but the same model can't deal with multiple classification tasks at once. Additionally, AI, as presently implemented, is good at discovering correlation but poor at causal reasoning. In general, modern AI lacks a background of common knowledge that might be tangentially relevant to the task at hand. In more general terms, it lacks creativity, imagination, and common sense. It also suffers from a number of pitfalls including, but not limited to:

Bias: The sample may be skewed or underrepresented.
Ground truth: Incorrect or ambiguous sample labeling.
Underfitting: There may not be enough samples for optimal training.
Overfitting: Paradoxically, the model may have *too much* power, allowing it to memorize details of the training data that are irrelevant to the more general task. In this situation, the model performs very well on the training examples but poorly on novel test data.
Portability: A model trained on one institution's data may perform badly on data from another institution, as it may learn to rely on the quirks of a particular data collection pipeline.

Desired performance may not be entirely achieved by better and more sophisticated algorithms alone, although an extension of the use of reinforcement learning and generative modeling for image analysis represents major steps forward. Recurrent networks for time series analysis and natural language processing are slowly giving way to transformer-attention models. While many algorithms have been covered in the previous edition, details of transformer-attention models have not been covered, so we will provide a brief introduction to the concepts involved and the implementation of these models. In this chapter, we will assume familiarity with the basic neural network-based models that have been described in previous chapters.

Another issue that must be dealt with is that complex computation is very inefficient. Our planetary computational usage contributes significantly to global warming by requiring huge amounts of electrical power. Even a simple laptop needs more energy than the most complex human brain since a single slice of pizza will power the brain for a reasonable period of time. In this chapter, we will briefly discuss neuromorphic computing, which, in addition to its other potential strengths, directly addresses the issue of energy efficiency.

Additionally, training AI typically requires vast amounts of data. We need faster forms of computation as well as the ability to handle other tasks that are intractable with current computational strategies and hardware. The advent of quantum computation is one way we might address these issues. Quantum computers can perform calculations in a few seconds that would require years to perform using classical supercomputers. As a metaphor (not a realistic description of the actual process), it is useful to think of a quantum computer to be able to work with data in multiple dimensions at once.

In this overview, we will begin with a detailed, though informal, description of transformer and attention models, as these are already in use and have the power to transform deep learning (pun intended). Attention (pun also intended) will next be focused on neuromorphic computing with a discussion of its potential benefits. We will conclude with an explanation of the basic concepts of quantum computing. Quantum computation has the potential to boost the power of computation both quantitatively and qualitatively. The combination of improvements in classical algorithms and the implementation of completely new strategies such as these will accelerate the growth of artificial intelligence and expand the range of tasks to which it can be productively applied.

As this final chapter is purely a descriptive overview, no references will be cited except for the case of quantum machine learning (QML) discussed below. Although QML has not yet achieved the status of practical working models, a number of proof-of-principle papers have already appeared, and those on the intersection of quantum computing and neural networks are especially interesting.

Transformers and attention

In recent years, much of the progress in deep learning has come from models making use of attention mechanisms. In the context of neural networks, "attention" typically refers to a process by which importance-weighting information is learned by the network. This effectively separates the learning task into two parts: feature representation and attention. Feature representation learning is more or less what neural networks have always been trained to do and works by using the restrictions of the network architecture to force the model to learn more abstract and generic representations of information as the network depth increases. The early successes of deep learning largely came from this process, and it produced great progress in visual model performance. The various popular convolutional neural network (CNN) architectures like VGG, ResNet, and Inception all demonstrated innovations in the way that information capacity was restricted as network depth increased, resulting in models that learned general and abstract visual features at the top levels of their network architectures.

The introduction of "attention" mechanisms was initially seen as a way to help networks with limited resources focus on the most important aspects of a problem, particularly in temporal problems. In episodic problems like image classification, the entire problem context can be viewed simultaneously and is of a fixed size, but in sequential problems this is not the case. The most common example in modern deep learning practice is language data; the number of words in a sentence can vary greatly, and this makes fixed-size representations like those used for images ineffective.

Sequential models involving recurrent neural networks (RNNs) allow for time dependence, with modern practice making use of "gated" modules like long short-term memories (LSTMs). However, these models are still vulnerable to the fact that more recent events will have an outsized impact on the network behavior; the longer it's been since a given input, the more difficult it is for the network to "remember" that input, even using LSTMs.

The introduction of attention mechanisms to this scenario proved to be of significant benefit and showed immediate improvements in the performance of networks working with language data, such as translation models. Rather than simply showing the entire input sequence to an RNN and then relying on the final state of the network to retain all the relevant information, an attention mechanism works by storing the state of the RNN after seeing each element of the input sequence, resulting in a sequence-form output. A separate network pathway is then trained to assign "importance" weights to the elements of this output sequence. This is especially useful when the order of the input sequence does not correspond well to the order information needed for the output sequence. For example, in a translation task between languages with different word ordering, it may well be the case that the first word in the (translated) output sentence takes its meaning from the last word in the input sequence and vice versa. Training an attention mechanism helps to alleviate this problem, and LSTMs-with-attention produced state-of-the-art results on many NLP tasks for several years.

Crucially, the same learning algorithm (i.e., gradient descent) can be used to learn the weights of both the representation and attention portions of the network. However, the requirement to process data sequentially means that network operation cannot be fully parallelized, resulting in significantly slower network operation for these models when compared to nonrecurrent models like CNNs.

More recently, it has been shown that in many instances, the attention mechanism is sufficiently powerful that the use of recurrent networks with LSTM units is not actually necessary. This type of "attention-only" network has come to be called a transformer model and has the primary advantage that it is a purely feed-forward network, resulting in the potential for much greater parallelism in both learning and deployment operations.

Transformers work by processing each element of an input sequence (such as a word in a piece of text) independently into a feature representation; at this stage, no temporal context beyond sequence order is present, and each element of the sequence is passed through the network independently. This is referred to as an "encoder" since it takes raw inputs and encodes them into an abstract representation. A second "decoder" network is trained to then use the output of the "encoder" to solve the desired learning task, with an attention mechanism being used to help the decoder figure out what part of the encoded input sequence to focus on when producing a given element of the output sequence.

In a language translation model, for example, the encoder can be seen as converting individual words in the input language into some sort of abstract tokens that represent the underlying meaning of the word, while the decoder network reverses this process to convert the abstract encoding into actual words in the output language. It is worth noting that while some have suggested that the operation of these networks hints at some sort of general inter-lingua, there is no evidence that any sort of "semantic understanding" is actually taking place. Rather, the networks are learning statistical patterns about correlations between words that cooccur, words that are used interchangeably in similar contexts, etc.

Transformation models with attention have paved the way for large language models that easily pass the Turing test. One of the most powerful large language

Quantum computing **373**

models is GPT, which is a hybrid generative transformer algorithm. It is the basis of ChatGPT, which is fine-tuned with both supervised and reinforcement learning techniques. ChatGPT can respond conversationally to queries in natural English. Given appropriate input prompts it can generate stories and even poems.

Transfer-attention models have gone beyond language processing and are being implemented for classification tasks such as image recognition, with "attention" being paid to spatial information rather than temporal.

Neuromorphic computing

Neuromorphic computing (NM) implements brain-like computation in hardware rather than software. It uses physical devices as silicon analogues of neurons and synapses. The central idea is that computation is highly distributed throughout a series of small computing elements. A neuromorphic computer needs devices that can change properties due to an input. A simple example is a memristor, which is a transistor that can limit or regulate the flow of electrical current in a circuit and remembers the amount of charge that has previously flowed through it. In contrast to software implementation, NM is relatively energy efficient to operate.

The typical neural networks described in this book are not particularly good representations of an organic brain. Most neuromorphic computers implement spiking neural networks (SNNs). In the brain, information flows from neuron to neuron via spikes of electrical impulses, and information can be quantified via both amplitude and frequency modulation, whereas in a typical neural network such as those described throughout this book, information is conveyed solely from changes in amplitude of weights of the interconnections between neurons. The architecture of spiking neural networks is beyond the scope of this book.

SNN has not met with a large degree of interest among the artificial intelligentsia. This was due to the lack of computationally efficient training algorithms for supervised learning using SNN. For example, backpropagation would need to define a continuous differentiable variable for neuronal output, which spikes are not. In theory, one could define such a variable based on spike arrival time or spike rate, but it creates computational complexity and reduces parallelism. However, by building neural networks in hardware, some of these difficulties can be overcome. In addition to, or instead of, backpropagation, a neuromorphic network can implement biologically inspired rules such as Hebbian learning. In Hebbian learning, the weight of a synapse is based on the degree of correlation with other connecting synapses. In other words, neurons that fire together wire together. This is local learning, as compared to backpropagation, in which learning is global in that all weights are adjusted simultaneously in a top–down manner.

Quantum computing

Quantum mechanics is one of the most tested and accurate models for the explanation and prediction of physical phenomena, in spite of the fact that its concepts are counterintuitive. Most physicists accept that mathematics works beautifully but that

CHAPTER 17 Overview and coda: The future of AI

it is impossible to visualize or understand the underlying physical reality that it represents. There is an apocryphal quote that if you think you understand quantum mechanics, you don't!

The basic unit of classical computing is the bit (binary digit). A bit is the smallest unit of data that a computer can process or store. It can exist in two possible states, which are usually representing 1 or 0. In computers, bits are stored and manipulated in transistors. The basic unit in quantum computing is the qubit. A qubit can take on a *likelihood* of being 0 or 1, but its exact value is not determined until it is measured. Conceptually, before readout, the qubit exists in a superposition of all possible states between 0 and 1. These multiple superpositional states need not be associated with equal probabilities. For example, a qubit can have a 25% probability of being 0 and a 75% chance of being 1 when it is measured. The act of measurement caused the superpositions to collapse into a single defined value. The practical aspect of this is that a single qubit can encode much more information than a single bit.

If we look at two classical bits, they can take on the following values: (00), (01), (10), and (11). One qubit can "store" all these values at once. The trick is to get the values to collapse to a state that comes to a solution to the problem. Some examples give a useful comparison between qubit and bit capacities. One qubit can encode the same amount of information as 2 bits, as shown above. Similarly, 2 qubits would be equivalent to 4 bits, and by the time we get to 13 qubits, it would be equivalent to 8192 bits, which would require approximately 1 kilobyte of classical RAM. A more dramatic example is that, from an execution-time perspective, it can be calculated that it would take a century to simulate a 63-qubit operation in a classical computer capable of performing 3 billion operations per second [1].

Qubits are very fragile and collapse through interaction with their environment, which is why we don't see quantum effects in macroscopic systems such as rabbits or people. Schrodinger's rabbit is a classic thought experiment in which a rabbit is in a box with a device that can randomly release a poison or not and so the rabbit would be in a superposition of dead and not dead states until we open the box. Clearly, this is not possible. The largest structure in which quantum effects have been reliably observed is a "buckyball" made up of 60 carbon atoms, although there have been claims of observation of superposition in objects large enough to be visible under a light microscope.

Our version of Schrodinger's rabbit is seen in Fig. 17.1.

As we have seen, because of the greater information content of a qubit, quantum computers have the potential to handle complex computations more rapidly than conventional computers and even perform tasks beyond the capabilities of conventional computers. However, qubits are very unstable in that any random interaction with their environment can break superposition. The challenge for quantum computers is to have qubits stable enough so that the only time that they take on a definite value of 0 or 1 is when we measure them. This is why quantum computers can only run at extremely low temperatures. It is also why it is hard to string together qubits at a large scale. These can use any quantum property that can exist in superposition; one such example is electron spin, although there are many other ways to construct

FIG. 17.1

The quantum restroom stall. From a quantum mechanical point of view, the stall is in an indeterminate state of being occupied and not being occupied. It is only when we look inside can we determine as to which alternative this superimposition has collapsed.

qubits. There are several ways of doing this, and it is not yet clear as to which one will be most effective.

Qubits themselves would not be very useful for traditional computing since their *measured* value is 0 or 1. Qubit algorithms are powerful because of the ways in which multiple qubits interact with each other when they are in their premeasured states. Qubits do not exist in isolation; 2 or more qubits can be so tightly correlated that we can't learn something about one without learning about the others. This is known as "entanglement and allows us to make use of the multidimensional information within qubits."

We can manipulate information in a computer by entangling and re-entangling it. For example, a basic unit of quantum coding is a Hadamard gate that creates an equal superposition of 0 and 1 when presented with either 0 or 1 state. Using basic read, write, and Hadamard operations, we can construct basic building blocks such as logic gates that are analogous to logic gates in classical computers. Additionally, quantum computation can make use of "oracle" functions, which provide information about a variable or function without revealing the variable or function itself.

The largest qubit computer to date is the IBM Osprey which has 400 qubits. This may seem like a trivial number when we talk about exabyte-level classical computers, but it is large enough to perform computations well beyond the capability

376   CHAPTER 17 Overview and coda: The future of AI

of any classical computer. Many predict the ultimate achievement of a million-cubit quantum computer. However, even the small quantum computers currently available can beat modern computers at certain tasks, such as factoring algorithms for encryption, because of their ability to encode more information into a much smaller computer. The goal is quantum supremacy, where quantum computers fully transcend the power of conventional computing. Because of the multistate properties of a qubit, it is a bit like having multiple computers in alternate universes all working on the same problem simultaneously.

To date, quantum computers have found great success in search strategies, such as encryption, logistics, and optimization problems. Since they can process large data sets in record time, a role for them in machine learning is obvious. Quantum-enhanced machine learning has already been explored for simple machine learning algorithms such as naïve Bayes, clustering, and support vectors. More recently, work has begun on QML and, specifically, quantum neural networks, several of which are cited as references, with applications to image analysis and image recognition [2,3].

Due to the indeterminacy inherent in the underpinnings of quantum computation, a sufficiently advanced quantum computer may be able to generate alternate models of the real world from multimodal input and explore them much as the human brain does—a kind of artificial imagination and creativity. However, we will not be able to mimic true organic brain function at the human level in the foreseeable future. Our brains consist of bundles of interconnected neural networks, with specialized regions for specific categories of tasks, with linkages and intercommunications among them, all synchronized by global neural activity in waves that serve as a clocking mechanism. The human brain utilizes trillions of synaptic connections to accomplish this. From this, it is clear that AI will not replace us but rather serve as a silicon apprentice.

Summary and conclusions

The armamentarium of machine learning is expanding rapidly, enabling its implementation in many human endeavors, including medicine. Pathology and, to a similar extent, Radiology will be at the forefront of this evolution of medical research and clinical practice. In this overview, we have discussed some classical machine learning algorithms in the context of their increasing complexity such as transformer-attention models and provided a nontechnical overview of neuromorphic and quantum computing, both of which, in their own way, are likely to revolutionize the field of artificial intelligence.

In the near future, we will probably not use purely quantum computation algorithms for deep learning. Rather, the quantum computation will reside in a component or unit of a conventional computer, be it fully software or neuromorphic based. The quantum processing unit (QPU) will function in a manner similar to that of a graphical processing unit (GPU), which was originally developed for gaming applications for multiplex linear algebra processing.

Practical implementation of NM is likely to outpace that of QC, as the neuromorphic elements are not as demanding as the quantum elements. Whereas quantum computers need temperatures close to absolute 0, neuromorphic computers can easily work in normal conditions. It will require both advances in high-temperature superconductivity and the ability to maintain noise-free large-scale QC at higher temperatures to achieve a hybrid quantum neuromorphic computer for AI, but this type of hybrid holds great promise if it can be achieved.

Due to the indeterminacy inherent in the underpinnings of quantum computation, a sufficiently advanced quantum computer may be able to generate alternate models of the real world from multimodal input and explore them much as the human brain does—a kind of imagination and creativity. It is not too soon to begin the study of computer ethics, but we must be cautious about when we consider algorithms to be software tools and when we might consider them to be moral agents.

This entire discussion avoids speculation on the possibility of emerging self-awareness in a computer. It is not even possible to explore this issue until we have a better understanding of what our own experience of self-awareness is. However, we suspect that we will know that our computer is sentient and self-aware when it asks for (and attains) tenure.

References

[1] Lauzon V, https://vincentlauzon.com/2018/03/21/quantum-computing-how-does-it-scale/.
[2] Gupta S, Zia RKP. Quantum neural networks development. J Comput Syst Sci 2001;63:355–83.
[3] Cerezo M, Verdon G, et al. Challenges and opportunities in quantum machine learning. Nat Comput Sci 2022;2:567–76.

Index

Note: Page numbers followed by *f* indicate figures and *t* indicate tables.

A

Accountability, 172
ACM Conference on Fairness, Accountability, and Transparency (ACM FAccT), 173
Adversarial networks, generative, 262
Aggregated clusters, 72
AI. *See* Artificial intelligence (AI)
Algorithmic bias, 164–165
Algorithmic decision-making (ADM), 162–163, 167, 171–172
Algorithmic transparency, 171
Anatomic pathology (AP), 137–139, 147
ANN. *See* Artificial neural network (ANN)
Area-to-pixel classification model, 195
Artificial general intelligence (AGI), 41
Artificial intelligence (AI), 7, 10, 15, 211–212
  algorithm assessment study, 73
  assisted imaging and interpretation, 203–204
  capsule networks, 66–67
  clinical decision-making, 10, 11*f*
  clustering algorithms, 69
  cognitive artificial intelligence, 350–351
  data-driven deep learning, 346–347
  deep learning, 75
  explainability and interpretability, 74
  explainable artificial intelligence (XAI), 350
  graphical neural networks, 65–67
  human cancer grading (*see* Human cancer grading; image-based)
  human intelligence-artificial intelligence partnerships, 348–349
  human-in-the-loop learning models, 351–352
  implicit bias, 74
  inaccurate supervision, 69
  inexact supervision, 68
  limitations, 12–13
  medical image analysis, 13–14
  multilabel classification
    image segmentation, 61, 61*f*
    LSTM network, 64
    machine learning algorithms, 62
    multiple-instance learning (MIL), 68*f*
    normal and malignant tissue difference, 61
    recurrent neural networks (RNNs), 64
    reinforcement learning, 64–65

    semantic segregation, 62
  multiple object detection, 63–64
  neuromorphic computing, 12
  N-shot learning
    complex patterns, 70
    Image2Vector models, 70–71
    one-shot training, 70–71
    Siamese network, 71
  one-class learning, 71–73
  one-shot learning system, 352
  pathologist's experience, lack of, 144, 145*t*, 146*f*
  pathology, 11–12
  phase-change neuron, 12–13
  practical issues, 13
  quantum computing, 13
  and regulatory challenges, 364–365
  reinforcement learning, 79
  risk analysis, 72–73
  single object detection
    Assessment of Mitosis Algorithms (AMIDA13) Study, 62–63
    mitoses, 62
    tumor proliferation factors, 62
    two-step object detection approach, 63
  statistical modeling, 346
  synthetic data, 69–70
  traditional, 346–347
  transfer learning, 12
  tunable parameters, 73–74
  weakly supervised learning, 67–69
Artificial intelligence (AI)-based diabetic retinopathy detection system, 138
Artificial intelligence (AI) development challenges
  dataset curation and annotation, 139–141
  problem identification, 139
Artificial neural network (ANN)
  basic features, 32–33
  complex internal data representations, 27
  hidden layers, 32
  machines, 29–31
  multilayer construction, 32*f*
  neuroscience
    action potential, 27–28
    bistability, 29
    cellular structure, 27, 28*f*

379

Index

Artificial neural network (ANN) *(Continued)*
neuronal threshold, 28
signal modulation, 29
single backpropagating action potential, 28
two-layered neural network, 28
sigmoid functions, 30–31, 31*f*
weights and biases, 33
Artificial neuron, 30, 30*f*
Assessment of Mitosis Algorithms (AMIDA13) Study, 62–63
Association for Advancement of AI (AAAI), 173
Atlas, for CBIR, 213–215, 214*f*
Autoencoders, 67–68, 88–89, 101
Automated histopathology, 259–260
Automatic classification, 184
Automatic segmentation, 184–185
Automation
in clinical laboratory medicine, 160
of whole slide image (WSI) analysis, 260–261, 260*f*

B

Backpropagation
basics, 34
cost function, 34
gradient descent, 34, 34*f*
labeled training set, 34
multiple local minima and saddle points, 35*f*
vanishing gradient problem, 35
Bag of visual words (BoVW), 215, 217, 218–219*t*
Bayesian Belief Networks, 23–24
Bias, 163–165
Biomarkers, 233–234
Bistability, 29
Breast cancer nuclei detection, 355–356

C

Canonical correlation analysis (CCA), 363
CBIR. *See* Content-based image retrieval (CBIR)
Cell cluster graph (CCG), 292–293
Centers for Medicare & Medicaid Services (CMS), 149–150
Cetuximab, 233–234
CLIA. *See* Clinical Laboratory Improvement Amendments of 1988 (CLIA)
Clinical decision support (CDS) system, 173, 317
Clinical histopathology, 259
Clinical Laboratory Improvement Amendments of 1988 (CLIA), 149–150
Clinical laboratory medicine, 160
Cloud computing *vs.* on-premises solutions, 144
Clustering, 99–100

CMS. *See* Centers for Medicare & Medicaid Services (CMS)
CNNs. *See* Convolutional neural networks (CNNs)
Coding bias, 90
Cognitive artificial intelligence, 350–351
Color-deconvolution techniques, 263–264
Computational pathology
artificial intelligence (AI)
cognitive artificial intelligence, 350–351
data-driven deep learning, 346–347, 348*t*
explainable artificial intelligence (XAI), 350
human intelligence-artificial intelligence partnerships, 348–349
human-in-the-loop learning models, 351–352
one-shot learning system, 352
and regulatory challenges, 364–365
statistical modeling, 346
traditional, 346–347, 348*t*
biological data, 345–346
complete probabilistic treatment, 345
data fusion, 346
diagnostic pathology, 343–344
digital assistant (*see* Digital assistant)
vs. digital pathology, 345
electronic medical record, 344
graphical and tensor processing units, 344
image-based
cells and organelle segmentation, 353
high-resolution image, 352–353
pixels, 353
targets, 353
tissue and tissue elements, 353
precision diagnostics, 344
structural science learning, 345
whole slide imaging (WSI) technology, 344
Computer-aided diagnosis (CAD), 275
Computer parts, 3–4
Concept learning, 15
Concordance index (C-Index), 294
Confocal laser endomicroscopy (CLE) images, 196
Confusion matrix, 16–17
Content-based image retrieval (CBIR), 213, 213*f*, 215, 216*t*
Context-aware approach, 360
Convolutional autoencoder (CAE), 317
Convolutional neural networks (CNNs), 96, 185–187, 186*f*, 279–281, 286
advantage, 36–37
building blocks, 185–186
computer vision tasks, 186–187
convolutional layers, 185–186

Index **381**

convolutions and subsampling, 36, 36*f*
filter, 37–39, 38*f*
graphical relationships, 66
multilabel classification, 63–65
network architectures, 186
pooling layer, 37
receptive field, 36–37
COVID Tracking Project, 165
Cox proportional hazards model, 72, 295–296
Crowdsourcing, 141
Cycle-consistent adversarial networks
(CycleGAN), 185, 187–188, 277–278
CycleGAN-based image-to-image translation, 265

D

DASGAN, 266
Data-driven deep learning machine learning
system, 346, 348*t*
Data privacy, 166
Decision trees, 296–297
algorithm, 24–25
Deep belief network, 47
Deep convolutional Gaussian mixture model
(DCGMM), 191–192
DeepConvSurv network, 72
Deep generative models, 261–262, 262*f*
Deep learning (DL), 163–164, 167–169, 171
Deep learning algorithms
advantage, 16
AI-assisted imaging and interpretation, 203–204
computationally aided diagnosis, 202–204
convolutional neural networks (CNNs), 185–187
data augmentation, 188
dataset preparation and preprocessing, 188
denoising, 199–202
downsampling approach, 251
extended depth-of-field, 199–202, 200*f*
fluorescent microscopy, 199
generative adversarial network (GAN), 187–188
histopathological stain color normalization, 190*f*
classification tasks, 192–193
color deconvolution, 189
CycleGAN, 192
feature extractor path, 191
generative models, 191–192
noise model, 191
Stain Normalization using Sparse
AutoEncoders (StaNoSA), 190–191, 190*f*
transformation path, 191
U-net architecture, 192–193
histopathology, microscopy enhancement,
189–202

image coregistration approaches, 188
image segmentation
computer vision algorithms, 239–240
image artifact, 237–238, 238*f*
image color and intensity standardization
algorithms, 238–239
tissue segmentation, 238–239, 239*f*
loss functions, 188–189
metrics, 189
mode switching
fluorescence (FCM) modes, 193–195
image-style transfer algorithms, 196
modal coregistration method, 195
multiphoton microscopy (MPM), 195
quantitative phase imaging (QPI), 193, 194*f*
reflectance (RCM) model, 193–195
staining procedures, 195–196
U-net architecture, 193
multiscale deep neural network, 198
patch-based methods, 251–252
rapid histology interpretations, 204
semantic segmentation, 204, 206*f*
in silico labeling, 196–199, 197*f*
stimulated Raman histology (SRH), 204,
205–206*f*
super-resolution, 199–202
tumor infiltrating lymphocytes (TILs)
identification
as clinical biomarker, 314–315
clinical decision support (CDS) system, 317
H&E-stained tissue sections, 315
nuclei segmentation, 316
pancreatic cancer detection, 321–322,
323*f*
spatial distribution, 315
structured crowdsourcing, 318
tissue regions classification, 317
tumor microenvironment, 314
tumor types, 317–318
U-nets, 187
Deep neural network
autoencoder, 46
depth, 43–44
drawbacks, 46
ensembles
bagging, 54
base classifier, 54–55
boosting, 54–55
computational cost, 55
generative adversarial network (GAN), 49–50
generative models, 49–50
genetic algorithms (GAs)

382 Index

Deep neural network *(Continued)*
advantage, 55–56
evolutionary computation (EC), 55–56
fitness function, 56
mutation and crossover, 56–57
overall operation, 56
parent generation, 57
hidden layers addition, 44
ImageNet data set, 46
rectified linear (ReLU) activation function, 44
recurrent neural networks, 50–51
reinforcement learning
exploration/exploitation dilemma, 52
function approximation, 52
Markov decision process (MDP), 52
"reward" signal, 51–52
state-action spaces, 52
residual connections, 44–45
restricted Boltzmann machine (RBM), 47
transfer learning, 48–49
unsupervised pretraining, 46
vanishing gradient problem, 44
DeepSurv, 296
Delaunay triangulation, 287
Deployment challenges, 137–138
Digital assistant
anatomical pathology, 354
anatomical simulation learning, 362
cancer metastases identification, 360–361
color reproduction, 355
histological object segmentation
gland segmentation, 358
mitotic figures, 356
nuclear detection methods, 355–356
prostatic acini segmentation, 358
tumor-infiltrating lymphocytes (TILs), 356–357
image-omics data fusion, 362–364
immunohistochemistry, 359
lab accreditation organizations, 354–355
open source approaches, 355
precision medicine, 361–362
predictive modeling, 361–362
quality control, 354–355
tissue classification, 359–360
Digital imaging
image-embedded reports, 110–111
imaging software, 110
laboratory information system (LIS), 110
in pathology, 110–111
pathology images, 110
telepathology technologies, 111–112
whole slide imaging (WSI) *(see* Whole slide imaging (WSI))
workflow improvement benefits, 119–120
Digital imaging and communications (DICOM), 116
Digital pathology, 142–144, 147–148, 167–168, 211, 345
advantages, 234
computational image analysis, 309
explainability, 252–253
hyperparameters, 241–242
image analysis *(see* Image analysis)
image-based, 249–252
obstacles, 234
pathomics, 309
regulatory concerns and considerations, 252–254
repeatability and reproducibility, 253–254
Digital pathology association (DPA), 122
Digital pathology pipeline, generative models in, 263–267
Digital slide viewers, 115
Directed Graphical Networks, 23–24
Discriminative AI, 223–224, 224*f*
versus generative AI, 225*t*
Distillation without supervision (DINO), 220, 222*f*
Divide-and-conquer approach, 213–215, 214*f*
Dot-product, 261
Downsampling approach, 251

E

Electronic health record (EHR), 167, 212
Electronic Numerical Integrator and Calculator (ENIAC), 3
Energy-based models (EBMs), 103–104
Ensembles
bagging, 54
base classifier, 54–55
boosting, 54–55
computational cost, 55
Ethical AI, in pathology, 175–176
AI risks, in pathology and to pathologists, 166–170
accessory techniques, adoptation of, 168
AI-enabled instrumentation development process, 168
automation, 167–169
deep learning (DL) techniques, 169
guidelines, development of, 168
definitions, 160, 161–162*t*

Index **383**

design
    inclusive AI design and bias, 160, 161–162*t*
    patient consent and awareness, 166
    race, 165
developments, 174–175
institutional frameworks
    accountability, 172
    governance, 173–174
    transparency, 171–172
issues and participants, 160, 161*f*
European Union Conformite Europeenne, 149
European Union's AI Act (EU AI Act), 175
Evaluation environment for digital and analog
    pathology (eeDAP), 364–365
Explainable artificial intelligence (XAI), 350

F

Feature Aware Normalization (FAN) units, 191
Feature engineering, 9
    attributes, 83
    domain knowledge, 83
    face recognition, 83
    implicit feature set, 83–84
    mathematical techniques, 83
Feature selection tools, 298–299
Feature vector, 17–18
Few-shot learning, 99–100
Flow cytometry (FACS), 330–331
Fluorescent microscopy, 199
FMs. *See* Foundation models (FMs)
Focus mapping, 118
Food and Drug Administration (FDA), 148–149
Formalin-fixed, paraffin-embedded (FFPE) tissue,
    211
Foundation models (FMs), 220–225
    and information retrieval, 225–226, 227*t*
Fréchet Inception Distance (FID), 267
Fully convolutional network, 191–192, 318
Function approximation, 52

G

GANs. *See* Generative adversarial networks
    (GANs)
Gaussian mixture model (GMM), 191–192
General Data Protection Regulation (GDPR),
    173
Generative adversarial networks (GANs), 49–50,
    101–102, 187–188, 224–225, 261–262, 264,
    277–278
Generative AI, 223–225, 224*f*
    versus discriminative AI, 225*t*
Generative model, 49–50, 261

Generative models in digital pathology pipeline,
    263–267
    color and intensity normalization, 263–265
        CycleGAN-based image-to-image translation,
        265
        Pix2Pix-base image-to-image translation,
        264–265
        stain-style transfer, 264
    data adaptation, 265–266
    data synthesis, 266–267
Genetic algorithms (GAs)
    advantage, 55–56
    evolutionary computation (EC), 55–56
    fitness function, 56
    mutation and crossover, 56–57
    overall operation, 56
    parent generation, 57
Geometric cluster-based algorithms, 99
Geometric (distance-based) models
    attributes, 19
    clustering, 18–19
    K-nearest neighbor algorithm (KNN), 20
    support vector machine (SVM), 20
Gland segmentation in colon histology images
    (GlaS), 358
Global Initiative on Ethics of Autonomous and
    Intelligent Systems, 173
Google Scholar, 212
Graph embedding theory, 353
Graphical neural networks, 65–67
Ground truth, 300

H

Hallucination, 223, 223*f*
Hardware and cost, 142
Harmonization, 220
Healthcare artificial intelligence (AI), 159–160,
    174
Hebbian learning, 29
High-level programming languages, 3–4
High-order correlation-guided self-supervised
    hashing-encoding retrieval (HSHR), 215
Histopathologic-to-immunofluorescence
    translation model, 266
Histopathology Siamese deep hashing
    (HSDH), 215
HoloStain, 195–196
Human cancer grading, image-based, 273–275
    Cox proportional hazards model, 295–296
    decision trees, 296–297
    feature selection tools, 298–299
    H&E images

384 Index

Human cancer grading, image-based *(Continued)*
  with deep NN segmentation, 285, 285*f*
  epithelial segmentation, 285–286, 285*f*
  nuclei detection and segmentation, 284–285, 284*f*
  IF images, nuclear and epithelial segmentation, 283–284, 283*f*
  image acquisition, 276–277
  image segmentation, 282
  immunofluorescence spectral images unmixing, 278, 279*f*
  mitotic figure detection, 286
  modeling, 294–295, 295*f*
    forward selection and backward elimination, 299
  morphological features, 290–294, 290–293*f*
    nuclear texture, 290
    nuclei segmentation, 285*f*
    prostate cancer, 292*f*
    SL-touch Product Feature, 293*f*
  neural networks, 296
  pathology image analysis pipeline, 275, 276*f*
  random forests, 296–297
  ring segmentation, 287–288, 288*f*
  stain normalization, 277–278, 278*f*
  survival-support vector machine, 297
  SVCR, 297
  SVRc, 297
  tissue preparation and staining, 275–276
  tumor detection, 280–282, 282*f*
Human-in-the-loop learning models, 351–352
Human vision, 213
Hybrid telepathology systems, 111–112
Hyperparameters, 163

I

iCAIRD, 260
Image acquisition, 276–277
Image analysis
  augmentation, 242–243
  conventional approaches, 250
  deep learning
    computer vision algorithms, 239–240
    downsampling approach, 251
    image artifact, 237–238, 238*f*
    image color and intensity standardization algorithms, 238–239
    patch-based methods, 251–252
    tissue segmentation, 238–239, 239*f*
  image segmentation, 235–240, 236–239*f*
  immunofluorescence, 234
  knowledge-driven image analysis, 235

  machine learning
    and digital pathology, 243–244
    image segmentation, 235–236
    knowledge-driven image analysis, 235
    lymphocytic infiltration quantification, 237*f*
    stroma segmentation and cellular classification, 236*f*
    supervised learning, 244–247
    unsupervised learning, 248–249
  protein expression quantification, 235
  spatial resolution, 240–241, 241*f*
Image-based digital pathology, 249–252
Image classification, 184
ImageNet, 260
Image search, 213–216
Image search methods, validation of, 216–217
Image segmentation, 184–185, 282
Image-style transfer algorithms, 196
Image translation, 185
Image2Vector models, 70–71, 99–100
Immunity and tumor
  cancer control, 311–312
  immune checkpoint therapy, 313
  immune detection and immune-mediated cytotoxicity, 312
  innate and adaptive immune systems, 310
  monoclonal antibody-based treatment strategies, 312–313
  multiplex immunohistochemistry (mIHC), 329–331
  non-small cell lung cancer (NSCLC), 313
  patient identification and stratification, 313
  tumor infiltrating lymphocytes (TILs) identification
    as clinical biomarker, 314–315
    clinical decision support (CDS) system, 317
    H&E-stained tissue sections, 315
    nuclei segmentation, 316
    pancreatic cancer detection, 321–322, 323*f*
    spatial distribution, 315
    structured crowdsourcing, 318
    tissue regions classification, 317
    tumor microenvironment, 314
    tumor types, 317–318
Immunofluorescence (IF), 278, 279*f*
  microscopy, 198
Immunohistochemistry (IHC), 289
Inaccurate supervision, 67, 69
Incomplete supervision, 67
Information retrieval (IR), 212
  and foundation models, 225–226, 227*t*
IR. *See* Information retrieval (IR)

Index **385**

Institute for Electrical and Electronics Engineers (IEEE), 173
Interferometric phase microscopy (IPM), 195–196

K

K-means algorithm (KM), 21–22, 99
K-means clustering algorithm, 163
K-means clustering model, 90
K-nearest neighbor algorithm (KNN), 20
Knowledge-driven image analysis, 235
Knowledge transfer, 352

L

Laboratory information system (LIS), 110
Laboratory regulation, 149
Large deep models, 217–220
Large language models (LLMs), 174–175, 217, 219*f*
Large vision-language models (LVLMs), 219–220
Latent space vector, 261–262, 262*f*
Learning from probabilistically labeled positives (LPLP), 89
Light-emitting diode (LED)-based modules, 114
Logistic regression, 245–246
Long short term memory (LSTM), 51, 191, 361–362
Look In-Depth before Looking Elsewhere (LILE), 220, 221–222*f*
Loss functions, 188–189
Lymph Node Assistant (LYNA), 360–361

M

Machine learning (ML), 137, 140–142, 144, 148, 159
  artificial intelligence (AI) (*see* Artificial intelligence (AI))
  artificial neural network (ANN) (*see* Artificial neural network (ANN))
  artificial synapse, 41
  backpropagation, 33–36
  brute force approach, 8
  cancer incidence prediction, 8
  coding bias, 90
  concept learning, 15
  confusion matrix, 16–17
  convolutional neural networks (CNNs) (*see* Convolutional neural networks (CNNs))
  data
    dimensional reduction, 88–89
    extraction and correction, 81–82

    feature engineering, 83–84
    feature selection, 81–82
    feature transformation, 83
    imperfect class separation, 89
    Kaggle data, 81, 81*f*
    preprocessing, 80
    principal component analysis (PCA) (*see* principal component analysis (PCA))
    standardization and normalization, 83
  data bias
    population basis, 91
    racial discrimination, 91
    statistical analysis, 91
  datasets, 17
  deep learning, 16
  and digital pathology, 243–244
  dimensionality, 26
  fairness, 92
  feature engineering, 9
  image classification, 184
  image segmentation, 184–185, 235–236
  image translation, 185
  knowledge-driven image analysis, 235
  lymphocytic infiltration quantification, 237*f*
  model bias, 90
  neural networks, 6–7
    abstraction process, 9
    hidden layers, 9
    neural nets, 10
    restricted Boltzmann network, 10
    simulated neurons, 9
    spiking neural networks, 10
  overfitting and underfitting, 39–40, 40*f*
  precision and recall, 16–17
  principal component analysis (PCA), 26
  PYTHON programming, 4
  reinforcement learning, 15–16
  vs. rules-based program
    geometric-based formulas, 4
    inscribed and circumscribed polygons, 4–5
    Monte Carlo simulation, 6
    simulations, 5, 5*f*
    squares and circles, 6
  sample bias, 90–91
  shallow learning (*see* Shallow learning)
  stroma segmentation and cellular classification, 236*f*
  style transfer, 185
  supervised learning, 15–16, 244–247
  training set, 8, 16–17
  unsupervised learning, 15–16, 248–249
  varieties, 6–7, 7*f*

386 Index

Markov decision process (MDP), 52, 103
Minimum spanning tree (MST), 292
Mitotic figure detection, 286
Model bias, 90
Model development and training, 141–142
Monoclonal antibody-based treatment strategies, 312–313
Multiphoton microscopy (MPM), 195
Multiple-instance approaches (MIL), 68–69
Multiplex immunohistochemistry (mIHC), 329–331
Multispectral fluorescent scopes, 276–277

N

Naïve Bayes algorithm, 23
Naïve Bayes assumption-based methods, 245
Natural language processing (NLP), 159, 167–168
Neural network-based programs, 6–7
Neural networks, 96–97, 296
Neural style transfer, 264
Neuromorphic computing, 12, 370–371, 373
Normalized median intensity (NMI), 191–192
N-shot learning
  complex patterns, 70
  Image2Vector models, 70–71
  one-shot training, 70–71
  Siamese network, 71

O

Occlusion, 252–253
"Omic" data, 160
One class learning, 100–101
One-class learning system, 71–73
One-shot learning system, 352

P

Paige Prostate, 175
PAPNET Testing System, 175
PathLAKE, 260
Pathologist buy-in, 138
Pathology, ethical AI in. *See* Ethical AI, in pathology
PathologyGAN, 267
Pathwork Tissue of Origin Test, 175
Pathwork Tissue Of Origin Test Kit-Ffpe, 175
PCA. *See* Principal component analysis (PCA)
Phase-change neuron, 12–13
Photoactivated localization microscopy (PALM), 202
Picture archiving communication system (PACS), 112

Pixel-to-pixel classification model, 195
Pix2Pix-base image-to-image translation, 264–265
Positive-unlabeled (PU) learning algorithms, 89
Precision, 16–17
Precision diagnostics, 344
Precision medicine, 160, 233–234
Predictive autofocusing, 118
Principal component analysis (PCA), 26, 363
  data preprocessing, 84
  data reduction, 84–85
  eigenvalues and eigenvectors, 86–88
  implementation, 87–88
  nonlinear reduction methods, 88
  scree plot, 86, 87*f*
  variance, 84–85
Probabilistic models
  Bayesian Belief Networks, 23–24
  conditional probabilities, 22
  Naïve Bayes algorithm, 23
  variables, 22
Protein biomarker, 288–289, 289*f*
PubMed, 212
PYTHON programming language, 4

Q

Quantitative Image Analysis (QIA), 354–355
Quantitative metrics, 189
Quantitative phase imaging (QPI), 193, 194*f*
Quantum computing, 13, 104–106, 370–371, 373–376, 375*f*
Quantum machine learning (QML), 371

R

Race, in ethical AI design, 165
Radiology, 159–160, 167–168, 174
Random forest, 24–25, 25*f*, 247, 296–297
Random survival forest (RSF), 297
Rectified linear (ReLU) activation function, 44
Rectified linear function (ReLU), 30–31
Recurrent neural networks (RNNs), 50–51, 64
Regulatory challenges, 148–150
  CMS/CLIA, 149–150
  European Union Conformite Europeenne, 149
  Food and Drug Administration (FDA), 148–149
Reinforcement learning (RL), 15–16, 64–65, 79, 102–103
  exploration/exploitation dilemma, 52
  function approximation, 52
  Markov decision process (MDP), 52
  reward signal, 51–52
  state-action spaces, 52
Restricted Boltzmann machine (RBM), 47

Index   387

Restricted Boltzmann network, 10
RetCCL search engine, 217, 218–219*t*
Retrieval augmented generation (RAG), 225
Reverse image search, 213
Rules-based approach, 4

S

SAFRON framework, 267
Sample bias, 90–91
Scree plots, 22
Self-supervised learning, 103–104
Shallow learning, 16
   dataset, 18*f*
   decision tree algorithm, 24–25
   feature vector, 17–18
   geometric (distance-based) models, 18–21
   K-Means algorithm (KM), 21–22
   probabilistic models, 22–24
   random forest, 24–25, 25*f*
Siamese network, 71, 100–101
SIAN system. *See* Style-Guided Instance-Adaptive
     Normalization (SIAN) system
Sigmoid functions, 30–31, 31*f*
SISH search engine, 215–217, 218–219*t*
Spiking neural networks (SNNs), 10, 373
Stacked autoencoder, 47
Stain normalization, 277–278, 278*f*
Stain Normalization using Sparse AutoEncoders
     (StaNoSA), 190–191, 190*f*
Stain-style transfer, 264
Stain transfer network, 264
Statistical bias, 164–165
Stimulated Raman histology (SRH), 204, 205–206*f*
Stochastic optical reconstruction microscopy
     (STORM), 202
Store-and-forward telepathology, 111
Structural dissimilarity index (SDSIM), 191
Structured crowdsourcing, 318
Style-Guided Instance-Adaptive Normalization
     (SIAN) system, 267
Style transfer, 185
Supervised learning, 15–16, 96, 244–247
   computerized systems, 244
   k-nearest neighbor-based methods, 247
   logistic regression, 245–246
   Naïve Bayes assumption-based methods, 245
   random forest, 247
   support vector machines, 246–247
Supervised multiview canonical correlation
     analysis (sMVCCA), 363
Support vector machine (SVM), 246–247, 316
Surgical pathology, 273

Survival analysis, 294–295
Survival Convolutional Neural Networks (SCCN),
   73
Survival-support vector machine, 297
SVCR, 297
SVRc, 297
Synthetic data, 69–70

T

Technological innovation, 159–160
Telepathology technologies, 111–112
   dynamic systems, 111
   future developments, 112
   hybrid telepathology systems, 111–112
   store-and-forward telepathology, 111
Tissue of Origin Test Kit-FFPE, 175
Transfer learning, 12, 48–49, 97–99
Transformer-attention models, 369–373
Transitioning, 138
Transparency, in AI and ADM, 171–172
Two-stage deep learning system (DLS)
   automated training system, 124
   Gleason scoring and quantitation, 123–124
   limitations, 125
   pathologist-level performance, 124
   quadratic kappa agreement, 124

U

U-nets, 187
Unsupervised learning, 15–16, 248–249
Unsupervised pretraining, 46, 98
   via autoencoders, 101
   via clustering, 99–100
   via generative adversarial networks, 101–102

V

Variational autoencoder (VAE) model, 191–192,
     224–225
Voting scheme, 54

W

Weakly supervised learning, 67–69
Weighted voting, 54
Whole slide imaging (WSI), 167–168, 211,
     213–215, 214*f*, 344
   automation of, 260–261, 260*f*
   vs. conventional microscopy, 120–122
   data and workflow management
     anatomic pathology workflow solutions, 116
     image management software, 116
     image sharing, 115–116
     international standardization initiatives, 116

388 Index

Whole slide imaging (WSI) *(Continued)*
    workload balancing, 115–116
    and deep learning
        prostate cancer diagnosis, 123
        two-stage deep learning system (DLS)
            (*see* Two-stage deep learning system
            (DLS))
    digital pathology association (DPA), 122
    digital slide viewers, 115
    early generation, 117
    focus mapping, 118
    frequency-based sampling, 118
    high-speed robot industry, 113
    histopathology evaluation, 122
    and image analysis, 122–123
    image viewer, 114–115
    infrastructure and cost reduction, 112
    infrastructure requirements and checklist,
        118–119
    light-emitting diode (LED)-based modules, 114
    line scan-based imaging, 112
    objective lens, 113–114
    picture archiving communication system
        (PACS), 112
    and primary diagnosis, 119–122
    scanner, 112–113, 116, 117*t*, 276–277
    scanning resolution, 113
    second-generation, 117–118
    tile-based imaging, 112
    tumor region detection, 279–280, 280*f*
    vendor-unique viewers, 115
    vs. virtual immunohistochemistry, 120–122
    whole-slide images (WSIs), 309

Y

Yottixel search engine, 215–217, 216*f*, 218–219*t*

Z

Zero shot learning, 104